EDWARD TENNER

WHY THINGS BITE BACK

Edward Tenner, former executive editor for physical science and history at Princeton University Press, holds a visiting research appointment in the Department of Geological and Geophysical Sciences at Princeton University. He received an A.B. from Princeton and a Ph.D. in history from the University of Chicago and has held visiting research positions at Rutgers University and the Institute for Advanced Study. In 1991–92 he was a John Simon Guggenheim Memorial Fellow and in 1995–96 a Fellow of the Woodrow Wilson International Center for Scholars.

ALSO BY **EDWARD TENNER**

Tech Speak

WHY THINGS BITE BACK

WHY THINGS BITE BACK

TECHNOLOGY AND THE REVENGE OF UNINTENDED CONSEQUENCES

EDWARD TENNER

VINTAGE BOOKS
A DIVISION OF RANDOM HOUSE, INC. NEW YORK

FIRST VINTAGE BOOKS EDITION, JULY 1997

 Published in the United States by Vintage Books, a division of Random House, Inc., New York, and simultaneously in Canada by Random House of Canada Limited, Toronto. Originally published in the United States in hardcover by Alfred A. Knopf, Inc., New York, in 1996.

Grateful acknowledgment is made to *Alfred A. Knopf, Inc.,* and *Peters Fraser & Dunlop Group Ltd.* for permission to reprint an excerpt from "Lord Finchley," from *Cautionary Verses* by Hilaire Belloc, copyright © 1931 by Hilaire Belloc, copyright renewed 1959 by Eleanor Jebb Belloc, Elizabeth Belloc, and Hilary Belloc. Rights in Canada administered by Peters Fraser & Dunlop Group Ltd., London, from *Complete Verse.* Reprinted by permission.

The Library of Congress has cataloged the Knopf edition as follows:
Tenner, Edward.
Why things bite back: technology and the revenge of unintended consequences / Edward Tenner.
p. cm.
Includes bibliographical references and index.
ISBN 0-679-42563-2
1. Technology—Social aspects. 2. Technology—Economic aspects.
I. Title.
T14.5.T459 1996
303.48'3—dc20 95-38036
CIP

Vintage ISBN: 0-679-74756-7

Book design by Mia Risberg

Random House Web address: http://www.randomhouse.com/

Printed in the United States of America
10 9 8 7 6 5 4 3

For My Mother and Brothers

Contents

PREFACE

This book began as it is ending—on paper. It started ten years ago when I saw something curious happening as personal computers appeared on one desk after another at my office, an academic publishing house. Futurism was thriving. Its most successful practitioner, Alvin Toffler, had declared in his best-selling *The Third Wave* that "making paper copies of anything is a primitive use of [electronic word processing] machines and violates their very spirit." Yet the paper recycling bins seemed always to be brimming with printouts. And even after the office was networked and electronic mail had replaced hard-copy memos, the paper deluge continued. I was not around to see it, but a former colleague explained. People were using electronic mail, including the Internet, in place of telephone calls. But they wisely did not trust the permanence of electronic records. Besides, many files circulated among three or four departments; cryptic cross-references to electronic files would not do. Networking had actually multiplied paper use. When branches of Staples and OfficeMax opened near Princeton, the first items in the customers' view (and in the catalogues) were five-thousand-sheet cases of paper for photocopiers, laser printers, and fax machines.

Use of paper has continued to soar. It is as though paper is taking its revenge on the futurists—not that any futurist has ever lost business because of a wrong prediction. And I wrote an essay titled "The Paradoxical Proliferation of Paper" that in turn made me look at the strange consequences of nearly everything, results that seemed to contradict every reasonable scenario. Paper seemed to have an existence of its own that defied the human will to control it. Things seemed to be fighting back.

Futurists had not reckoned with the perversity of ordinary objects and systems. Neither libertarians trusting in infallible market processes nor Greens prophesying starvation and environmental collapse had imagined the real future. But it was also impossible to refute either point of view. For the first, hardships and disasters are mere challenges to a boundless human ingenuity; for the second, any gain in living standards is just another charge against a dread future reckoning. Humankind is either on its way to the stars or hurtling out of a high-rise window to the street and mumbling, "So far, so good."

I am not trying to resolve this debate. The poet Paul Valéry was surely right when he wrote in 1944:

> Unpredictability in every field is the result of the conquest of the whole of the present world by scientific power. This invasion by active knowledge tends to transform man's environment and man himself—to what extent, with what risks, what deviations from the basic conditions of existence and of the preservation of life we simply do not know. Life has become, in short, the object of an experiment of which we can say only one thing—that it tends to estrange us more and more from what we were, or what we think we are, and that it is leading us . . . we do not know and can by no means imagine where.[1]

Valéry in 1944 was evoking the errors that the best thinkers of 1890 would have made in trying to foresee the next fifty years. In fact these errors were not just hypothetical. Valéry had no way of knowing it, but in the early 1890s a U.S. newspaper syndicate had surveyed seventy-four eminent American men and women about life in 1993, and it published the results for the Chicago World's Columbian Exposition of 1893. Dave Walter has compiled their responses in a fascinating book, and they show the perils of technological extrapolation. What is remarkable is not any omission of nuclear weapons or microelectronics; many public figures of the late nineteenth century foresaw both new weapons of mass destruction and new forms of electrically powered global communication. The striking oversight is the rise of mass motoring. Nobody understood the chain of technological, commercial, social, and political events that would surround the internal-combustion vehicle which Karl Benz had patented in 1886. For that matter, even the physics professors of Harvard a century ago failed to recognize the impending transformation of their own field; their 1880s Jefferson Laboratory, still in use, was designed especially to advance the physics of the future—terrestrial magnetism. But then, as a wise musician once said, "If I knew the jazz of the future, I'd play it."[2]

I do not know the jazz of the future, either. This is not a book of prediction. It is a new look at the obvious. It is obvious that technology makes many things better. By technology I mean humankind's modification of its biological and physical surroundings. It is also obvious that we are still unhappy about those surroundings, more discontented than when they were inferior. I consider four main areas: health and medicine; the environment; the office; and sports. I describe real technological gains in coping with human problems but also the frustrations that have accompanied those gains.

The technological dream of a self-correcting world is an illusion no less than John von Neumann's 1955 prediction of energy too cheap to meter by 1980. The social goal of a new Athens, of machine-supported leisure, has proved a noble mirage. Technology demands more, not less, human work to function. And it introduces more subtle and insidious problems to replace acute ones. Nor are the acute ones ever completely eliminated; in fact, unless we exercise constant care and alertness, they have a way of coming back with new strength. We are on a treadmill that we can no longer dismount from. We cannot turn back to a wholesome past, if only because the past, while sometimes more decorous, was far messier than we realize or perhaps can realize. A whole book could be written about mud, and another about dust.[3]

We are unhappy, I suggest, for two reasons. First, in controlling the catastrophic problems we are exposing ourselves to more elusive chronic ones that are even harder to address. And second, our greater safety demands more and more vigilance. Occasionally we really do simplify things. Programming a videocassette recorder was once an almost comic exercise; now ingenious mathematical routines built into the machines let us tape television shows by punching in a handful of numbers printed in the newspapers. Far more often, though, the apparent simplicity of technology conceals underlying problems that then become far more difficult to diagnose and treat, whether they are in automobiles or in personal computers. My experience leads me to doubt whether the electrical system of a car, once it starts to malfunction, can ever truly be repaired. And a stylish and even "friendly" computer interface may conceal a fatal tangle of corrupted files, not to mention latent bugs. Chronic problems almost by definition demand maintenance rather than solution; while the need for vigilance and care becomes itself a chronic irritation.

I see no way around our predicament; neither the quest for a spartan "sustainable" existence nor reliance on the forces of the marketplace will release us. I am writing not for policymakers but for people stumbling through a Rube Goldberg world and trying to make sense of it. And I am arguing not against change, but for a modest, tentative, and skeptical acceptance of it.

An award from the John Simon Guggenheim Foundation made this project possible. I am indebted to the foundation for its generosity, flexibility, and understanding. A grant from the Exxon Education Foundation for another project also enabled me to continue my career as an independent writer. Special thanks are due to the program officer who initiated the grant, Richard R. Johnson, and to the president of the foundation, Edward F. Ahnert.

Years ago, William H. McNeill opened my eyes to new ways of bringing science and technology into history. Thomas P. Hughes and John McPhee also have given me indispensable encouragement.

I began the project as a visitor in the School of Social Science at the Institute for Advanced Study. I am grateful to Joan W. Scott, executive officer at the time, and to the other faculty of the school for their patience and comments. Albert O. Hirschman of the School of Social Science and Freeman Dyson of the School of Natural Science were especially helpful. At the Rutgers Center for Historical Analysis, John R. Gillis and Victoria de Grazia introduced me to another vigorous academic community. And most recently, Robert O. Phinney, John Suppe, and S. George Philander helped make my affiliation with the Department of Geological and Geophysical Sciences at Princeton University an equally productive one.

Dozens of scientists, scholars, and government officials answered my questions patiently and responded generously to requests for reprints. I should single out a few for special thanks. In the Princeton and Rutgers communities I had excellent counsel from Enoch Durbin, Freeman Dyson, Robert Freidin, Gerald L. Geison, John R. Gillis, Stanley N. Katz, Jackson Lears, and Michael S. Mahoney.

John T. Bethell and Christopher Reed of *Harvard Magazine*, Michelle Preston and later James I. Merritt III of the *Princeton Alumni Weekly*, and Jay Tolson and Steven Lagerfeld of the *Wilson Quarterly* all encouraged me to develop the ideas that led to this project. Joseph W. Dauben's encyclopedic knowledge and superb editorial eye have helped me overcome many a hurdle. Barbara Freidin's enthusiasm for the project and her organizational skills made it possible to put in order the reams of research materials accumulated in this project. I have also learned much from conversations with Charles L. Creesy, Michael Volk, Russell C. Maulitz, M.D., Stephen S. Morse, Stephen J. Pyne, Don C. Schmitz, Jesse Ausubel, Joseph Keller, J. Nadine Gelberg, and Daniel Headrick. James Andrew Secord's inspired phrase "Ever since Frankenstein" has become a chapter title with his kind permission.

Because I have written this book for nonspecialists and have only been able to touch on vast literatures like those of risk assessment, natural hazards, and conservation biology, it is especially important that none of these teachers, colleagues, friends, and consultants should be blamed for any error or omission.

Ashbel Green's editorial guidance and encouragement, and the acumen and thoroughness of Alfred A. Knopf's production department, have been invaluable. So has the understanding and skill of my literary agent, Peter L. Ginsberg, and his colleagues at Curtis Brown Ltd.

Edward Tenner

Princeton, New Jersey, May 1995

WHY THINGS BITE BACK

1

Ever Since Frankenstein

One of industrial and postindustrial humanity's perennial nightmares is the machine that passes from stubbornness to rebellion. As Rod Serling's 1961 short story "A Thing About Machines" begins, a fastidious, acerbic food writer named Bartlett Finchley is insulting a long-suffering television repairman. The man is fixing a set that Finchley pummeled after it failed to "work properly." Finchley mishandles the devices around him with the same malice he shows toward people. When his departing secretary wishes him defeat in "this mortal combat between you and the appliances," his electric typewriter begins spontaneously to produce the words "Get out of here, Finchley." A Mexican dancer looks straight at him out of a television program, repeating the message. Soon a mechanical mutiny erupts. His electric razor nearly mauls his face. The chimes of his clock ring uncontrollably. His car slips its emergency brake and rolls down the driveway, almost striking a child on a bicycle.[1]

After the police come and go, the appliances drive Finchley from the house. The car—forerunner of Stephen King's Christine—accelerates spontaneously, pursuing him around the neighborhood. Forced into his own swimming pool, he drowns as the engine seems to "let out a deep roar like some triumphant shout." The story ends with a cemetery caretaker puzzling over why his power lawn mower has tugged him off its path to strike Finchley's tombstone. The machines have retaliated where the stoic human victims of Finchley's corrosive wit had simply thrown up their hands.

Finchley's nemesis was technological vengeance, American-style. His namesake, the ill-fated early do-it-yourselfer Lord Finchley in Hilaire Belloc's verse, is electrocuted after breaching noblesse oblige by rummaging around in his fusebox rather than defer to a tradesman:

Lord Finchley tried to mend the Electric Light Himself.
It struck him dead: And serve him right!
It is the business of the wealthy man
To give employment to the artisan.[2]

Bartlett Finchley, who does hire people but insults them, dies to atone for his affronts to machines. The human beings in the story are too polite or too afraid of his acid tongue to retaliate. Not so the objects.

Serling was mocking not technology but sarcasm and arrogance. Yet he was also tapping into the misgivings of industrial society, doubts that have grown recently to proportions even greater than those of the 1960s. What if Serling had lived to the 1990s to present Mr. Finchley in a "smart" house? With heating, cooling, timekeeping, security, television, and telephone coordinated by a master computer, Finchley's aggrieved machines could have choreographed their retaliation under the diabolical intelligence of a central processing unit. Serling recognized that, without being Finchleys, we all have at times "a thing about machines."

Serling's Finchley, waspish cosmopolitan caricature that he is, ridicules his all-American suburban neighbors. But he also expresses very American misgivings about machines. America may be the home of *Popular Mechanics* and Tomorrowland, but it is also the birthplace of lemon laws. New Jersey and Pennsylvania, for example, officially define a lemon as a new car with a "substantial" defect that cannot be repaired after three attempts, or is out of service for a total of twenty to thirty days.[3] Americans believe that things can be simply no good, that things can be contrary by nature.

In fact, many objects in America, as elsewhere, *are* designed badly and even dangerously. Horror stories are no mere inventions of consumer lobbyists and plaintiffs' attorneys. But the elusive, irreparable problem is different. In the mid-1980s, more than twenty-five years after Serling's story was published, dozens of people claimed their Audi sedans had accelerated suddenly and spontaneously from a standing position, occasionally threatening startled drivers who had stepped out of their cars. Engineers could not duplicate the problem. The National Highway Traffic Safety Administration eventually concluded that people had been hitting the accelerator when they thought they were simply braking. Judges dismissed the lawsuits. But the myth of the killer machine—already present in early-1900s warnings that the automobile "may run amok"—would not die.[4]

Even computer professionals aren't immune from demon theories. Some specialists acknowledge, as one put it, that certain problems are "in the area of metaphysics," and that "strange things happen in electronics for which there is no reason." He reports that at one company, a system would crash during critical client preparations whenever a certain engineering manager appeared. The problem was solved by excluding him from its room whenever an important demonstration was being prepared.[5]

In vain do statisticians point out how likely weird coincidences really are. It would be, mathematically speaking, grounds for suspicion of human interference if *nobody* ever won a major lottery twice. Nor have psychologists persuaded many people that perceptions of bad luck can be highly selective. (We are more likely to remember missing a subway or elevator car than arriving just in time.) Mediterranean peoples fear the evil eye, which may belong to an otherwise decent but hapless *person*. Americans are more likely to attribute the problem not to a person but to a thing, because they expect so much of objects.

From Revenge to the Revenge Effect

Bartlett Finchley provoked the technology around him into murderous rebellion. But for most people, the problem is not open malice but repeated small episodes of frustration. The damages, real or perceived, are sometimes more psychological than material. In fact, sometimes the safer and better off we are, the more threatened we feel. Our discontent with experts who promise progress depends on our sense of entitlement to progress. Indignation swells when change fails to bring promised improvement. We are alarmed when our factories close, yet our magazines celebrate the "smashing success" of the early-nineteenth-century Luddites. Articles recall admiringly the stocking workers of the English Midlands who broke the power looms that were depressing prices. A psychologist publishes "Notes Toward a Neo-Luddite Manifesto," and a historian of technology praises the machine-wreckers' conviction that sometimes progress is worth stopping.[6]

The indignation of nineteenth-century producers has yielded to the irritation of late-twentieth-century consumers. Why are the lines at automatic cash dispensers (so much for the paperless society) longer in the evening than those at tellers' windows used to be during banking hours? Why do helmets and other protective gear help make football more dangerous than rugby? Why do filter-tip cigarettes usually fail to reduce nicotine intake? Why has yesterday's miracle vine become today's weed from hell? And why have today's paperback prices overtaken yesterday's clothbound prices? Why has the leisure society gone the way of the leisure suit?

The real revenge is not what we do intentionally against one another. It is the tendency of the world around us to get even, to twist our cleverness against us. Or it is our own unconscious twisting against ourselves. Either way, wherever we turn we face the ironic unintended consequences of mechanical, chemical, biological, and medical ingenuity—revenge effects, they might be called.

Noticing revenge effects doesn't mean denying that life generally has improved in the West over the centuries. The English essayist Paul Jennings, in a celebrated early send-up of Heidegger and Sartre titled "Report on Resistentialism," discovered the new Parisian insight that "*les choses sont contre nous*," that "man's increase in [an] illusory domination over Things has been matched, *pari passu,* by the increasing hostility (and greater force) of the Things arrayed against him." We have more "Things" to worry about because we have replaced a small number of them ("the lack of satisfactory illumination at night, the primitive hole in the roof blowing the smoke back and letting the rain in") with "far more opportunities for battle-losing against Things—can-openers, collar studs, chests of drawers, open manholes, shoelaces. . . ."[7]

The paradoxical behavior of objects isn't always negative. Some things last longer if used regularly. Automobiles that have gone hundreds of thousands of miles are no miracle. Cars are built for prolonged highway cruising. Occasional short-distance driving wrecks engines, and long-term storage deforms tires. The Library of Congress has for years sponsored chamber concerts not only for audience enjoyment but to help preserve its priceless early-modern stringed instruments by letting master musicians play them. Years of disuse, even at optimal temperature and humidity, would damage the wood. Even electronic devices act as though they want to be used. The cones of high-fidelity speakers last longer if used regularly. A computer that is left off for weeks or months often needs time for its capacitors to work properly again and for mechanical connections to readjust to higher temperatures. Hard-disk drives act as though they prefer whirring around the clock to starting up and shutting down. In fact, many computer specialists favor never turning a computer off during the working day, or even after hours. At least one laser printer is now sold without an on/off switch; instead, after a certain period in idleness it automatically goes into an energy saving sleep mode, from which it is revived by any print command.

Still, most unintended consequences are unpleasantly rather than pleasantly surprising. We usually discover even the positive effects only after negative experience—for example, realizing that repeated heating and cooling damages electronic components. And whenever we try to take advantage of some new technology, we may discover that it induces behavior which appears to cancel out the very reason for using it. The electronic gear that lets people work at home doesn't necessarily free them from the office; urgent network messages and faxes may arrive at all hours, tying them more closely to business than before.

Anatomy of Revenge

A revenge effect is not the same thing as a side effect. If a cancer chemotherapy treatment causes baldness, that is not a revenge effect; but if it induces another, equally lethal cancer, that is a revenge effect. If an experimental hair-growing drug were shown to raise the likelihood of cancer, it would be banned; but its risk would be a side effect rather than a revenge effect. On the other hand, if it turned out to accelerate hair loss under certain conditions, that would be a revenge effect. A revenge effect also is not just a trade-off. If legally required safety features raise airline fares, that is a trade-off. But suppose, say, requiring separate seats (with child restraints) for infants, and charging a child's fare for them, would lead many families to drive rather than fly. More children could in principle die from transportation accidents than if the airlines had continued to permit parents to hold babies on their laps. This outcome would be a revenge effect.

Security is another window on revenge effects. Power door locks, now standard on most cars, increase the sense of safety. But they have helped triple or quadruple the number of drivers locked out over the last two decades—costing $400 million a year and exposing stranded drivers to the very criminals the locks were supposed to defeat. Advanced alarm systems also are now standard equipment on many luxury cars and popular options on even moderately priced models. It is true that most owners don't mind occasional incidents. They'd rather have false positives than false negatives. But squirrel exploration and other transient events spook the systems so easily that the rest of us assume sirens to be screaming wolf. In cities where alarms appear most needed, hotheaded neighbors silence malfunctioning systems by trashing cars. Then the damages are a revenge effect. If legislatures, manufacturers, and insurance companies encourage installation of the alarms and frustrated automobile thieves turn to armed carjacking, there is not just an individual but a social revenge effect. At home, too, cheaper security systems are flooding police with false alarms, half of them caused by user errors. In Philadelphia, only 3,000 of 157,000 calls from automatic security systems over three years were real; by diverting the full-time equivalent of fifty-eight police officers for useless calls, the systems may have promoted crime elsewhere.[8]

How Revenge Effects Happen

Technology alone usually doesn't produce a revenge effect. Only when we anchor it in laws, regulations, customs, and habits does an irony reach its full potential. Take our ability to alter the landscape and relocate its inhabitants. Florida developers, legally required to relocate tortoises, appear to have inadvertently spread an undesirable microorganism to new populations. Officials believe the respiratory disease it causes has been killing the endangered and ecologically important gopher tortoise.[9]

Or consider shifting heat around a city. Urban centers are notoriously hotter in summer than the surrounding countryside. Air conditioning has made a normal business pace possible even through July and August, but at the cost of raising the ambient temperature on the street, making time spent outdoors even more unpleasant. (There is also the transfer of discomfort to nonclimatized quarters, but because it does not affect air-conditioned people, it is not a revenge effect.) Similarly, air-conditioned mass transit may be raising platform temperatures by as much as 10 degrees F. Someone who waits ten minutes for a ten-minute ride may be a net loser. And if the heat overloads and disables the train cooling units themselves, riders may be even worse off, sweltering behind windows sealed to prevent interference with the system when it does work. Like the tortoise relocations, this might be called a *rearranging effect*.

Think of the results of devices and systems that are supposed to free time for other things. "Machines should work. People should think"—the celebrated motto of IBM Corporation in the golden days of the mainframe era—may still be the finest brief summary of technological optimism. But it doesn't always work that way. The historian of technology Ruth Schwartz Cowan has shown in her book *More Work for Mother* that while vacuum cleaners, washing machines, and other "labor-saving" appliances did gradually improve the working-class standard of living, they saved no time for middle-class housewives. Women who had once sent soiled clothing to a commercial laundry began to do more and more washing at home. And as laundries and other services went out of business, fewer choices remained. We can call this a *repeating effect*: doing the same thing more often rather than gaining free time to do other things. Like other revenge effects, it changed when its social framework did; the rise of two-career families in the 1970s and 1980s sent housework time down again. But as we will see, there have been repeating effects at the office, too.

Few people prefer rotary to pushbutton telephones. There was never anything graceful about looking for the right opening, twirling the dial, and hitting a small metal object with one's finger. A button is less likely to be pressed mistakenly, and voice recognition technology actually produces fewer wrong numbers than either manual system. But the awkwardness of the mechanical switching had an advantage. There were limits to the size of numbers that people were expected to dial—and to remember. By now the time savings of punching rather than dialing have been more than consumed by elaborate systems built to take advantage of it. When a carrier access code and credit card number are added to the number itself, a single call may need thirty digits. A voice-mail system may then take over, demanding still more digits and waiting. Who has never fallen into an endless message loop? And just as last-number dialing appeared even on low-priced telephones to automate repeat calls, companies began to install equipment that could stack dozens of incoming lines for indefinite periods—often at the callers' expense. All these are *recomplicating effects.*

Sometimes a practice or device can multiply a problem. According to a study by the political scientist Theodore A. Postol, damage to Tel Aviv during the Gulf War may have actually increased after the United States deployed Patriot missiles as a shield against Scud attacks. More people were injured and more apartments damaged during the Patriot defense than before, though fewer Scuds were launched. Some of the Scuds the Patriots broke up might have landed without damage. According to Postol's calculations, a Patriot hitting a Scud at 5.5 kilometers altitude could produce debris extending over 5 kilometers. A spinning piece the size of a soft-drink can could break through a five-inch concrete slab. So, among other things, the Patriot might have transformed the Scud into smaller projectiles.[10] This hydralike response to technology is a *regenerating effect*. As we shall see, manipulating nature to promote "beneficial" organisms and suppress "pests" can produce regenerating effects even more spectacularly damaging than advanced conventional weaponry.

When innovation opens a new space, there is at first a euphoria of endless horizons. Somehow, though, a new frontier is never stable. Either people lose interest and it becomes a series of literal or metaphoric ghost towns, or it is soon as crowded as the space that people left. Consider the thirty to seventy thousand pieces of space debris cluttering the earth's orbit. Each measures a centimeter or more in diameter and can shatter a spacecraft, much as a Scud fragment could crash its way through an apartment house. Meanwhile, back on earth the density of marine litter deposited on a remote island in the Atlantic increased between 1984 and 1990 from 500 to over 2,200 items per kilometer. The electromagnetic spectrum that once appeared so ample is now so close to bursting with media and telecommunication channels that re-allocation may leave some users out in the cold or largely dependent on telephone lines. These are examples of *recongesting effects*.[11]

Reverse Revenge

Revenge effects happen because new structures, devices, and organisms react with real people in real situations in ways we could not foresee. There are occasional reverse revenge effects: unexpected benefits of technology adopted for another reason. (Like revenge effects themselves, reverse revenge effects are a rough but useful metaphor: in one case, for the way reality seems to strike back at our efforts, and in the other, for the equally unexpected ways in which we benefit from the complexity of the world's mechanisms.) Sometimes these are earlier devices that were good for us in ways we and our parents didn't realize. In retrospect, the key-pounding, carriage-returning, paper-feeding chores required by old-style manual typewriters had the reverse revenge effect of reducing the likelihood of carpal tunnel syndrome. Unfortunately the light touch and blazing speed of computer keyboard entry often turned out to cause unexpected pain. And what are we to make of stepping from office elevators into our cars and driving to health clubs to use treadmills (a feature of nineteenth-century prisons) and stair-climbing machines?

If older technology turns out to have hidden benefits, so has change. In some former arsenals in Colorado and eastern Germany, rare animal species abound because artillery shells and toxic waste have kept people out; it's the suburb that menaces diversity. My own town of Princeton became the charming place it is in part because the future Pennsylvania Railroad main line was rerouted three miles away from it during the Civil War. Western railroads—plus mechanization and falling grain prices—also helped begin to restore trees and wildlife to the surrounding landscape, which was already farmland by the time of the Battle of Princeton in 1777. Suburban sprawl now threatens the second-growth forests that were regained, but it was new technology that indirectly helped restore them.

Malignant Machinery

From the earliest days of the industrial age, the greatest artists and writers of the West have had their eyes on the recalcitrant and even malevolent machine. Human ingenuity turning against itself was not a new conceit. The perils of magic were well known. The medieval Jewish legend of the Golem had already inspired the story of a clay monster that was fashioned by Rabbi Löw of sixteenth-century Prague, only to turn against its creator. When Hamlet declares it "sport to have the enginer/ Hoist with his own petar," his metaphor was of a crude small bomb used to blow away a gate or part of a wall, sometimes taking its creator with it.[12]

If men and women before 1800 or so had any idea of a malevolent machine, nobody knows about it. No student of early European popular culture I have asked has found an example of it. The elites who did most of the writing had little contact with the mechanics of a household. Adam Smith, in his *Theory of Moral Sentiments*, treats early mechanical consumer goods as playthings. The men who worked with the most complex and perilous machines of medieval and early modern Europe, sailors and miners, do not seem to have endowed ships and equipment with the malicious independent will so familiar to twentieth-century people. There were spirits *in* the mines—in America, the Tommy Knockers and the Old Man—who demanded respectful treatment and could help and rescue as well as bedevil. But these creatures were almost literally ghosts in the machine, not features of the shafts, pumps, or tools. Breaking a shipboard custom could threaten the safety of the vessel and crew, but not because the ship's design itself had hidden dangerous and refractory powers.

NATURE'S REVENGE

In traditional lore, when nature took its revenge it was to punish sin. After the newly rich miners of a Polish town began to wear silver buckles on their shoes and throw bread in the street to protect them from the dirt, their mine was flooded and they had to go without bread. Pride, arrogance, exploitation, and avarice, not technological overreaching, brought disaster. These stories were entirely consistent with the early modern idea that nature itself would expose and defeat evil conduct.[13]

Even a scientific figure as important as the eighteenth-century botanist Linnaeus collected dozens of anecdotes to this effect in a secret notebook for his son, published only after the Second World War as *Nemesis Divina*. Linnaeus saw no opposition between the natural and the divine order. The powers of nature itself punished wrongdoers who had kept their secrets from their fellow men and women. Like calamities of the miners' lore, each of the natural disasters in Linnaeus avenged some moral transgression. And like the miners' stories, Linnaeus's parables—which read today like a mixture of Kierkegaard and the *National Enquirer*—were not for a broad reading public but for a small, intimate circle.

The Real Frankenstein

It was Mary Shelley's *Frankenstein* that first connected Promethean technology with unintended havoc. The theme riveted audiences over a century before the most rudimentary organ transplantation was medically possible. The literary historian Steven Early Forry has shown how rapidly Shelley's story spread across the stages of London and Paris within a few years of its publication in 1818. In fact, the stock figures of mad scientist, cretinous assistant, and brutal monster probably owe more to these early stage versions than to Shelley's original text.[14]

The Victor Frankenstein of the novel was no doctor—of either medicine or philosophy. But his project was a scientific and technological experiment, and he had left his studies at the University of Ingolstadt after a successful career in which he "had made some discoveries in the improvement of chemical instruments, which procured me great esteem and admiration in the university." He was a genteel amateur in the eighteenth-century mode, created just before the rise of nineteenth-century academic and industrial science.[15]

Mary Shelley based her story on actual "galvanic" experiments wherein corpses were animated electrically. Was she warning against the conquest of nature by science, against male appropriation of life-giving power? A growing number of critics think so, perhaps reading certain twentieth-century attitudes into Shelley's story. Victor Frankenstein was reassembling and reviving life, not growing it. They are certainly right, though, in seeing in the book a revenge of nature against practitioners of a technology that surpasses their understanding.[16]

Frankenstein's fateful error was to consider everything but the sum of the parts he had assembled. The "limbs were in proportion, and I had selected his features as beautiful." The hair was "of a lustrous black, and flowing; his teeth of a pearly whiteness." But he had failed to understand the body as a *system.* Thus the "yellow skin scarcely covered the work of muscles and arteries beneath," and the "watery eyes . . . seemed almost of the same color as the dun white sockets in which they were set."[17]

Mary Shelley was pointing to a dilemma of all science-based technology—at a time when science was only starting to influence technological practice. How can we understand a system before we try to change it? Disaster inspires so much of our understanding. Victor Frankenstein had pursued years of steady research and development, punctuated by readings of the latest journals. "I prepared myself for a multitude of reverses; my operations might be incessantly baffled, and at last my work might be imperfect; yet, when I considered the improvement which every day takes place in science and mathematics, I was encouraged to hope my present attempts would at least lay the foundations of future success." And in fact Frankenstein's last words, after an unconvincing plea against the life of science, are: "Yet why do I say this? I have myself been blasted in these hopes, yet others may succeed."[18]

From Tool to System: Transforming Revenge

Mary Shelley wrote prophetically at the dawn of technological systems thinking. She does not treat the monster as a machine, but neither is it human despite its articulate and moving speech. Still less is it an animal. Neither its creator nor any other person in the story gives it a name of its own. It is a kind of system, though, a creature with unintended emotions, including rage and a passion for vengeance against its creator.

A machine can't appear to have a will of its own unless it is a system, not just a device. It needs parts that interact in unexpected and sometimes unstable and unwanted ways. A flat tire is not a system problem. A failure of battery charging may well be one. Any one of a number of parts of the automobile's circuit, or interactions among them, may be responsible. An individual part may be warranted for thirty days; no electrical system repair is likely to come with any meaningful guarantee. Industrial society did not begin the deceptive sale of an inferior product or an unhealthy animal. Horse traders once had the reputation enjoyed by used-car salespeople today. But there is a difference. The complexity of mechanical systems makes it impossible to test for all possible malfunctions and makes it inevitable that in actual use, some great flaws will appear that were hidden from designers.

From Use to Management

Technology before Mary Shelley's time did not come in systems. Well into the nineteenth century, artisanal tools and farm implements were extensions of the user's mind and body. In Central Europe and no doubt elsewhere, a scythe, for example, was custom-proportioned to the cultivator's body as a suit of clothes might be. Even a large, bureaucratically supervised enterprise like an arsenal or print shop was a complex of craftsmen rather than a factory in the nineteenth- or twentieth-century sense.[19]

As the museum curator James R. Blackaby has pointed out, the link between person and instrument was changing in America on the eve of industrialization. A rough, low bench called a shaving horse was common in colonial America. With a foot-operated clamp, it involved the whole body of the operator. The finer workbench, then already long used by professional artisans, began to displace the shaving horse on the farm in the nineteenth century.[20]

The workbench changed the relationship between the operator's body and the tools. It is a well-finished, solid table for anchoring the material with pegs and vises. The operator usually stands. And above all, the tools have more of the operator's intelligence and skill built into them. Adzes and drawknives demand experience and judgment. Planes have to be mastered, too. Yet once a job is set up right, they make constant judgment and adjustment unnecessary. Even an inexperienced woodworker can cut an elaborate piece by setting up the work and the plane blade properly. Most molding and grooving planes are constructed to cut only to a preadjusted depth. Skill is concentrated more in the conception, setup, and beginning of work than in each successive detail of its execution.

We cease to be tool *users* and, in Blackaby's phrase, begin to be tool *managers*. We direct and control processes that take place rather than shape them. Blackaby has contrasted the leather-cased ivory slide rule presented to him as a college freshman by his father with the electronic calculator he has since come to use. One requires human judgment, experience, and the constant exercise of skill; the other simply executes the operations it is programmed to do.

The calculator is in principle accurate to several more decimal places than the slide rule, but it is so only as long as its solid-state and mechanical parts are interacting properly. And there may be no clue when they are not. I once discovered—at tax time, of course—that an electronic printing calculator I had bought was starting to get its sums wrong. The problem was almost certainly in the print wheel advance mechanisms, but the tapes showed no sign of it. They looked impeccable until I saw that the numbers did not add up. Unlike a mechanical adding machine, my calculator did not merely malfunction; it gave dangerously wrong readings without a clue. The precision of the managed tool has a price. It may be less robust and, as it becomes more complex, less predictable.

Systems and the Birth of the Bug

A printing four-function calculator is both one of the simplest and one of the most advanced examples of a special kind of managed technology: the system. As the historian of technology Thomas P. Hughes has suggested, America's great contribution to world technology was the idea of a system, a set of matched, standardized, interacting components linked to a broad market.

The bug, that perverse and elusive malfunctioning of hardware and later of software, was born in the nineteenth century. It was already accepted shop slang as early as 1878, when Thomas Edison described his style of invention in a letter to a European representative: "The first step is an intuition and it comes with a burst, then difficulties arise—this thing gives out and then that—'Bugs'—as such little faults and difficulties are called—show themselves, and months of intense watching, study and labor are requisite before commercial success—or failure—is certainly reached."[21]

Edison implies that this use of "bug" had not begun in his laboratory but was already standard jargon. The expression seems to have originated as telegraphers' slang. Western Union and other telegraph companies, with their associated branch offices, formed America's first high-technology system. About the time of Edison's letter, Western Union had over twelve thousand stations, and it was their condition that probably helped inspire the metaphor. City offices were filthy, and clerks exchanged verse about the gymnastics of insects cavorting in the cloakrooms. When, in 1945, a moth in a relay crashed the Mark II electromechanical calculator that the Navy was running at Harvard—it can still be seen taped in the original logbook—the bug metaphor had already been around for at least seventy-five years. Anything can break. Only a system can have a bug.[22]

In the late nineteenth century, it was not just mechanical and electrical systems that began to show unintended and unwelcome properties. The literal foundations of the city were beginning to be unpredictable. When workers filled in Boston's Back Bay in the thirty years from the late 1850s through most of the 1880s, they supported the new brick townhouses with the trunks of spruce trees. The trunks were driven upside down into the firm marine clay of the tidal flats that underlay the sand and gravel brought by rounds of railroad trains from the hills of Needham. The builders thought this method would preserve the trunks from decay by keeping them submerged just below sea level in the dense soil.[23]

Decades of railroad, subway, and sewer construction turned the soil beneath the Back Bay into a system of unpredictable complexity. The new construction dammed and channeled groundwater, exposing some pilings to oxygen and thus to fungi and bacteria that began to rot the wood. By the mid-1980s, owners of townhouses on one street faced bills as high as $150,000 to $200,000 to remove the tops of the fragile old piles and reinforce the foundation with concrete and steel.

None of the engineers concerned appear to have violated the good practice of their time. No single construction project was responsible in itself. The problem did not appear to be any single contractor's greed or shoddy work. As in communications systems and microcomputer software, the interaction of acceptable components could produce an unacceptable result.

All this is no argument against advanced technology. To the contrary, only closed-circuit television in the 1980s revealed the submerged fragility of the system. For the first time, it was possible to discover and correct (at late-twentieth-century prices, admittedly) nineteenth- and early-twentieth-century mistakes.

System Effects

The best framework for understanding the emerging systems of the late nineteenth century comes from the diagnosis of the sociologist Charles Perrow. Perrow has argued that certain technologies are so inherently unsafe that what is called "operator error" is actually made inevitable by the way in which parts of a system are related.

Perrow has classified systems as tightly and loosely coupled. In human terms, even thousands of people on a crowded beach form a loosely coupled system. If a bodybuilding bully kicks sand in some weakling's face, or even if two bullies start to bully each other, the limited personal space around the bathers will usually suffice to confine the problem. There is open access at a number of points, and of course a smooth transition between sea and sand. There is risk to each swimmer in the ocean, but (apart from attempted amateur rescues) little chance that one swimmer's mishap will spread to dozens of others. Even if a storm approaches or a shark is spotted, an orderly closing of the beach by lifeguards is usually possible.[24]

Now imagine the same crowd packed in a stadium, surrounded by gates, turnstiles, wire mesh, and other control devices. Some of these are part of normal ticket-access routines; others have been added to keep disturbances from spreading. But in installing these new barriers, the management has turned the place into a much more tightly coupled system. The barriers serve to keep troublemakers off the field. Unfortunately they also make it more likely that a single problem will be tragically amplified. The fall of a single person can panic a crowd, part of which is then crushed against some obstacle. This is a tightly coupled system.

Perrow's argument is that many late-twentieth-century systems are not only tightly coupled but complex. Components have multiple links that can affect each other unexpectedly, as when an airline coffeemaker heats concealed wires and turns a routine short circuit into a forced landing and near-crash. Complexity makes it impossible for anyone to understand how the system might act; tight coupling spreads problems once they begin. To take another example from the airline industry: fatigue cracks in aging aircraft may not stop at the internal tear straps at which the panels are joined. Small cracks, each hard to detect and apparently harmless alone, may consolidate as a crack large enough to cause rapid loss of pressure.[25]

The Rise and Fall of Prometheanism

The Back Bay contractors were beginning a system without realizing it. Neither they nor the first buyers of the land they filled and houses they put up could have foreseen the interactions they were creating. (Conservative Beacon Hill residents discouraged their children from setting up households on "made land" because it was infra-dig, not because it was inferior digging.) It was the following generations that saw a level of technological transformation unequaled ever since. Electrification of industry helped create what the historian and critic Lewis Mumford was to call a "neotechnic" era of power grids in place of steam engines and new alloys and materials alongside steel and other conventional ones. Mumford urged a new political and social order to decentralize work from grimy urban factories to smaller, electrically powered workshops dotting the countryside. The era's most celebrated social critics faulted not technology but entrenched finance and management; Thorstein Veblen urged a national industrial "network" of mechanical processes, overseen not by industrialists and bankers but by councils of engineers.[26]

The Apex of Optimism

Americans from 1880 to 1929 were probably more optimistic about the electrical, mechanical, and chemical transformation of society than any other people has ever been. Neither the sinking of the *Titanic* in 1912 nor the devastation of the First World War could destroy their confidence. Just as Veblen advocated rule by "soviets" of technical experts, Lenin and Stalin extolled American scientific management, industrial complexes, and electric grids. And there was reason for this prestige. Even the pioneer of artificial intelligence John McCarthy, a firm believer in the transforming power of the computer, pointed out in 1983 that television and computers had until then prompted only modest changes in people's lives compared with the lighting, transportation, and communications revolutions of the 1890–1920 era.[27]

Even the era's satires of technology were affectionate. In the 1880s the French comic illustrator Albert Robida produced what have turned out to be stunningly accurate visions of twentieth-century technology as nightmares and absurdities, complete with chemical warfare, flat-screen television, and test-tube babies. Contrast the vision of Rube Goldberg, the constructor of bizarre and delightful thought experiments in intentionally needless linkages. Goldberg's work is a tribute to the pure joy of system construction—to what the historian Daniel J. Boorstin has called "complicated ways of simplifying everyday life." A Rube Goldberg contraption, ridiculous as it is, is also reassuring. Not only is the purpose nearly always benign; the system is, in Perrow's terms, tightly coupled but unidirectional. The chickens, cats, or whatever wait patiently for their cues. And the consequences of a Rube Goldberg technology are definitely intended.[28]

Misgivings

Official America held fast to technological optimism throughout the Depression and the Second World War. The Tennessee Valley Authority, the Hoover Dam, the streamlined defense plants, the large-scale production of penicillin—each seemed to show that despite the troubles of the economy, rational planning could conquer almost any task. Even critics of technology like the early Lewis Mumford believed that properly implemented, it could promote a more humane life. Science and technology appeared to be benign alternatives to the greed and irrationality that were thought to have brought about the Depression. While atomic weapons turned Mumford and others against this vision, even these terrible instruments had done what they were intended to do. They had (apparently) saved countless thousands of lives by compelling Japanese surrender.[29]

Still, the complexity of wartime systems was already bringing home to troops and civilians how many things could malfunction. A writer in the London *Observer* confided in 1942 that the behavior of aircraft "couldn't always be explained by . . . laws of aerodynamics. And so, lacking a Devil, the young fliers . . . invented a whole hierarchy of devils. They called them Gremlins. . . ."[30]

Strangely enough, it wasn't aviators but engineers and aircraft manufacturers who did most to spread the notion of the recalcitrant machine. Captain Edward Murphy, Jr., of Edwards Air Force Base, an engineer, believed in technological improvement. Murphy's boss, Major John Paul Stapp, a biophysicist and medical doctor, was his own crash dummy for harrowing tests of high-deceleration stress. He had just exceeded his old record of thirty-one times the force of gravity on the rocket sled, but nobody could say by how much—the gauges hadn't worked. Murphy found that a technician had installed each of them backward. He drew this lesson: "If there's more than one way to do a job and one of those ways will end in disaster, then somebody will do it that way."[31]

At a later press conference Stapp referred to "Murphy's Law," which he expressed in the classically succinct form "If anything can go wrong, it will." Soon aircraft companies began to advertise their products as exempt from Murphy's Law, and the term passed into technological folklore. Murphy originally was calling only for tightened vigilance—and implicitly for redesigned sensors that could be attached only correctly. (In the consumer world, this form of precautionary design had long been called "foolproofing"; the *Oxford English Dictionary* records "fool-proof" in a 1902 book on the automobile, and the word seems to have spread along with the new consumer technologies of the 1920s.)

Stapp went on to test human endurance on himself for five more years. In his last rocket-sled test in December 1954 he decelerated from 632 miles an hour to zero in 1.4 seconds. Magazines called him the world's bravest man. Stapp then began a successful campaign for mandated automotive seat belts. Volvo began to put belts in all its cars only three years later, and the Stapp Car Crash Conference remains a major annual event.[32]

Murphy and Stapp had proved their point. Murphy's Law is not a fatalistic, defeatist principle. It's a call for alertness and adaptation. But Murphy's and Stapp's work made another point, far from the optimism of early motoring. They were showing the power of innovation to master acute, sudden, catastrophic problems—including those that other new technologies had created.

Learning from Disaster

The rocket sleds and safety belts of John Paul Stapp represented two sides of the same technological coin: a tendency to multiply and amplify hazards, and an ability to reduce and control them. The two aren't really contradictory. It is both a sad and a happy fact of engineering history that disasters have been powerful instruments of change. Designers learn from failure. Industrial society did not invent grand works of engineering, and it was not the first to know design failure. What it did do was develop powerful techniques for learning from the experience of past disasters. It is extremely rare today for an apartment house in North America, Europe, or Japan to fall down. Ancient Rome had large apartment buildings, too, but while its public baths, bridges, and aqueducts have lasted for two thousand years, its big residential blocks collapsed with appalling regularity. Not one is left in modern Rome, even as a ruin.[33]

Not every technological catastrophe is, strictly speaking, a revenge effect. The *Exxon Valdez* oil spill, the release of radioactive material at Three Mile Island, and the *Challenger* explosion, to name only three of the most celebrated recent disasters of advanced technology, are system-related "normal accidents" in Perrow's sense, but only one of them may be regarded as an indirect result of trying to make things safer. The meltdown of one of the Chernobyl reactors, because it occurred during an override of safety systems to test an improved emergency procedure, was in part a revenge effect.

When a safety system encourages enough additional risk-taking that it helps cause accidents, that is a revenge effect. While the *Titanic*'s owners never actually claimed their ship was unsinkable, the crew's and passengers' overconfidence in her advanced construction proved fateful. The Iroquois Theatre in Chicago was deemed so fireproof that it opened before its sprinkler system was ready to operate. It had no firefighting equipment. When it burned during a performance only a few months after its first night in 1903, over six hundred people lost their lives in what remains the largest American disaster of its kind. (To this day, authorities in England and Australia have criticized smoke alarms for making people less vigilant in preventing fires.)[34]

Maintenance Compulsion

The importance of past tragedy (whether of natural or human origin) for safety suggests a positive corollary of Murphy's Law. It is that sometimes things can go right only by first going very wrong. The *Titanic* sinking soon led to the founding of the International Ice Patrol, to legally mandated iceberg reporting, and ultimately to aerial and satellite surveillance, advanced radar systems, and iceberg-mounted radio transmitters. The London smog that killed over four thousand people in December 1952 built public opinion for the Clean Air Act of 1956 and hastened the end of coal fires and the spread of electric heating. Neither remedy was necessarily permanent or universal; icebergs now threaten environmentally catastrophic collisions with drilling platforms, and photochemical smog menaces health in cities around the world. But in both cases, a modern disaster brought impressive gains against a long-established problem.[35]

A visible catastrophe has a positive value. Avoiding one is a powerful incentive to do things right. The economist Albert O. Hirschman recognized this feature of technology in the phrase "maintenance compulsion." He points out that Venezuela, with a poor road system, had air routes with a good safety record. Airline crashes are discrete, well-publicized, and indisputable events. Defective maintenance and operation show up quickly and tragically. Individual automobile accidents rarely approach the scale of air catastrophes. One popular but systematic compilation of world disasters lists twenty-four major air accidents since 1908 but only five automobile crashes. One of these was a professional racing accident and another related to a bridge collapse. Of the two automobile pile-ups that made the list, one had three fatalities among fifty-three cars involved, and the other had twelve among eighty-three.[36]

Because each air accident is so serious, vigilance at each point in the system—not only maintenance, but design, training, and control—is intense. (Of course, there are other reasons. We are willing to take more risks when we feel in control, as when driving, than when giving responsibility to professionals. And powerful and influential people fly a lot, usually on public carriers, reinforcing political interest in air safety.) Remarkably, the growing complexity of aircraft and dependence on automatic systems have improved rather than compromised safety. Aircraft and spacecraft design recognized and worked around revenge effects. Designers built in more than one way of doing things. The system could go on even if a part failed. The record of aviation safety shows the power of potential disaster to catalyze change. From 1970–78 to 1986–88, passenger fatalities per million enplanements in the United States declined from 0.42 to 0.18; serious injuries to passengers plunged from 0.25 to 0.07—all this after the deregulation of the 1980s. Aviation safety does not show that our fear of catastrophe is unfounded. To the contrary, it underscores how important fear has been for improvement.[37]

Long-term, intractable, progressive, degenerative problems—those we call chronic—have always existed alongside sudden, intense, episodic ones—those we call acute or in extreme cases catastrophic. And each kind of problem can provoke the other. A small shock can cause the collapse of structures weakened by slow corrosion. A sudden temperature inversion producing the London smog strikes especially at people with chronic respiratory diseases. Slow climate change can increase the likelihood of devastating floods, and possibly of tropical storms as well. Chronic conditions may have acute episodes. Acute shocks may have chronic consequences.

Until the later twentieth century, acute problems dominated consciousness. New measuring and imaging technologies of science made it far easier to localize problems and to treat them with apparent precision. For physicians, for scientists and engineers, the instruments of localizing—the stethoscope, the microscope, the X-ray machine—represented authority and trust. These professionals recognized longer-term, slower-acting problems that did not respond to specific treatments. But they, and most of the public, understandably concentrated their attention on what they were able to do rather than on what they were not.

The instruments and concepts of the late twentieth century have let us shift our attention to the cumulative impact of the sometimes imperceptibly gradual. Measuring and imaging technologies let us detect earlier stages of long-duration problems—mechanical, chemical, or biological. We can measure substances in concentrations once too minute to detect. We can capture patterns once inaccessible or hopelessly blurred. We can not only measure present conditions but extrapolate future ones. All these abilities raise gradual processes to a new level of concern. They make them as real, and as catastrophic in their implications, as high-impact events.

The ability to project the catastrophic impact of the chronic came slowly. The debate on atmospheric testing of nuclear weapons in the 1950s was probably the watershed. The testing of the early hydrogen bomb (as opposed to the largely localized effects of the atomic weapons dropped on Hiroshima and Nagasaki in 1945) may have been the first technology ever to have an immediate and measurable global environmental impact. The turn of opinion against testing showed that the certainty of steady, invisible, and not immediately hazardous processes—the accumulation of strontium 90 in human bones, plus cumulative genetic damage—could be as frightening as the much smaller chance of a direct nuclear confrontation.

The debate on fallout also showed the limits of concern over chronic, cumulative problems. Fallout was frightening in a way that medical X-rays and smoking were not as yet, not only because it was an entirely involuntary risk that especially affected children, but because it was linked with the ultimate catastrophe, nuclear war. It was a foretaste of the unthinkable.

Contrast the slow acknowledgment of global warming. The theory behind it is nearly a hundred years old, first described by the Swedish geochemist Svante Arrhenius as early as 1896. It took the satellite and computer techniques of the last thirty years finally to confirm his analysis. Still, as the historian of science Spencer Weart has pointed out, while there were grounds for deep concern as early as 1960, the greenhouse effect did not become a major scientific and lay issue until the late 1980s. The risk from thermo-nuclear weapons had an almost built-in maintenance compulsion. The deferred consequences of climate change did not.[38]

As with the change from nuclear winter to solar summer, we usually (though not always) come late to the chronic. And many revenge effects amount to a conversion of sudden impacts that we perceive immediately into long-term problems far more difficult to remedy.

As the twentieth century ends, the very devices that helped diagnose, treat, and prevent acute and catastrophic conditions become causes and portents of chronic ones. X-rays have not been abandoned; they are even used against a new form of acute threat, skyjackings. Dosages are lower and more controlled, and there is little doubt that, carefully used, X-rays on balance save lives. But they still represent some long-term danger as well as more immediate safety; one cumulative side effect of medical X-rays is thought to be a small but significant number of new cases of cancer each year. Asbestos promised protection from fire and collision. It retained heat in nineteenth-century railroad boilers. It stopped and still stops railroad trains when used in brake shoes. Theater owners proudly advertised its presence on the curtains that shielded audiences from that archetypical nineteenth-century tragedy, the backstage fire. Yet by the 1980s, asbestos had turned out to be a cause of slow death from a form of cancer called mesothelioma. This mineral had become so symbolically frightening that it was and still is being removed at immense expense from buildings where it would often be harmless if properly immobilized. And eliminating it for fear of chronic risks also brings back acute ones. Since federal rules forced asbestos out of truck brake-drum linings in the 1980s, thousands of drums have shattered every year after developing cracks. Within two months on roads in the Washington, D.C., area alone, a twenty-seven-pound chunk of brake drum moving at an estimated one hundred miles per hour killed one passenger, and a smaller fragment struck a two-year-old through a windshield. The problem was probably negligent maintenance rather than the absence of asbestos, but that is part of the point: working without asbestos demands more vigilance.[39]

Fire protection may sometimes expose us to unexpected cancer risks. Solvent tanks in the semiconductor industry of Silicon Valley, buried years ago in compliance with local legislation to reduce fire risks, now are thought to be leaking carcinogens into the local water supply. The PCB insulation in electrical equipment, also mandated by codes to replace dangerously flammable mineral oil, appears to be carcinogenic. Chlorofluorocarbons (CFCs), which helped make refrigerators universal household appliances by replacing potentially explosive chemicals, have tended to deplete the ozone layer precisely because they are so stable in the lower atmosphere. They are stable enough to reach the stratosphere, where they break down to form chlorine, which reacts with, and thus destroys, ozone. The halon gas used in another safety technology, fire extinguishers, may cause up to a sixth of long-term loss of upper-atmosphere ozone. Halon is three to ten times as harmful to the ozone layer as an equal volume of CFCs. Preventing fires and explosions on earth thus produces an increase of ultraviolet radiation that in turn raises the risk of skin cancer—another exchange of catastrophic for chronic hazards.[40]

Most of the residential smoke detectors that have saved so many people from the acute danger of fire emit small amounts of ionizing radiation, also a cancer risk (if a smaller risk than that of not having a detector). United States legislators attempted to protect children from burns by requiring fireproofing of pajamas, only to learn that the chemical of choice, TRIS, was carcinogenic. Conversely, mirex, the carcinogenic pesticide applied in the ill-fated fire ant campaigns of the 1960s (see Chapter 7 below), is still in use as a flame retardant. (Even keeping floods at bay has many of the same perils as combating fires; waterlogged sandbags from the Mississippi floods of 1993 had such concentrations of pesticides, industrial waste, and sewage residues that authorities warned against direct skin contact.)[41]

Since our knowledge of the relationship between dose and response in the origin of cancer and other chronic illness is so imperfect, these hazards make us more nervous than our ancestors were, even though we are far safer. The catastrophic risks of the nineteenth century at least had visible outcomes. A train completed its run unless it derailed. A steamship arrived safely unless a collision, a storm, or a boiler explosion sank it. The hazards of steady, long-term, repeated exposure are usually statistical: more sickness or death than would have happened in the absence of the risk factor. A radiological researcher, Eric J. Hall, told the *Washington Post* that out of 100,000 survivors of the Hiroshima and Nagasaki atomic explosions that he and colleagues had studied, "20,000 are going to die of cancer anyway. We are looking at the difference between 20,000 [who would normally get cancer] and 20,400. It is not a big effect, and it is hard to see." The bombs caused, beyond doubt, hundreds of additional deaths in the long run, but it is impossible to say which of the survivors who later died of cancer would otherwise have died of other causes.[42]

Technology's Revenge Revisited

Looking back over the last two hundred years, we can see a pattern. The nineteenth and early twentieth centuries were an age of crisis, a time when people were awed by technological scale and intensity, when people would come at great expense to world's fairs to ogle steam engines, and when artists painted new furnaces and forges in romantically outsized dimensions. Even Krupp's huge cannon attracted admirers from the very lands at which they were soon to be aimed. The combination of scale and the complexity of technological systems guaranteed that catastrophes happened far more often than they had in previous centuries. High-speed printing presses helped fix them in public consciousness. But these same catastrophes also were catalysts for technological and legal changes that have reduced their impact on human life while increasing their material cost. As early as the 1850s, the carnage of the Crimean War stimulated innovations in nursing and humanitarian relief that brought long-term benefits for civilian health as well as military medicine.

But something else was happening as disasters were coming under control in the West. The very means of preventing them sometimes created the risk of even larger ones in the future. And, even more significant, the gradual, long-term, dispersed problem proved far less tractable than the sudden, shocking one. As we shall see, the steady seepage of petroleum products from small industrial, residential, and service-station tanks became a more serious problem than any of the great oil spills.

Catastrophes still happen. And safety technology is useless if consumer behavior, building codes, and inspections don't ensure that it is applied. But more and more of the risks that disturb us most are not towering infernos or shattered aircraft, still less the evil robot of Fritz Lang's *Metropolis* or the vengeful appliances of Rod Serling's Bartlett Finchley. The old disasters were spectacular; like the final hiss as the *Titanic*'s boilers were extinguished and the upended ship was swallowed by the sea, they made for ghastly theater. The new ones are diffuse, silent processes that continue almost invisibly and usually too late. (Even the Chernobyl meltdown left only limited outward signs of damage.) Five years, a decade, twenty years may pass between cause and manifestation. And the cause is usually not a single event but the cumulative effect of many small doses.

Classic disasters were deterministic. Cause and effect were linked. An exploding boiler killed those it killed, and spared those it spared. Late-twentieth-century disasters are expressed as deviations from a baseline of "normal" background tragedy. The truth is not in immediate view. It emerges from the statistical inference of trained professionals; to see it, laypeople must learn at least the basics of their language. The old disasters were localized and sudden. New ones may be global and gradual, from radioactive isotopes in milk in the 1950s to climate change in the 1990s.

Our control of the acute has indirectly promoted chronic problems. Medical researchers have recognized this trend for years and have been shifting their efforts to chronic diseases—though so far not with the same results they have had with injury, infection, and acute illness. Our ability to transport animals and plants among continents, deliberately and accidentally, has on balance been decreasing rather than promoting species diversity. But the invaders have also failed to be as catastrophic to trees and crops as some had feared. Like many chronic illnesses, they have become manageable nuisances, neither conquerable nor fatal, but demanding time-consuming vigilance. Our efforts to modify our environment have also produced chronic problems: the comforts of home have helped produce the annoyances of allergies, suppressing forest fires has helped make them a greater threat, and protecting the shoreline is helping to erode it. In the office, not viruses or system failures but repeated low-grade problems are the thieves of productivity. On the road, mass motoring tends to make driving safer but also slower. And technology not only tends to turn leisure hours into work; it also promotes new and usually chronic categories of injury as it deals with old ones.

Just as we have spread the cost of savings and loan failures laterally across the taxpaying public and forward to new generations, we have resolved problems by broadening their base in space and time. But this is a hopeful as well as a frustrating sign. It is auspicious because it seems to take a sense of urgency to force new thinking. As the conclusion of this book will suggest, it is part of the nature of radically new ideas that they are not the kind of ideas we thought they would be.

2

Medicine: Conquest of the Catastrophic

An account of technology's frustrations can start anywhere, but sooner or later it leads to medicine. People in the United States and other industrial countries have never been healthier—or more anxious about illness. In the 1990s people describe themselves as less healthy than they did in the 1970s, although most medical indicators have been pointing up, not down, with medicine deserving much of the credit. True, concern and even some fear are justified. AIDS and resurging tuberculosis are all too real. Still, medicine is more effective and usually less invasive or painful than it was even a generation ago, but neither refinements of medical technology nor more cautious living habits have brought peace of mind.

Escalating costs are but one reason for unhappiness. Ivan Illich's memorable 1976 polemic, *Medical Nemesis*, found evidence even in mainstream professional journals that "medical bureaucracy creates ill-health by increasing stress, by multiplying disabling dependence, by generating new painful needs, by lowering the levels of tolerance for discomfort or pain, by reducing the leeway that people are wont to concede to an individual when he suffers, and by abolishing even the right to self-care." Few other critics of medicine have gone so far, yet most have wondered at the paradox, so succinctly put by the political scientist Aaron Wildavsky, that we are "doing better but feeling worse." Writers within the medical profession, too, have commented on this irony. The psychiatrist Arthur J. Barsky has pointed out that while medicine is now able to do much more for people—according to one survey in an internal medicine journal, the proportion of treatable major illnesses has increased from under 10 percent to over 50 percent since the turn of the century—medical treatment has also focused our attention on symptoms and dangers. People get better medical care than ever. They know more about diet and exercise. They smoke and drink less. By any objective measure, the American middle class is healthier. Yet it is also more concerned about being sick.[1]

Is anxiety about health only a mental revenge effect? Have the public and private insurance plans of the industrial world eroded responsibility and rewarded sickness? Have men and women grown neurotic and unappreciative of their medical tutelage, reading fatal portents in every twitch or sniffle, seduced by false hopes of eternal youth and painless ease? Some people do meet this description, having fallen into a health obsession of their own making. Others have learned from the medical system to focus on their symptoms, amplifying them. Behind this worry and discomfort, though, there is a reason—a revenge effect that has to do with one of the most complex systems we know, our own bodies. Medical knowledge has a profound strength but a corresponding weakness. Supported by communication and transportation, it has a superb record of coping with disasters: treating traumatic injury, rehabilitating the bodies of survivors, controlling potentially deadly epidemics of infectious disease. Medical knowledge has also helped engineers save literally millions of lives by improving technologies from water and sewer systems to automobiles.

These very accomplishments have had major unintended effects. Thanks to advanced technology, many procedures, while speedier and less invasive, are not always easier to perform. The demands on surgeons' craftsmanship as well as knowledge may be greater. Technological systems also multiply the opportunities for miscalculation and for infection. Above all, as we shall see in the next chapter, the improvement of overall health has made chronic illness more important. Sometimes difficult-to-treat conditions have been the price of survival. Sometimes longer life has meant a sicker life.

Before we look at the revenge effects of medicine, it is important to see why medical technology has become as meaningful—and as costly—as it is.

Health Without Medicine?

While many may think of the industrial age as a medical disaster, as a dark fountain of effluents and pollutants, as a purgatory of consumptive operatives in satanic mills, it has also had remarkably positive, unintended consequences, reverse revenge effects. Medical practice has been less important and economic growth more important in increasing the human life span than most people realize. True, the great population centers have looked unhealthy and have been unhealthy in many ways. And as the economist and philosopher Amartya Sen has recently pointed out, developing countries today can achieve a remarkably high level of public health even before building an industrial base. Still, the improvement of public health over the last century and a half is as much a consequence of increasing income as it is of scientific medicine. We still don't understand which of the benefits of economic growth really matter to health, and how. But we do know that lower rates of death and serious illness have accompanied growth. We are aware, conversely, that illness and mortality grow again when living standards decline.[2]

The historical epidemiologist and physician Thomas McKeown assembled powerful evidence that deaths from infectious disease began to decline as early as the eighteenth century. He discovered that airborne disease was in retreat decades before effective vaccines and treatments were available.

Tuberculosis in England decreased steadily beginning in the 1830s. Eighty-six percent of the drop in the tuberculosis death rate took place before streptomycin was introduced in 1947. Sixty-eight percent of the reduction in bronchitis, pneumonia, and influenza deaths, 90 percent of the decline of whooping cough, and 70 percent of the decline of scarlet fever and diphtheria occurred before the sulfa drugs appeared in the 1930s. Death rates from water- and food-borne diseases (cholera, diarrhea, dysentery, nonrespiratory tuberculosis, typhoid, and typhus) also lessened decades before they could be treated, mainly thanks to improved water supplies. There were exceptions. Vaccination has brought the smallpox virus to the brink of extinction. More than nine-tenths of the decline in mortality from ear, nose, and throat infections followed the introduction of antibiotics. Still, when McKeown considered the impact of specific measures on all airborne diseases, he found that only 25 percent of the fever deaths occurred after vaccines and therapies were introduced.[3]

McKeown drew his conclusions mainly from English and Welsh statistics, but American data seem to confirm them. John B. McKinlay, a sociologist, and Sonja M. McKinlay, a mathematician, have studied the decline in mortality from infectious diseases and other causes in the United States since 1900. They, too, found that in most cases effective therapies and vaccines appeared only after most of the easing of the death rate for a given disease had been achieved. In fact, when the proportion of the gross national product devoted to medical care began its sharp ascent in the late 1950s, "*nearly all (92 percent) of the modern decline in mortality this century had already occurred*" (emphasis in original). Critics of McKeown have pointed out the limits of such statistical arguments, which ignore how medical knowledge and medical initiatives worked in the public arena—how physicians supported better urban sanitation, for example, and how special hospitals for tuberculosis patients by the late nineteenth century slowed the spread of the disease. The statistics also neglect a physician's ability to comfort and to relieve pain, and the undisputed power of the placebo effect. Furthermore, they slight the more recent gains that clearly are due to antibiotics and other medication, and to diagnostic technology. Even so, the role of medical therapeutics in long-term health remains uncertain.[4]

If medicine has done less in combating infectious disease than is usually assumed, what *has* been responsible for declining mortality? McKeown discounts alterations in either microorganisms or people: Too many diseases or people would have had to change at virtually the same time. In any case, by the eighteenth century, mortality was too low to support the idea that there could have been any natural selection for superior immunity to infection. Nor do better living conditions in themselves explain lower mortality. (Declining living standards in early-nineteenth-century England did not bring a corresponding large increase in mortality.) Certainly clean water and effective sewage systems curbed cholera and other water- and food-borne diseases. But many improvements in food handling, especially widespread pasteurization of milk, were not available to most people until the early twentieth century, and by then gains were already substantial. McKeown insists that nutrition is the only remaining explanation for most of the lower death rates for infectious diseases, although he acknowledges a dearth of positive evidence for his argument. Historical information on eating habits is sketchy, and statistical data are too meager to suggest how changes in diet could have improved health as McKeown believes they did. Eating habits depend on culture as much as on income and availability of food; skeletons from the wealthy ancient Greek colony of Metaponto suggest widespread malnutrition and disease. In Victorian England, even amply fed and well-off parents did not realize how much food growing children need; pupils went hungry at the best public schools.[5]

Another physician-historian, Leonard A. Sagan, has challenged the nutrition hypothesis, with logic as rigorous as McKeown's. Absent actual famine, wartime rationing has generally improved health in the twentieth century, putting nations on a forced high-carbohydrate, lower-calorie diet. Grains have replaced saturated fats during food shortages. (The most serious food demonstrations in Berlin during the First World War were not about bread but the disappearance of butter.) Sending food to contemporary developing countries does not always improve death rates, except where mothers and young children have been seriously undernourished (and even for these groups only modestly).[6]

In the twentieth century, good nutrition does not appear to have enhanced survival rates in epidemics, including the influenza pandemic of 1918. Sagan argues that something about nineteenth- and twentieth-century society has made people more resilient and self-confident. Many disease microorganisms are present in individuals who never develop symptoms; differences in immune response must be the cause. Those who feel in control of their future apparently have stronger immune systems. Literacy and education are better indicators of state of health than is income. In fact, the economist Donald S. Kenkel has discovered that college graduates (regardless of how much they know about health) are significantly less likely to smoke and slightly more likely to exercise than people (regardless of education) who know most about health and behavior. In other words, schooling in general is somehow incrementally more effective in promoting healthful living than specific knowledge about health, and not simply because better-educated people are better paid.[7]

Conversely, poverty seems to engender a self-destructive fatalism. In Glasgow, Scotland, where in late 1992 one in five residents was unemployed and three out of four received some public assistance, 83 percent of middle-aged men were regular smokers. Worse yet, one so-called greengrocer's vegetable offerings were limited to "marrowfat processed peas." It is no wonder that cases of scurvy and rickets are still reported, while doctors speculate that depressed immune systems may be responsible for lung cancer rates among smokers in Scotland that are twice those of American smokers. At home, cooking skills often stop at "dumping chips in the Masterfry."[8]

We are left with an uncomfortable conclusion about health, and especially about longevity. For all the contributions of medical technology, other things have meant more. The real mechanisms of improved health are so entangled in other good things—the economy, education, environmental quality—that we still don't understand just what has happened. We know, as Aaron Wildavsky put it so well, that richer is better. But because we don't understand how and why, it is difficult rationally to say exactly what sorts of public and private applications of money will do most for health. The statistics don't necessarily imply that funds should be shifted from medicine to education, though they may suggest that raiding education programs to pay for health care may have unexpected revenge effects on health itself. Good health turns out to be a positive by-product of the pursuit of other things. This is a point to which we will return later in examining the implications of revenge effects for the future of medicine.

The Power of Localizing

If the contribution of medicine to longevity and health is harder to measure than we once supposed, it is still real enough. The reason for the profession's success in the last 150 years is no mystery. It was a set of new physical and mental tools that let physicians intervene with an exactness that had once been rare. The view of the organism that prevailed before the nineteenth century was in many ways an attractive intellectual system. Traditional medicine was like some of today's alternative therapies in viewing the patient as a whole person, but in practice it was not necessarily gentle or humane and had powerful revenge effects of its own. Well into the eighteenth century, most physicians thought of health as a proper balance of the body's humors (black bile, yellow bile, blood, and phlegm), along with other substances. The often frightful cures visited on the patients of eminent physicians were efforts to reestablish humoral balance. Most people were fortunate in being unable to afford treatment, but the elites who could suffered counterproductive agonies. Doctors and patients both tied pain to gain. "Gentle purges and slight phlebotomies [bloodlettings] are not my favourites," declared Samuel Johnson, "they are Popgun batteries, which lose time and effect nothing." Dr. Benjamin Rush, a signer of the American Declaration of Independence, believed with equal firmness in massive bloodletting and purging—for himself and for his patients. He helped promote calomel, a toxic purgative of mercurous chloride. Physicians treated George Washington himself for "quinsy" (severe streptococcal disease) with large quantities of mercury in his last days. Even medical theorists who tried to localize diseases in particular organs could not escape the tendency to build systems. And bloodletting remained central to the idea of balance in these. As late as 1833, France imported 42 million leeches in a single year for this purpose.[9] (Physicians have begun to apply therapeutic leeches again—but on a far smaller scale and with a different theoretical basis.)

Treating the whole person in this way sometimes had horrific immediate results, but they were not side effects, even when the patient later appeared to improve because of such treatment. The modern idea of the side effect depends on a concept that barely existed in the early nineteenth century: a focused attack on the site of a problem. This idea, now so commonplace that few physicians or laypeople think twice about it, was once heretical. As early as the seventeenth century the English clinician Thomas Sydenham had awakened interest in diseases as distinctive entities, but the eighteenth century lacked the science and technology to draw meaningful distinctions. The discomfort and even danger from bleeding, purging, and mercury compounds were no mere painful inconveniences but part of the healing process, proof that the cure was working. Many people administered these cures to themselves and their families. And whatever the results were called, nineteenth-century remedies could be as harrowing as aggressive twentieth-century chemotherapy. The virtuoso Niccolò Paganini, treated with mercury for suspected syphilis, lost his teeth to the spread of an infection. He suffered severe coughing spells, failing eyesight, and a collapse of his confidence and ambition—all due to the mercury treatment.[10]

Instruments of Localization

Nineteenth- and twentieth-century doctors gained their power, prestige, and wealth by persuading patients to accept more and more localized diagnoses and treatments. They developed new techniques to isolate and attack diseases, mainly acute ones. Movements that resisted the trend, from Christian Science to chiropractic, promoted their own visions of healing the whole person, but they remained marginal. Localization did not always produce better results than the older medical alternatives it replaced—at least at first. The stethoscope, first used in an elementary form by the French physician René Laennec in 1816, required special training to yield more meaningful information than old-fashioned auscultation. But the idea of the stethoscope was as important as the results it yielded. It pinpointed conditions that were once considered more diffusely. The inventor of a binaural stethoscope wrote in 1851: "As certain nebulae are resolved into stars by powerful telescopes, so specific chest sounds, obscure from their lowness, may be determined. . . ."[11]

In the same year the German physician and physicist Hermann von Helmholtz introduced that other canonical medical instrument the ophthalmoscope. By letting a trained observer see the detailed structure of the retina, it also permitted an earlier and more accurate detection of illness. And a hundred years ago the discovery of X-radiation by the German physicist Wilhelm Roentgen appeared to do for the entire body what the ophthalmoscope had done for the eye. In 1918, the American physician James B. Herrick showed that an electrocardiogram, especially in the presence of certain symptoms, could identify the existence and location of a coronary artery occlusion—a condition more common than doctors had realized. What the stethoscope, ophthalmoscope, and thermometer did for examination, the microscope and related devices like the microtome did for laboratory analysis. Compound microscopes had existed since the seventeenth century, but distortion limited their utility. Only around 1830 was there a breakthrough leading to a new way of correcting aberrations in microscope lenses. It soon led to instruments of unprecedented power, "to resolve," as one reviewer put it, "the solid products of disease into the simplicity of their elementary components."[12]

The technological transformation of medicine was due both to the deployment of new devices and to their cultural consequences. The new instruments revolutionized the ways doctors saw, heard, and thought—and in turn changed the attitude of patients toward their physicians as well as their own bodies. Listening to chest sounds and interpreting X-rays demand practice in finding or creating patterns where naive observers would find ambiguity. Patients wanted—and still want—this ability to specify and localize. Some doctors abuse technological enthusiasm for their own profit, but it is demand that lets them do it. Even before 1830, patients suspected doctors who had not adopted the stethoscope. X-rays were an international sensation of popular culture in the late 1890s. People with foreign objects lodged in their bodies, even those who suffered no pain from them, now asked to have them extracted—a procedure that sometimes had fatal complications. Localization, too, could and can have revenge effects.[13]

Localizing Surgery

Once acute illness had been identified specifically, it could be targeted. In the 1840s the American dentists Horace Green and William Morton demonstrated the surgical possibilities of nitrous oxide and ether, two gases previously best known as recreational drugs. As the number of general and local anesthetics grew, opportunities for surgery multiplied. Antisepsis and asepsis reduced its risks sharply. But critics of medicine point out a revenge effect of anesthesia. By making the immediate surgical intervention less painful, it has encouraged surgery and (if such things are quantifiable) possibly increased the sum of medically induced pain, especially postoperatively. This argument takes no account of the pain, not to mention death and debility, that naturally resulted and would still result from *not* having surgery. Much surgery has been what Lewis Thomas has called "halfway technology," prolonging life and relieving discomfort at great cost without treating the cause of pathology itself.[14]

On the other hand, localization makes many kinds of surgery far less invasive; indeed, imaging often renders surgery unnecessary. In the mid-nineteenth century, one hallmark of surgical skill was the speed of an amputation; in the late twentieth, it is the fineness of an incision. When the editor of the scientific journal *Nature*, John Maddox, had a cartilage removed from his knee in the early 1960s, it was open surgery. Recovery took several months. His surgeon performed a repeat operation in 1991 with arthroscopy, in which a light source, a miniature camera, and a video recorder permit the whole procedure to be completed through three small holes. Maddox spent a single day in the hospital, then took a taxi to work.[15]

Localizing Pharmacy

Like diagnostic instruments, drugs targeted for particular ailments are a late-nineteenth-century innovation following the germ theory of disease. In the early nineteenth century, doctors classified drugs by their effect on the whole body. Terms like "cathartic," "diuretic," and "narcotic" survive from this vocabulary. Dosage depended not only on the drug's usual effects but on the patient's constitution and even the local climate. Prescribing a single drug uniquely for a certain condition was the mark of a quack. Specific drugs were slow to arrive. At first, specificity was a matter of degree. Progressively, new chemical knowledge produced increasingly effective substances from familiar natural remedies. In 1804, Armand Séguin obtained morphine from opium, and by the 1840s doctors were able to inject it with hypodermic needles. In 1822, quinine was derived from the cruder chinchona bark, a giant step in the treatment of malaria. These new drugs were more focused versions of the old. It was chemical synthesis in the late nineteenth century that made specific medicine possible.[16]

It was only a hundred years ago, in 1890, that Paul Ehrlich advanced the side-chain theory: that a drug can act by binding to specific kinds of cells, as a key fits a lock, and neutralizing them. The next two decades saw specific treatments for diphtheria, tetanus, and syphilis. In 1910, Ehrlich was able to introduce Salvarsan, a drug that identified and attacked syphilis spirochetes while sparing the body's own cells—the first "magic bullet," in Ehrlich's own phrase. The sulfa drugs of the 1930s, penicillin and streptomycin in the 1940s, the polio vaccines of the 1950s, and the other celebrated drugs of the middle years of the twentieth century all excited doctors and patients alike as few other scientific discoveries. Meanwhile, governments in the United States and elsewhere were suppressing the remaining patent-medicine "tonics," discredited vestiges of the time when treating the whole body was the norm.[17]

Technology has been developing for over 150 years toward just this kind of localization. New generations of instruments have changed the nature of medical training. Although the goal of established medicine—perhaps even more in America than elsewhere—has been a measured, speedy, and precise application of a pharmaceutical or device, the ideal of precision is far from realized. Nonetheless, it is what distinguishes our medicine from the treatments of 150 years ago. Then patients expected medication to restore the balance of their humors; now they want not only a precise diagnosis of their condition but an equally precise remedy for it. Half the doctors interviewed in a survey by the American Medical Association in 1989 believed their patients were requesting unnecessary treatment. Were the patients indeed "worried sick"? Or were they instead desperate for relief from conditions for which there is no localized therapy—the chronic conditions we will see in the next chapter?[18]

Mastering the Urgent

If medicine showed how the understanding of specific mechanisms could produce effective therapies for acute illness, the development of surgery showed that the more acute and serious an emergency, the more impressive the rescue and rehabilitation that were possible. The rise of emergency treatment has demonstrated once more the power of a positive revenge effect. As new munitions, vehicles, and industries made both military and civilian injuries more severe, the techniques of emergency medicine responded. Technology showed its power to deal with the sudden and the life-threatening. The history of military medicine is worth studying in detail, because it is the best-documented long-term record of the ability of medicine to treat casualties. It also shows one of the happiest unexpected benefits of technology: how military carnage eventually came to the aid of civilian surgery.

At least until the time of the American Civil War, military surgery had been a craft, sometimes skillfully exercised, but little improved over the centuries. The military historian Richard Holmes has written that "a Macedonian phalangist, treated for a sword-cut received when Alexander beat the Persians at Arbela in 331 B.C., probably had a better chance of avoiding gangrene than a British soldier whose similar wound was quickly bandaged by his regimental surgeon before Sebastopol in 1854." Well into the nineteenth century, infections rather than wounds themselves were the real hazards of battle. During the American War of Independence, American troops suffered a 2 percent battle casualty rate; meanwhile, 75 percent of those treated in hospitals did not survive. Disease, most of it unrelated to wounds, caused nine out of ten fatalities.[19]

Still, the toll on the battlefield was increasing. In the middle decades of the nineteenth century, what William H. McNeill has called the Industrial Revolution of War brought unprecedented combat deaths. Supply and troop trains, smokeless propellants, breech-loading rifles, armor-piercing artillery, and vast conscript armies began to define war as it was known through most of the twentieth century. One result was staggering numbers of wounded as well as dead troops.[20]

The new technology changed the hazards of warfare more in the generation of the 1850s and 1860s than in the previous 150 years. Although epidemics were still more dangerous than enemy action, the minié bullet and rifled barrel widely used by the French and Russians made infantry fire seven times deadlier and produced wounds far more extensive than older ammunition of comparable size. In the Crimean War the Russian army lost a higher proportion of its troops per year in battle, and the French lost a higher proportion to disease, than in any previously recorded war. The death rate per thousand per year was over 253 for the French, 161 for the British, and 119 for the Russians.[21]

Propelled by the new Springfield .58 rifle, the minié bullet proved especially devastating in the American Civil War. The new firepower was massed, just as the inaccurate and slow-reloading muskets of the seventeenth and eighteenth centuries had been. Nearly 30,000 of the over 174,000 gunshot wounds to the arms and legs of Union troops led to amputation. The U.S. Patent Office recorded dozens of new designs to meet the demand for artificial limbs. The new intensity of casualties inspired impressive gains in treatment. Surgeons learned more effective use of ligatures. The mortality rate from wound infection fell from up to 60 percent early in the war to 3 percent at its end.

Infectious disease still caused most military deaths. (One estimate is that 110,000 Union and 94,000 Confederate troops died of wounds, whereas 250,000 and 164,000 succumbed to disease.)[22]

Although military medicine and sanitation do not seem to have been any more effective than their civilian counterparts, antiseptic technique and other surgical innovations began to change that. The First and Second World Wars, despite vastly greater total fatalities, actually had fewer deaths (combat- and disease-related) per thousand troop-years than the Crimean War and U.S. Civil War. By the end of the First World War, for example, better treatment improved the death rate to 8 percent of the wounded, down from 20 percent in the Crimean War and 13.3 percent in the Civil War. Yet in other ways wounds became more severe. Higher-velocity ammunition and fragmentation weapons multiplied the chances of disfiguring injury, but this also had a reverse revenge effect. The First World War was a milestone for the emerging specialties of orthopedic surgery and plastic and reconstructive surgery for civilian as well as military patients. The Second World War revealed how effective the medical response to urgent problems had become. Under wartime pressure the production of penicillin increased in a single year from just enough for a hundred patients to the billions of units needed for Allied casualties after D-Day. In the Second World War, and even more in the Korean War and Vietnam War, increasingly rapid responses cut death rates. The Mobile Army Surgical Hospital (MASH) units of Korea saw the median wait of wounded soldiers for treatment reduced to 1.5 hours (55 percent of wounded troops were hospitalized on the same day they were wounded).[23]

American field medicine in Vietnam did even more for emergency treatment, civilian as well as combat. Only decades earlier, a soldier wounded in New Guinea needed sixteen native bearers for transportation, with the nearest military hospital in Australia. The average soldier needed medical assistance for malaria four times a year. In the Vietnam jungles, however, helicopters with medically trained crews were able to bring most seriously wounded soldiers to surgery less than two hours after the event. The flight itself took no more than half an hour. Only 2.5 percent of those arriving alive at the hospital failed to survive—and an unprecedented 87 percent of the hospitalized wounded were able to return to duty. A wounded soldier in the Second World War had a 71 percent chance of survival; in Korea, 74 percent; and in Vietnam, 81 percent.[24]

From the horse-drawn ambulances of the Napoleonic Wars to the litter jeeps of the Second World War and the medevac Huey helicopters of Vietnam, medical technology has been most dramatically effective when injury has been most severe and has needed the most prompt attention. Moreover, the results have not been of military importance alone. Warfare has always been a laboratory for extreme situations. The military has been able to mobilize resources, as in the development and commercial production of penicillin in the Second World War, at a pace that civilian market economies cannot match. But it has galvanized research and practice for certain problems: injuries and infections of young men and women carefully screened for good health. Military medicine has been care at its most acute.

The Emergence of Emergency: Civilian Accidents and Epidemics

Civilian medicine has shown the same attention to urgency, the same alternation of destruction and repair. Like military treatment, civilian practice was forced to respond to nineteenth- and early-twentieth-century technology, which has tended to expose people in unprecedented concentrations to sudden and violent forces: train (and later automobile and aircraft) crashes, theater fires, industrial explosions and accidents. By the mid-nineteenth century, a charitable appeal in Philadelphia drew attention to the rising number of industrial mishaps. Accident admissions to the city's Pennsylvania Hospital, for example, increased from 140 to 400 annually in the twenty years after 1827. To these, the twentieth century added many more—including automotive accidents.[25]

With new antiseptic procedures, X-rays and other imaging, blood typing, and antibiotics, emergency medicine has flourished in the last hundred years. Its story, unlike that of military medicine, remains to be told—partly because it was a neglected and usually low-status specialty until its rapid rise in prestige during the 1970s and 1980s. Late in the twentieth century, the ability of emergency medical crews to respond to disaster reached levels hard to imagine only decades earlier. Spinal-cord injuries that once meant irreversible paralysis now can be treated with the synthetic steroid methylprednisolone (MP) soon after an accident to reduce inflammation and cell damage from free radicals. In 1993 the former New York Jets lineman Dennis Byrd was able to walk again less than a year after breaking his neck.[26]

Another prominent example of recovery from catastrophe shows even more dramatically how well hospital-based medicine can respond to extreme conditions. In January 1993 a gunman shot five people at close range outside the Langley, Virginia, headquarters of the Central Intelligence Agency. Two died immediately, and another was treated for minor injuries. The fate of other victims was revealing. One sixty-year-old CIA employee was flown by helicopter to a local hospital, where an operating team had been alerted. A bullet, after passing through the bone and blood vessels of one arm, was lodged in his chest. He had lost so much blood that, as the head of the hospital trauma unit put it, the man "had one foot on a banana peel and the other on the edge of a precipice." In Vietnam, where the hospital's trauma service director had served, the injury would have meant amputation or even death. Treatment required almost twelve hours of surgery. The hospital, calling for donors of the patient's unusual O-negative blood type, promptly found 450 people who contributed blood for him and others. It was not only the skills of surgeons, anesthesiologists, and nurses that saved the patient's life and arm; it was speedy evacuation and communication. New weaponry multiplies emergencies and makes them more severe, but faster evacuation and better life-support systems rise to the occasion. Yes, it is tragic that these systems and skills are necessary. Many people would gladly trade some of them for a society that needed them less often. The point is, bloodshed has had an unexpected and unintended positive side.[27]

Just as warfare, transportation, and industry were increasing the number of casualties, surging world trade and population movements were accelerating another acute problem: the national and international circulation of infectious disease. They made it easier to define the situation as urgent when actually it was still relatively small. In 1875, in the midst of worldwide concern over the spread of rabies, the death rate in Great Britain was still two per million of population. But published statistics revealed that the rate had risen more than sixfold, from 0.3 per million, in only fifteen years. Expanded shipping was transferring rabid animals to and from the New World and Asia. Rabid dogs were even more frightening in growing cities than they had been in the countryside. Louis Pasteur's controversial trials of human rabies vaccine in 1885 and his successful vaccination of thousands of exposed individuals showed how well the specificity of medicine could be mobilized in a crisis. Pasteur had identified the virus responsible for rabies, as well as the sites of the disease, the brain and nervous system. He became an international hero by providing a specific response to an urgent situation.

Even in the absence of vaccines, medical knowledge and communication technologies have helped to prevent the worst consequences of infection. In the early twentieth century there was still no effective treatment for bubonic plague. There was also a great reservoir of the plague bacillus in Asia, where political and military upheavals in China helped to increase its range. The plague, as William H. McNeill has pointed out, could have devastated North America and Europe. It killed 6 million people in India alone in the decade after its arrival in Bombay in 1898. (Between 1346 and 1350 alone, 20 million out of a total of 100 million Europeans were lost in the Great Dying or Great Pestilence, as it was originally called.) Earlier, slower ocean voyages protected overseas populations by exhausting the supply of hosts: uninfected fleas, rats, and people. Late-nineteenth-century steamers, with their higher speeds and greater capacity, made the establishment of new plague reservoirs more likely. Worse yet, the United States, South America, and other regions have burrowing rodents that can form permanent reservoirs.[28]

Against great odds, international cooperation checked the disaster in most parts of the world. The plague bacillus, *Yersinia pestis*, was discovered in 1894 by Japanese and French bacteriologists working independently. As scientists studied the spread of the bacillus, they learned enough about its mechanisms to permit effective quarantine. But while such public health measures stopped the spread of the disease, they did not do so for the bacillus, which increased its range well into the twentieth century. Ranchers in the American West inadvertently gave it a free ride when they tried to eradicate prairie dog colonies by introducing sick animals. The result was a memorable revenge effect of the resurging variety: there were just as many prairie dogs, but now *infected* ones. In 1940, thirty-four species of burrowing rodents and thirty-five species of fleas in the United States had become plague carriers, thanks in part to the ranchers' efforts.[29]

In spite of the spread of the plague, the remarkable fact is how little dread was left. In 1992 alone, at least ten cases were reported in the United States. One man in Arizona died from pneumonic plague after handling an infected domesticated cat he had removed from the crawl space of a house. The measures taken show the changes to an ancient threat. Authorities dusted the house and nearby rodent burrows with an insecticide against fleas. They dusted the cats and dogs in the house. They advised the owners to continue periodic dusting. The medical professionals who had treated the man tested negative. Of these, only two nurses requested and received preventive treatment with tetracycline. The Centers for Disease Control did issue a warning to veterinarians that animals on the West Coast and in the Southwest were at risk, but the level of alarm was notably low compared to fears that risk factors for other chronic illness raised—including asbestos and toxic waste exposure.[30]

What changed the plague from the most dreaded disease of the West to merely a localized hazard? The immediate answer is that there are now fewer reservoirs of rodents interacting with concentrated human populations. A long-term explanation probably is that a less virulent form of the plague bacillus has spread over the centuries, immunizing both animal and human hosts against the deadlier form. Fortunately the immunity of much of the world at the turn of the last century did not have to be tested. Bacteriologists of the early 1890s were finally able to identify the bacterium responsible for the plague and the mechanism of transmission, and this specific knowledge made intervention effective against catastrophe. Once the problem was localized, both therapy and prevention could also be targeted.

Of course, there are limits to our protection, especially against rapidly mutating viral infections. A generation after Western nations were able to raise effective barriers to the plague, pandemic influenza struck in 1918, causing up to 40 million deaths worldwide, nearly 200,000 of which were in the United States. Today's communications and biomedical technology make it possible to formulate a new influenza vaccine every year after determining the most likely new RNA mutations that would render previous immunizations ineffective. It is true that mass immunization has risks of its own; in 1976 a thousand people developed the paralytic disease Guillain-Barré syndrome after an American program of mass immunization against a swine flu epidemic that was anticipated to have the potential of the 1918 disaster. This does not change the fact that medical technology, on balance, has been a powerful force against biological catastrophe. "Better a vaccine without an epidemic than an epidemic without a vaccine," as one vaccination advocate put it later. Just as boiler explosions, railroad crashes, automobile accidents, lethal munitions, and other nineteenth- and early-twentieth-century innovations motivated new treatments for the victims of catastrophe, the international spread of infectious disease by improved transportation helped produce more effective means of prevention and treatment.

Paradoxes of Vigilance and Craft

Critics of medicine have found in cases like swine flu a distressing increase in the rate of iatrogenic—physician-caused—disease, of side effects worse than the symptoms if not the actual diseases they are intended to treat. Some might consider these the main revenge effects of medicine, and iatrogenic diseases are in fact an important problem. But are they true revenge effects? It is worth keeping in mind that the concept of a side effect depends on the ideal of modern medicine that we have just traced: targeted, localized, measured interventions with effects that can be verified with controls. In practice, doctors have still not studied the outcomes of many medical procedures rigorously. Some accepted ones, on closer scrutiny, undoubtedly will turn out to be ineffective or even counterproductive. Some controversial treatments will be vindicated. But almost everyone agrees on the principle of weighing benefits against risks in statistically sound trials. And this is a relatively new concept in medical history.

Medical writers have long recognized malpractice. Many recognize that improper drugs can cause or worsen disease. But the localization of medicine has brought a different notion of medically caused pain and harm. When medicine was supposed to treat the whole body, disruption could signify effectiveness. Medicine was expected to be bitter and painful. It was quacks who introduced sugared pills, as the medical historians Roy and Dorothy Porter have pointed out. Patients like Samuel Johnson and Paganini expressed the logic of premodern medicine in their positive view of the suffering their treatment brought.[31]

Medical Technology and Its Discontents

The real revenge effects of medical localization and faith in objective measurement are more subtle than futile procedures. As the physician and medical historian Stanley Joel Reiser has noted, there is a hierarchy of medical evidence, with tests and imaging results at the top, the physician's direct visual and aural examination in the middle, and the patient's account of illness at the bottom. Doctors and patients alike have sought both objective and localized diagnosis, as well as specific treatment. Yet neither professionals nor laypeople have been satisfied. Doctors want the prestige and higher fees that come with advanced procedures but not the government regulation and insurance bureaucracy that expensive treatment entails. And they fear, probably rightly, that computerized norms will rob them of professional discretion and initiative. Patients sometimes demand unproven technology or therapies from a conservative or skeptical profession, but they also resent impersonality. And they believe, correctly, that in learning to interpret technologically produced data, doctors have failed to remember the skills of looking, touching, and listening. When all tests are negative, in the presence of painful symptoms, is the problem a virus or other chronic disease not yet detectable, or is the complaint psychological? An overreliance on tests can also defeat medical common sense. A Stanford premedical student wrote movingly of four weeks of horrific tests before doctors in the university hospital finally acknowledged that a ruptured retrocecal appendix was the cause of his agonizing stomach pains. A retired physician and family friend had recognized the symptoms at once, but the young doctors trusted tests above traditional judgment.[32]

It isn't hard to draw up lists of drugs, treatments, and procedures that many or most physicians consider doubtful or bad. Cases fill weighty textbooks like Robert H. Moser's *Diseases of Medical Progress*, general-interest treatments like Illich's *Medical Nemesis*, and Diana Dutton's *Worse Than the Disease*, not to mention the tabloid press. It takes a high degree of medical *and* statistical knowledge to evaluate the merits of most cases. Learning from mistakes is inevitable, although too much professional and popular skepticism about new therapies could obviously result in corresponding revenge effects. Just as many people seem to demand aggressive treatment as fear it. Probably the most fundamental revenge effects of contemporary medicine are systematic tendencies, not the dead ends and errors of therapeutics. The problem of today's medicine, and main revenge effect of new therapies, is that contrary to our expectations of technology, the more advanced it becomes, the more it demands in vigilance and craftsmanship. Elsewhere we expect—generally incorrectly, as we shall see—that more advanced technology will bring more safeguards against things going wrong, less need for attention, and fewer workers with craft knowledge. In medicine the increased potential hazards of diagnostic and therapeutic equipment, complex procedures, and the possible interactions of drugs require an unusual degree of attention. The proof is the surprising frequency of serious errors in medical practice. A group of Harvard researchers studied over thirty thousand randomly selected records from fifty-one acute-care hospitals in New York State from 1984. Internists and surgeons from outside the hospitals coded treatments and outcomes. The study concluded that 3.7 percent of all hospital stays include at least one "adverse event." Of these, more than half resulted from some physician error. Medical negligence injured one patient in a hundred.[33]

Judged by the risks of everyday life, a 3.7 percent mishap and 1.0 percent incompetence rate are high. And the more people are attracted to the medical system, the more casualties even a low rate of substandard care will cause. The authors projected nearly 100,000 adverse effects, over 27,000 with negligence, among 2,670,000 patients discharged from New York hospitals. Over 6,300 people were permanently impaired, and over 13,400 died, all needlessly. At a symposium in 1992, Lucian L. Leape, one of the project leaders, estimated that almost one American hospital patient in twenty-five, or 1.3 million each year, suffers an adverse effect and that one in four hundred, or 100,000, dies. This does not include unpredictable events like allergic reactions or risks of properly followed established procedures like cancer chemotherapy. If Leape's estimate is correct, avoidable medical injuries take twice as many lives as highway accidents.[34]

Here is a repeating effect: Improvements in treatment encourage more people to undergo procedures. More advanced technology may produce better results when all goes well, but because there are more interacting systems, more things can go wrong—a recomplicating effect. As their numbers multiply and as the population ages, even a small rate of error and malpractice can produce shocking casualties. A single hospital might dispense over 2.8 million drug doses each year (based on Leape's example wherein each of 24,000 patients received ten medications twice a day during an average six-day stay). Even an error rate of less than 0.2 percent would result in five thousand pharmaceutical mistakes and perhaps five hundred adverse events. The point is not only that there is much more malpractice than ever reaches the courts—probably in other Western countries as well as the United States—but that avoiding serious error requires greater attention on the part of physicians, surgeons, nurses, and technicians, and increasingly of computer programmers and software developers. Software can injure and kill not only by failing to signal dangerous conditions but by triggering the delivery of too much or too little medication, or by sending too many false alarms. Glitches in computer-controlled radiation dosage can be fatal. Automated treatment may decrease catastrophic risk to patients, but it imposes a need for more safeguards, more attention, and more stress. Technologically intense medicine is, to use Charles Perrow's term again, a tightly coupled system, and often a complex one. With proper training, supervision, and peer review, it can provide a high level of benefit, but it also increases the chances for problems like unwanted interactions of one medication with another.

BURDENS OF CRAFT

Consider laparoscopy, a form of surgery that appears much neater and less traumatic than conventional procedures. Laparoscopic surgeons manipulate fiber-optic light sources and cameras and insert miniaturized instruments into the body through a few small incisions. Patients recover in days rather than weeks. The procedure, which originated in the late 1980s in community hospitals rather than in academic medical centers, became so popular that by 1991 it accounted for 400,000 of 600,000 gallbladder removals. Yet financially, it has disappointed health insurers. According to one study, although it cost 25 percent less than conventional techniques, the rate of gallbladder surgery increased by 50 percent in one health maintenance organization, raising total expenditures on gallbladder care by 11 percent. People whose discomfort was only mild, and who might otherwise have avoided the pain and lost income from surgery as long as possible, instead find the new-style procedure attractive.[35]

But the real revenge effect of laparoscopy has turned out to be a medical rather than a financial one. After all, both surgeons and HMO patients have benefited materially, even if the insurers have not. Unfortunately, for all its apparent tidiness, laparoscopy may be more hazardous than conventional surgery: a recomplicating effect. Complications may occur up to ten times more often than with traditional procedures, in as many as one case in fifty. A television image is narrower, grainier, and of course flatter than the immediate if messy sight of an open abdomen. Surgeons can't feel internal organs directly with their fingers. To perform what critics have called "Nintendo surgery" requires skills different from those for traditional procedures—different enough that some surgeons with the right spatial-motor abilities are proficient after a few supervised operations while others are said to need dozens or even hundreds. New York State demands at least fifteen supervised operations before permitting solo laparoscopy.[36]

Far from making work surer for the surgeon, the new equipment demands mental and physical contortions in virtual space. Will a future generation of surgeons, growing up with video games and other flattened mappings of the world, learn a new way of relating to the body, as their predecessors learned to interpret stethoscope sounds, microscopic images, and X-rays? Or will a gradual loss of some older tactile skills impoverish medicine? Since there will always be times when advanced procedures can't be used—a hospital may lack appropriate equipment, the procedures may pose unacceptable risks in a specific case—surgeons probably need training in both new and traditional techniques. Instead of simplifying the medical curriculum, speedier procedures may in the long run add to it: a recomplicating effect.

Stanley Joel Reiser has called attention to the dangers of imputing more validity to "objective" test reports and imaging than to a physician's own observations and the patient's own account of illness. We have already seen how the shift from tool use to tool management enhances our power over the physical world while reducing immediacy of understanding. The case of laparoscopy, at least so far, shows that the management of new tools can be a craft in its own right, and not an easy one to master.

Other technologies may be dangerous ways of gathering additional information. Catheterization is one of the most controversial. Inserting tubes in patients' bodies for monitoring bodily functions appeals to the physician's desire for information. Working with more precise information enhances professional prestige. Yet results are still inconclusive. Few specialists agree with the charge of Eugene D. Robin and Robert F. McCauley, professors of medicine at Stanford, that up to 100,000 patients have died from the use of pulmonary artery catheters, as many of the patients were at high risk. But the procedure remains contested. The risks include infections, perforations, arterial ruptures, thrombosis, and heart blocks. Many physicians have not learned to insert the catheter, calibrate it, or interpret the data gathered with it. The catheter, Robin and McCauley conclude, is "user friendly, even though it may be usee unfriendly." The point is not when, if ever, catheterization is worth the pain and risks. It is that this additional source of information makes decisions more difficult for physicians—if only because they need to decide whether or not to use the technique in the first place. It takes more skill to be a patient: to learn something about the procedure and to discuss it knowledgeably with the doctor.[37]

Burdens of Vigilance

At the other end of the spectrum, far removed from the skills of advanced surgery, are fatiguing, repetitive, and familiar low-technology duties that the pressure of working with new systems makes it easy—but dangerous—to ignore. Elementary sanitation is one. About 6 percent of all hospital patients are infected by microbes they encounter upon entering the hospital. Medical personnel transmit many of these infections. Some doctors and nurses may be asymptomatic carriers, transmitting bacteria by merely breathing or walking around. But a far simpler problem, inadequate hand-washing, probably causes more complications. A study at the University of Iowa Hospital showed how the pressures of work could displace even this most basic and valuable routine. Investigators studying hand-cleaning preparations found to their dismay that fewer than half of intensive-care personnel—even when aware of the observers—washed their hands at all. "Experts in infection control coax, cajole, threaten, and plead, but still their colleagues neglect to wash their hands," a *New England Journal of Medicine* editorial lamented. Doctors appear even more negligent than other staff about washing before examining patients—and not just in America. A professor of clinical hygiene in Freiburg, Germany, Dr. Frank Daschner, once infuriated his colleagues by declaring: "You can sit on any toilet seat without the least risk, but don't, whatever you do, shake hands with your doctor," and he stood by the proposition in 1990. Advanced technology, far from banishing the rituals of the past, makes them even more crucial by multiplying the possibilities of infection.[38]

Medicine also imposes another need for vigilance: the interpretation of life-critical test results is in the hands not only of medical specialists, but of inadequately trained and supervised technicians. Some tests can be automated and even digitized, but all too often the recognition of dangerous conditions depends not on science but on skills that vary with innate ability and practice. The Pap test for cancer has saved many women's lives, but when misinterpreted to give false reassurance, as it often is, it can be a risk to life. There are risks in false positive diagnosis, too. The twin burdens of new competence and added vigilance help explain why medical costs are so difficult to control. The costs in themselves are not revenge effects; they may or may not be worth the benefits received. But multiplying medications and treatments increases the risk of possible unwanted interactions.

Our discontent with medicine is not merely a psychological trick we are playing on ourselves. Nor is it, at the other extreme, a reaction against medically caused illness. The exploding worldwide demand for medical services shows, to the contrary, how much the power of technologically based medicine is appreciated. But we have to consider the nature and limits of this power more closely. Technology has benefited human longevity more by raising the standard of living than by raising the standards of healing, though this does not mean that medicine has not counted. Medicine was transformed beginning in the early nineteenth century when physicians turned from treating the whole system to identifying specific diseases with specific remedies. At the same time, the rising toll of warfare and industrial accidents was helping raise the level of military and emergency medicine. The threat of worldwide epidemics was mobilizing medical authorities in Europe and North America. Of course, catastrophic injury, acute illness, and infection had not really been mastered. The influenza pandemic of 1918 and the casualties of two world wars demonstrated this all too well. What the developers of medical technologies had left beyond doubt was that the search for more targeted treatment of acute conditions and for more precise surgical procedures was prolonging lives.

Many people would say that the revenge effects of the new techniques were injuries and deaths from unsuccessful procedures, the side effects of therapies and vaccines. But patients as well as physicians are ready to accept risks; in fact, as the study of Leape and his colleagues showed, the great majority of avoidable medical errors go unnoticed. The real revenge effect is on the practice of medicine itself. We expect technological refinement to reduce dependence on human attention, and it often does. Automobiles and even commercial jet aircraft have fewer gauges to be watched than they did a generation ago. Paved roads actually need less maintenance than dirt ones. But because the human body is a tightly coupled system, in which treatments can make parts interact in unexpected ways, advanced medicine usually requires more rather than less human attention. More and more care becomes intensive, potential complications multiply, and deviations can be fatal. For physicians, new technology requires more rather than less craft. For all health workers, it demands more rather than less vigilance. Medicine costs as much as it does not simply because machines are so expensive, but because successful medical technology multiplies the need for nontechnological services.

And what of the other sources of medical discontent that were mentioned at the beginning of this chapter? Many of them arise from another, and more troublesome, revenge effect of medical improvement: the shift from acute to chronic illness.

3

Medicine: Revenge of the Chronic

So far we have seen two kinds of revenge effects of medical technology. The first is that *non*medical technological change appears to have done more for human health than medical procedures; the second is that more advanced procedures and drugs demand a higher level of craft skills and impose a burden of greater attention—contrary to most expectations of technological change. But these hardly account for the health anxieties of people with excellent access to medical care. One explanation, advanced by Ivan Illich and other critics, is that physicians and other professionals are manipulating demand, promoting dependency on themselves. It was in this spirit that the Viennese writer Karl Kraus once defined psychoanalysis as "the disease of which it purports to be the cure." This line of attack has some merit, but cannot explain why so many people ask for more treatment than practitioners want to provide. Professionals dread the hard core of complainers, the "turkeys" who haunt emergency rooms.

Physicians do realize, however, that millions of people without easily diagnosed complaints are really in pain. And these patients' ailments have turned out to elude the strengths of medical technology. Whereas drugs and procedures target specific local problems, individual symptoms are often vague: headaches, fatigue, pain, digestive problems. And while X-ray and magnetic resonance devices at first seemed to have revealed the body's innermost secrets, now scans and tests all too often reveal nothing unusual. Surgery, having saved countless patients, now often, it seems, can do little. The illnesses we used to worry most about were visible, local, and relatively brief; now they are elusive, polymorphous, and open-ended. Medical technology can manage some chronic conditions, but sometimes only with serious side effects. It can deflect the acute phases of others, prolonging life. But after a century of brilliant successes with specific treatments, medicine today must come to terms with the very conception that it thought it could do without: the patient as a whole system.

The Rediscovery of Chronic Illness

Early in the twentieth century, hospital-based physicians avoided chronic cases as discouraging and intellectually unpromising. But as the sociologist and historian of medicine Daniel M. Fox has pointed out, interest in "incurable" disease began to grow in the 1920s and 1930s. With the apparent conquest of tuberculosis by the late 1940s, TB researchers shifted their work to chronic respiratory diseases. Funding increased steadily for these and for cancer, heart disease, and stroke in the 1950s and 1960s. At that very time, the microbiologist René Dubos, in his book *Mirage of Health*, stressed not only the limits of the germ theory and the importance of the environment, but the fact that medicines "are far more effective in the dramatic acute conditions which are relatively rare than in the countless chronic ailments that account for so much misery in everyday life." The epidemiologist Abdel R. Omran put this paradox in a global perspective in a 1971 article titled "The Epidemiologic Transition." He argued for a three-stage process of medical development: an Age of Pestilence and Famine, an Age of Receding Pandemics, and an Age of Degenerative and Man-Made Diseases, with respective average life expectancies at birth of twenty to forty years, thirty to fifty years, and over fifty years. The West entered the third stage in the 1920s. Omran saw the rise of chronic illness as a positive sign of the banishment of pestilence, famine, and pandemic. He was concerned more with the conditions of social and economic modernization than with the problems of chronic illness in the developed world.[1]

But a few years later, another article drew more pessimistic conclusions for industrial countries. In the influential analysis "The Failures of Success," the psychiatrist and epidemiologist Ernest M. Gruenberg recalled a paradox still cited in the 1935 edition of William Osler's classic textbook on medicine. Osler had written that "persons rarely die of the disease with which they suffer. Secondary *terminal* infections carry off many patients with incurable disease." Yet by the late 1930s, sulfa drugs were cutting the death rate from pneumonia in half, while penicillin and other antibiotics were to reduce it still more in the postwar period. The patients saved by new treatments were disproportionately people already weakened by incurable disease. Instead of improving health, medical research was inadvertently impairing it by increasing the prevalence of disability and chronic illness.[2]

Gruenberg was also struck by the increase of the proportion of children affected by what is now called Down's syndrome. One study revealed that it had doubled in the twenty years after 1929, and had doubled again to one in a thousand by 1958. Another suggested that among children from five to fourteen years old, the prevalence rate had doubled even between 1961 and 1974. Once, most children with Down's syndrome died of pneumonia before the age of six. By the 1970s, even the diseases of middle age were being successfully treated, and doctors were expecting the first generation of elderly Down's syndrome patients. Alzheimer's disease patients also seemed to be surviving twice as long. There was evidence that better treatments for pneumonia, heart attacks, and stroke were letting arteriosclerosis progress further, resulting in more chronic brain, heart, and kidney disease.

Gruenberg's point was to shift the goals of medical research away from eradicating infections to understanding the mechanisms of chronic and degenerative disease. The discovery that fluorine deficiency contributed to dental caries appeared to show how research could target a widespread chronic disease successfully. Gruenberg, a survivor of severe injuries from an automobile accident, recognized that many chronically ill and disabled people could still enjoy a high quality of life. His plea was not to stop saving lives but to abandon the illusion that reducing mortality rates was the only, or even the main, goal of the public health movement.

When medical and public health thinkers rediscovered chronic illness in the midst of postwar medical optimism, they generally passed over a difficult question. How much chronic illness had existed before Omran's epidemiological transition, undiagnosed or even unnamed? While chronic and degenerative illnesses are the cause of death more often than a hundred years ago, nobody knows how much more prevalent they really are. A social historian of eighteenth-century Germany recently mentioned in a seminar a case she had found in an archival document. A woman in intense pain from a bunion saw a physician. She also had cancer, but that was not her chief complaint. She believed, no doubt correctly, that it could not be cured. Meanwhile the bunion was keeping her from doing necessary work, and it *could* be treated.

In our own time, postmortem necropsy has revealed a much greater incidence of cancer, heart disease, and other "silent" conditions not listed on death certificates. The multitude of deaths from infectious disease in middle age a hundred years ago undoubtedly concealed many chronic conditions. Like the eighteenth-century cancer patient, the nineteenth-century quarryman suffering from silicosis, the coal miner from black lung, the cotton-mill worker from brown lung, and the asbestos worker from white lung usually had little to gain from reporting chronic occupational illness. Far from receiving workers' compensation or disability payments, a sick worker might become unemployed and unemployable.[3]

The success of the germ theory had the revenge effect of delaying recognition of diseases that develop slowly from exposure to environmental insult. Silicosis became a special problem as power tools increased the volume of dust to which workers were exposed.[4] Yet instead of pursuing this link, scientific medical researchers turned their backs on industrial and environmental factors to hunt for disease-specific germs. Until South African commissions established that silicosis was a disease in its own right, European and American medical authorities believed that phthisis, as it was called, was a form of tuberculosis. Even in our own time it is difficult to diagnose silicosis—X-rays are unreliable—and the authors of the standard history of the disease insist it remains widely underreported. Already in the late nineteenth century the observation of a 1950s miners' union official must have applied: "It's a grim joke in mining camps that miners never die of silicosis. That's because silicosis brings on tuberculosis or heart failure, or some kind of infection. You will find these other diseases on the records but without the silicosis they wouldn't have occurred."[5]

Even conditions like chronic fatigue syndrome (CFS) and carpal tunnel syndrome (CTS), recently named and linked to the new stresses of twentieth-century life, had precursors. English clerks of the early nineteenth century first reported a "writer's cramp" affecting the thumb, forefinger, and middle finger and arising in the same kind of stressful white-collar environment as today's CTS. (We know less about possible nineteenth-century blue-collar CTS, possibly because the processes of the time did not expose workers to the same risks.) The "vapors" of the same period, and the later-nineteenth-century diagnosis of neurasthenia, were applied to a cluster of symptoms similar to chronic fatigue syndrome: exhaustion, multiple pains, fevers, and depression. The similarity is attractive to psychiatrists and psychologists favoring psychological and social—as opposed to viral or other organic—explanations for CFS. But it is equally possible that a virus causing CFS has been active continuously or intermittently for a century or more, and that cultural change only led more patients to report it as a disease.[6]

Infectious diseases must have masked many hidden chronic conditions, which in turn may have affected the rate of infection. There are many reports of severe long-term diseases of the past, including the recurrent gastric distress and exhaustion that severely limited Charles Darwin's work schedule for most of his life, that we may never be able to identify. (In Darwin's case, historians and physicians have proposed a brain disease from a South American insect bite, the effects of long-term self-medication with a patent medicine, the stress of controversy and family tragedies, and, most recently, chronic fatigue syndrome.) Thus we probably will never be able to say how much more chronic illness exists now than did 100 or 150 years ago. But no matter how much of the shift to chronic illness is organic and how much is cultural or psychological, chronic illness is what we fear most.[7]

Whether new or just newly identified, chronic illnesses run counter to most of the strengths of technological medicine. Physicians do not want only to care—they want to cure. The director of a major New York hospital once acknowledged that in a setting for acute medical treatment, "chronic disease is an accusation," and a desire to do more for chronic patients conflicts with frustration in treating acute disease. Staff members are demoralized when "nothing can be done." The medical sociologist Anselm L. Strauss and his colleagues have listed seven characteristics that set chronic illnesses apart. They are *long-term* by nature; workers have to be prepared to help patients for years rather than days or weeks. They have an *uncertain* prognosis, with acute episodes alternating with periods of remission, while new treatments are always under evaluation. They demand relatively more *symptomatic relief*, which in turn can be costly. They are *multiple* diseases, in which the breakdown of one system can damage others—a characteristic of tight coupling, as Perrow has pointed out. They are *socially disruptive* for patients, who may not be able to keep their jobs or living arrangements. They need a variety of *social* as well as medical services. And they are *costly*.[8]

There is yet another characteristic, especially for chronically ill people without major disabilities: they encourage unorthodox treatments. Physicians are beginning to realize that a large number of their patients—fully a third, according to one survey—are seeing chiropractors, acupuncturists, massage and relaxation therapists, and other unconventional practitioners, usually to supplement standard medical care. Overwhelmingly these providers treat chronic complaints, including back pain, allergies, arthritis, headaches, and sleep disorders. The same American public otherwise internationally known for its love of intensive medical intervention is equally ready to look elsewhere when medical science and technology don't appear to work.[9]

Management of some chronic illnesses has improved enormously in the twentieth century. Diabetes is one. As early as 1923, the isolation of insulin made it possible to control a condition affecting millions. The Diabetes Control and Complications Trial, a ten-year American study begun in 1983, showed that tighter control of insulin dosage could reduce eye, kidney, and nerve complications dramatically. But technology once more has brought not a tidy solution but a burden of additional vigilance. Insulin therapy demands constant attention to maintaining proper dosage of the hormone in order to avoid circulation problems that can lead to blindness and amputations. To be certain that blood sugar does not exceed or drop below a narrow range, patients need to test their blood up to twelve times daily and inject repeated doses of insulin. This intensive self-monitoring requires hours of training, strict self-discipline, and medical supervision. Continuous insulin pumps can simplify injection but need watching, too. So in mainstream as well as alternative medicine, chronic disease reinforces the necessity of constant care, belying hopes that technology can release us from attention to open-ended, repetitive work.[10]

The Cost of Survival

Since it is so hard to say how new, or even how changed, a disease is, a better way to approach technology and the history of chronic illness is to see how improvements in care can actually help increase the number of people left with chronic conditions. Survival is, of course, nearly always better than the alternative. And the contributions that people with chronic conditions and disabilities make are immense. Calling the problems of survival revenge effects, then, does not imply that saving people is futile. It does mean that in solving one problem—reducing the casualties of catastrophes—we have created another that also demands better solutions.

Late-twentieth-century technology has given survivors of catastrophic injury an unprecedented chance to continue developing and using their abilities. Computers and peripherals can scan and read text, recognize spoken commands, and even synthesize speech. New materials help restore motor function and mobility. Unfortunately, technology is able to do much less about the chronic mental disturbances that can follow psychological trauma. The cultural historian Wolfgang Schivelbusch has shown how important railroad accidents became in creating and bringing to notice the chronic mental disturbance that the late twentieth century was to call post-traumatic stress disorder. Schivelbusch quotes an 1866 treatise by the medical writer William Camps. Camps believed that the severity of railroad accidents was assaulting the nervous system in an unprecedented way. Even where physical injuries were slight, there could be "such a *shock* to the system as for a time to shatter the whole constitution, and this, moreover, to such a degree, to such an extent, that the unfortunate sufferer may not altogether recover throughout the remainder of his life, which . . . may . . . be curtailed in its duration." Another early author, John Eric Erichsen, writing independently, described the sufferings of an accident survivor: deficiencies in memory, sexual function, and digestion; tiredness, insomnia, and nightmares. Improvements in control, signaling, tracks, rolling stock, and brakes could reduce the rate of physical accidents. Better emergency treatment could save even more survivors. And it is possible that as men and women became accustomed to motion at higher speeds, the perceived shock of accidents diminished, at least in the absence of serious physical injury. But the chronic psychological outcome of catastrophes has continued to trouble medical and legal writers. Like other forms of pain and suffering it presents the problem, uncomfortable to quantitative investigators, of being both real and unmeasurable. And it poses the dilemma, disturbing to legal analysts, of ignoring suffering or rewarding fraud.[11]

It was not in civilian accidents, though, that the most striking chronic consequences of the catastrophic appeared. The very efficiency of military medicine that, as we have seen, sent more and more wounded troops back to the battlefield also contributed to the changed nature of wartime casualties. Post-traumatic stress grew daily in medical and social importance in war after war. The Vietnam conflict ironically seemed to show that the psychological toll of warfare could be slashed. Like the mortality rate in military hospitals, the curve of battle-related mental illness appeared to be turning downward. In the First World War, an estimated 200,000 British soldiers were discharged for shell shock, despite early official prohibition of the diagnosis and often brutal treatment of those affected. In the Second World War, what we now call post-traumatic stress accounted for 23 percent of casualties; in the Korean War this was cut to 12 percent; and in Vietnam it was only one-tenth the Korean rate, or 1.2 percent. And a psychiatrist believed that only 5 percent of these were true combat fatigue; 40 percent were problems that could have begun after discharge from service.[12]

The Army believed it had learned from previous wars. It set limits to tours of duty. It returned wounded soldiers quickly to combat. Yet as the years went by after the Vietnam War ended, it became apparent how serious and difficult to treat were the war's psychological consequences. The most comprehensive recent study, by William Schlenger of the Research Triangle Institute, estimates that a third of all Vietnam combat veterans, a total of about 470,000, suffer from post-traumatic stress disorder (PTSD). Symptoms often did not appear until a year or more after discharge. As the psychologist Ghislaine Boulanger has pointed out, PTSD was a new diagnosis but the same syndrome as the war neuroses of earlier conflicts. Like them, it was directly related to the intensity of combat experience. Studies from the Second World War also showed that symptoms could last for decades. As in earlier war neuroses, symptoms included vivid reexperience of traumatic wartime events, usually in sleep but sometimes while awake. Boulanger, following the third edition of the American Psychiatric Association's *Diagnostic and Statistical Manual* (*DSM-III*), also lists "excessive autonomic arousal, hyperalertness, exaggerated startle reactions, difficulty falling asleep, and the feeling of being on the edge of losing control." It was the time lag between the traumatic event and the appearance of symptoms that delayed official recognition of the syndrome by American psychiatry.[13]

Despite decades of research on post-traumatic stress, not only among troops but among survivors of incest, rape, and other crimes, we still understand little about the circumstances in which the event produces the symptom. We cannot link the relatively long onset time and longer persistence of PTSD among Vietnam veterans to any single feature of the conflict. Nor can we say that the success of evacuation and treatment directly contributed to PTSD. Yet indirectly, it probably did. The Vietnam combat experience was even more stressful than the conventional battlefields of previous wars. The excellence of medical treatment not only increased the effectiveness of American forces but may have encouraged commanders to expose more troops to the frontline conditions that eventually led to PTSD, just as some users of civilian safety equipment take risks they otherwise would shun. The military medicine that had succeeded so brilliantly in dealing with the catastrophes of combat proved unable to anticipate or prevent its long-term, chronic consequences.

The long-term consequences of better emergency medicine can be agonizing in civilian life too. Organic brain injury has assumed tragic proportions over the last twenty years. In the 1970s the United States government began a far-reaching technological assault on the hundreds of thousands of annual deaths from trauma. It helped improve telephone response, rescue services, and hospital emergency rooms, often with skills developed during the Vietnam era. New imaging and brain surgery techniques spread. While shock trauma centers have been able to save two out of three serious head injury cases and to monitor the swelling of the brain that is responsible for massive damage, the net result of better medical treatment for emergency cases is the survival of unprecedented numbers of head-injured patients. Even with the best care, the brain needs more prompt attention than other injured organs. For the rest of the body, the grace period may be an hour or more; for the higher functions of the brain, it may be minutes.

Head injuries still account for millions of emergency room visits, and for 500,000 hospitalizations each year in the United States. Trauma to the head is still the largest single cause of death for Americans under twenty-four. But between the late 1970s and the late 1980s, according to the National Head Injury Foundation, the survival rate increased from 5 percent to 50 or 60 percent—an improvement that led to a "silent epidemic" of disability. The number of teenage head injury survivors alone has quadrupled. Of brain trauma survivors, up to 90,000 a year have severe disabilities.[14]

For reading, for speaking, for mobility, there are hundreds of devices that have helped people with physical limitations. But for many forms of brain damage there still are few products that restore independence. Hundreds of thousands of brain-injured people need both high- and low-technology care, both labor-intensive. Home care can become a full-time task. Normal swallowing and breathing may be impossible. Incontinence can prevent normal participation in society. Even when the injury does not fully disable a survivor, it can change personality in dramatic ways. A brain-damaged person may not recognize the parents or children who are providing care; the injury may have so impaired memory that the person will never reestablish relationships, will ask the same questions day after day. In a famous paper, "The Head-Injured Family," the neuropsychologist D. Neil Brooks has reviewed the disturbing conclusions of research on the lives of spouses, children, and others close to brain-injury survivors. Brain damage disrupts domestic and social life even more profoundly than spinal cord injuries do. The burden often grows heavier with time, not lighter as with many other disabilities. Depression and anxiety are common. Up to nine in ten feel trapped. The very efficacy of our response to sudden injuries has left us with an epidemic of prolonged suffering, not just of individuals but of whole families.[15]

Surviving Childhood Illness

Even medicine's greatest success stories in treating chronic illness have multiplied other chronic conditions. Survivors of acute conditions have long been familiar with chronic consequences. Post-polio muscular atrophy may eventually affect as many as a quarter of the 250,000 to 300,000 survivors of the American polio epidemic of the 1950s. It mimics the effects of aging and arthritis. Nerves signaling muscles, including muscles not apparently affected by the original disease, are destroyed. Though records go back more than a century, doctors were slow to acknowledge the reality of post-polio syndrome. Even in the early 1990s it was not clear whether muscle strength actually declines more in post-polio patients than in healthy people of the same age.[16]

The recent improvement of the rate of survival from childhood cancer is in some ways as impressive a story as the development of the vaccines that virtually eliminated poliomyelitis. Seven out of ten pediatric cancer patients live five years; only two of ten adult patients do. One estimate predicts over 200,000 American survivors of childhood cancer by the year 2000, or one adult in nine hundred. Yet in one study of childhood survivors, four out of ten identified physical or mental problems brought on by treatment. Successful therapies can still cause reproductive difficulties, cataracts, and heart disease. Radiation can damage bones and lung tissues, and its application in cases of brain tumors often turns out to impair learning ability and may even cause some mental retardation. For these children, there is a tenfold increase in the probability of a future cancer, and most of that increase is due to therapy rather than to any genetic tendency.[17]

The revenge effects of survival may go beyond the effects of trauma and childhood disease. There is the less clearly documented but real possibility that the decline of infectious disease—both from medical intervention and from better education and living conditions—has actually promoted chronic illness. This is the thesis of the historian James C. Riley, who has adopted the concept of "insult accumulation" from an essay by the medical physicist Hardin B. Jones. Jones argued, in effect, that any sickness is hazardous to your long-term health. In Riley's quotation and paraphrase: "Every disease episode does some damage to physiologic function. Diseases tend to facilitate the growth of disease states at all ages, including adulthood. Favored cohorts—those that have experienced less disease—enjoy longer life spans and greater vigor at every subsequent stage of life." As expanded by other medical theorists and by Riley, insult accumulation means that diseases, injuries, and risk factors (like cigarettes and alcohol) don't cause only overt and recognized damage. While infections may produce a beneficial and adaptive immune response, insult-accumulation theorists believe that lasting tissue damage may make the body less resistant to future stress by shifting resources to "repair and adaptation."[18]

Riley's hypothesis is difficult to test, but evolutionary theory argues for it. The most important single study was a review of three thousand autopsies performed by a San Francisco physician, William Ophüls, in the first quarter of this century. Ophüls's most important finding was that people with a history of infectious disease showed significantly more arterial and heart disease. Other conditions, he believed, hastened heart disease by damaging arterial walls. Lesions arising during an early disease episode could attract plaque and so lead to atherosclerosis. Riley analyzed the records of the sick funds of late-nineteenth-century England. He concluded that while more workers survived into their forties and fifties as time went on, in any given age range there was a higher rate of illness. People were living longer, but appeared to be sicker as they got older, at least until they reached sixty-five, when aging itself became a more important source of impairment than previous disease. Insult accumulation is the strongest general form of a revenge theory of medicine. It is also the most speculative because its factual base is so weak. As the case of silicosis shows, it is not easy to study slow-onset conditions even with modern imaging techniques.[19]

Even if insult accumulation is a valid hypothesis, avoiding injury and illness at all costs would almost certainly have revenge effects of its own. It would imply a highly protected and isolated childhood, one without the chance to acquire the antibodies so useful to the mature adult. As early as the 1920s and 1930s, epidemiologists were finding that the crusade for cleanliness at the beginning of the century, far from combating polio, was promoting it. When all infants acquired the virus in the first days of their lives, while still protected by antibodies from their mothers' blood, paralysis was almost unknown. Epidemics became most severe where standards of plumbing and cleanliness were highest. There, young people were first exposed to the virus long after the end of maternal immunity. German measles, too, turned into a serious adult disease only after fewer and fewer children were infected with it. More recently, the science writer Lynn Payer has pointed out that the more casual French attitude toward exposure to germs makes their effects less severe in later life. In countries where fewer people have had hepatitis A and toxoplasmosis in childhood, more people are seriously affected when they have the disease for the first time at a later age. We will not be able to judge the insult-accumulation hypothesis, or to draw conclusions for living habits and medical practice, until the science of long-term health reaches maturity.[20]

Are We Booby-Trapped?

Evolutionary theory suggests that we don't need the concept of insult accumulation to explain the increase of chronic and degenerative illness. Plants and animals in nature succumb to all kinds of natural hazards, accidents, and predators while in perfect health. Even without any disease, populations of a given age would thin out naturally as time went on, a fact stressed by the evolutionary biologist George C. Williams. Natural selection favors organisms that reproduce rapidly and early. Think of a gene that helps an animal multiply relatively early in life but increases its risk of cancer, say, years later. Those eventual deaths will retard the spread of the gene, it is true. But quite independently of the gene, few animals carrying it may survive to an age when it would promote the disease. The early gain would outweigh the later loss. Natural selection neglects the fitness of organisms that have passed their peak reproductive years. As the evolutionary biologist Steven Austad has put it: "Since most new mutations are harmful and since mutations with effects late in life tend not to be weeded out, late-acting harmful mutations can accumulate over time."[21]

In fact, we know of mechanisms that help early reproductive success at the cost of an individual's long-term health. The biologists Robert M. Sapolsky and Caleb E. Finch have pointed out that the mating season of Pacific salmon and marsupial mice exposes these animals to extraordinary levels of gluticortoids—a "hormonal death switch," as they call it. Yet this stress is a by-product of the intense reproductive activity that helps maintain these species. Some human genes may promote reproduction at an early age at the cost of later chronic illness. Sapolsky and Finch point to the higher-than-normal sexual activity and reproduction rate of men with the Huntington's disease gene. They cite research suggesting that prostate cancer is also a consequence of genes that promote seminal fluid production and thus greater fertility earlier in life. It is likely that more studies will show how the very genes that help our species reproduce in early life turn against it later. Even without insult accumulation, simply surviving accidents, injuries, and food shortages with improved technology gives us that much more time for the latent lethal genes to act: yet another way of running from catastrophic hazards to chronic risks.[22]

Resurging Infection

So far we have seen the rise of chronic illness helped by the success of technology in preventing and repairing the results of injury and infectious disease. But there is also a sense in which control of infectious disease, while still effective, has turned out to be a chronic problem in itself: long- rather than short-term, demanding vigilance, increasingly costly, and without a neat solution in sight. Where we once hoped to eradicate, we are now struggling to manage. Intensive use of drugs against some bacteria and viruses has helped promote the growth of resistant strains. At the same time, medical pessimists have turned out to be as wrong as the optimists. Even the greatest medical disaster of recent origin, HIV infection, is not the kind of plague that it was at first feared to be. Medication has helped transform it in many cases into a slower and longer-term condition that has more in common with cancer and other dreaded chronic illnesses than with the acute epidemics that once decimated Europe overnight.

Only a generation ago, we seemed to be on the verge of the chemical conquest of infection. In 1967, U.S. Surgeon General William H. Stewart declared it was "time to close the books on infectious disease." Twenty-five years later, the books are open again. People are beginning to fear infection once more. The streptococcal attack that killed the puppeteer Jim Henson in 1990 was one of tens of thousands of adult cases in the United States in the early 1990s. As far as we know, this resurgence had nothing to do with the antibiotics that had helped make the most feared complication of streptococcal infection, rheumatic fever, a rarity among children. In fact, specialists believe that prompt antibiotic treatment could have saved Henson's life. But resistance has become a serious problem in treating other infections.[23]

The natural selection of drug-resistant strains has been known since the early days of antibiotics. Penicillin-resistant bacteria were at first only a minor problem. Alexander Fleming discovered that use of penicillin would select mutant strains with walls resistant to the drug. He warned as early as 1945 that, freely taken, an oral form of the drug could breed resistant strains that could then infect other people—especially if patients stopped treatment before all bacteria were destroyed. Resistance turned out to be more serious than Fleming had expected. Treatment indeed selected for natural variants of bacteria that could not only resist but destroy penicillin. Resistant strains began to overwhelm hospitals in the fifties, sixties, and seventies. Then in the 1970s, resistant forms of the bacteria causing meningitis and gonorrhea began to appear.

When streptomycin was introduced against the tuberculosis bacillus, a dangerously rapid rate of mutation led scientists to develop a family of related antibiotics in the 1950s and 1960s, some of which are still widely used. Tetracyclines and other broad spectrum antibiotics were also introduced beginning in the late 1940s. In this case the alarming rise of resistance had one positive effect: promoting research into bacterial structure. Bacterial plasmids—minute, self-replicating pieces of DNA outside the chromosome—add valuable capabilities to their hosts. Plasmids conferring antibiotic resistance to their host bacteria naturally multiplied vigorously in the presence of penicillin and other antibiotics. Most seriously, in 1959 a bacterium appeared in Japan that resisted four antibiotics—and the same resistance was soon found in another bacterium. Plasmids called R factors were moving among bacterial species.

R factors have always been present. So have plasmids. Introduction of antibiotics created a powerful evolutionary pressure. The mechanisms of genetic exchange among bacteria, their plasmids, and associated viruses (phages) allow an unceasing transfer of resistance genes. Otherwise harmless bacteria in the human urogenital and respiratory tracts now carry resistance genes that they furnish to pathogenic bacteria. While there are also genes that express specific mechanisms for neutralizing the effects of particular antibiotics or for destroying them, microbiologists have found that resistance accumulates. Bacteria assemble their own arsenals of resistant genes as hackers swap access codes and passwords. We are awakening from the nineteenth-century dream of specificity. The boundaries between species and organisms are not as well defined as our ancestors believed. In only fifty years we have gone from the offense to the defense. And that should not be completely surprising, since the two human generations since the introduction of antibiotics have faced countless bacterial generations.

In hindsight, antibiotics and the smaller number of antiviral drugs are what lawyers call wasting assets. We deplete their value slightly each time we use them. Always present as natural substances in the soil and in some traditional remedies, they had never been released massively until the Second World War and its aftermath. Even conservative use would have promoted resistance genes, though of course more slowly. Because the world has become a single microbiological system, drugs can stay effective only if used conservatively not only where they were introduced but wherever they are available. The actions of every physician and every patient help determine the length of a medication's useful life. Alternatives usually cost much more than the drugs for which they have been substituted.

The startling wartime successes of penicillin created the dangerous myth of an antibiotic panacea. Even after the U.S. Food and Drug Administration began to require prescriptions in the mid-1950s, an antibiotic injection or prescription remained for many people the payoff of a medical encounter. They resisted the medical fact that antibiotics can do nothing against colds and other viral diseases. In many other countries, antibiotics are still sold legally over the counter to patients who may never get proper instructions about dosage or the importance of completing a course of treatment. Dr. Stuart B. Levy of Boston cites an Argentinian businessman who was cured of leukemia but died of an infection by the common bacterium *E. coli.* Ten years of self-medication had produced plasmids in his body that were resistant to every antibiotic used. Governments, too, have unintentionally promoted resurgence. Indonesian authorities have literally ladled out preventive doses of tetracycline to 100,000 Muslim pilgrims for a week at a time. Since the Mecca pilgrimage has historically been one of the great mixing bowls of microorganisms, it is especially disturbing to learn that half of all cholera bacilli in Africa are now resistant to tetracycline.[24]

Until recently, bacterial resistance was good for the drug business. Older medications lose protection against generic versions; resistant strains create opportunities for patenting new drugs. But the industry is discovering that new products, too, can have disappointingly short useful lives. Promotion of one new antibiotic, Ciprofloxacin, destroyed much of its greatest asset—inexpensive protection against hospital-based infections—within a few years of its U.S. introduction. One critic of industry promotion, Dr. Calvin Kunin, has compared resistance to the planned obsolescence of automobiles, but others fear that pharmaceutical companies have lost interest in developing new antibiotics even as the old ones lose their power. The U.S. National Institutes of Health, concentrating funds on AIDS, also have cut back. Only five new antimicrobials were approved in 1991 and only two in 1990.[25]

Steady application of antibiotics on the farm has also been crowding the commons. Most U.S. swine, poultry, cattle, and even farmed fish get regular doses of antibiotics. In the 1980s, the proportion of drug-resistant salmonella bacteria doubled from 16 to 32 percent. Animals and people have been exchanging antibiotic resistance genes. And people who are taking antibiotics become more vulnerable to salmonella bacteria because antibiotics have wiped out so many beneficial or neutral intestinal bacteria. Like a disturbed grassland, the body on antibiotics is primed for an invasion of weed species. Mitchell L. Cohen of the National Center for Infectious Diseases has warned that without prompt action, "the post-antimicrobial era may be rapidly approaching in which infectious disease wards housing untreatable conditions again will be seen."[26]

The return of tuberculosis in New York City in the 1990s shows how fragile our line of antibiotic defense can be. In the 1970s, TB hospitals seemed costly relics in an age of miracle cures. New York, New Jersey, and other states abolished their inpatient treatment centers, relying on outpatient treatment. When declining living standards, intravenous drug use, and AIDS began to help spread TB again, a revenge effect of drug therapy appeared. Like some other antibiotics, TB drugs need to be taken for a time after the symptoms of the disease are no longer apparent to the patient. A full course of treatment takes from six months to two years. Even infected (but not diseased) health professionals in at least one study failed to complete their course of treatment—another measure of what a burden vigilance can be. An interrupted course does not in itself promote resistant strains of the TB bacterium. But if the initial prescription is not appropriate, *and* if patients discontinue some pills before others, bacteria will be selected for multidrug resistance. Pieces of DNA called transposons hop easily from cell to cell, transferring resistance genes, which accumulate. Expert treatment can reduce the chance of developing multiple resistance, but it demands a carefully balanced combination of antibiotics taken together. The chief tuberculosis control official for the New York City Department of Health, Tom Frieden, has acknowledged that only "a couple of dozen" doctors in the city are capable of managing tuberculosis cases properly. In the early 1990s, about one in seven new tuberculosis infections resisted one or more drugs. In New York City, half of all cases have resisted treatment. For patients infected with bacilli resistant to the two leading antibiotics, isoniazid and rifampin, cure rates may be only 56 percent or less.[27]

Antibiotic resistance does not have to return us to pre-penicillin days, but it shows how a technology that seems decisive can turn into a protracted effort. Antibiotics once appeared to mark a decisive change in the human relationship to bacterial infections, just as vaccination effectively removed smallpox from the list of dread diseases. What they actually accomplished was still considerable, but different.

Antibiotics took a problem of acute illness and turned it into one of chronic attention. It is not a solvable problem, but it can be a manageable one if antibiotic resistance can be protected as a resource. In Tanzania, for example, authorities have limited the spread of drug-resistant strains of TB. In the long run, as we will see later, there are better ways to protect ourselves from microorganisms: to domesticate them.[28]

Antibiotic-resistant TB is not only a menace in its own right but a powerful ally of an even greater threat, the human immunodeficiency virus (HIV). Given the chance, this virus and TB bacteria can activate each other in a deadly synergy. Even more than the revival of tuberculosis, the emergence of new viruses like HIV has shaken the medical optimism that prevailed from the early antibiotic era through the 1970s. Acquired immunodeficiency syndrome (AIDS) has shown how a virus previously undetected in human beings can in less than a generation become one of the world's most serious public health problems. A few researchers claimed that AIDS itself is a revenge effect of African field tests for a polio vaccine, but this theory has been found to have too many missing links. How HIV became a virulent source of human disease is still unknown. Nor do we yet understand how diet, sexual habits, and living conditions affect the immune system's ability to neutralize HIV toxins and delay the onset of full-blown AIDS. The population biologists Roy M. Anderson and Robert M. May have concluded from molecular sequencing studies that human HIV may have been present in Africa for more than a century, possibly for more than two. Some researchers now think it could have been a human virus before it became a monkey virus. Just as virulent strains of the ubiquitous and normally harmless bacterium *E. coli* occasionally appear, the HIV virus may have coexisted harmlessly with human beings for a century or more. The answer will be hard to find, because infectious diseases with a doubling time of a few years can spread very slowly for decades and then extend explosively, leaving few traces of their history. In Africa, shifting populations, rural poverty, separation of male workers from families, and the growth of road networks in the twentieth century created ideal conditions for disseminating the virus. But it may also have been present in North America and Europe long before its supposed invasion from Africa.[29]

Whatever the mechanisms and geography of AIDS, it and other emerging viruses could well be a revenge effect of our victories against other infectious diseases. The physician and historian Mirko D. Grmek has proposed the concept of "pathocenosis." Disease organisms, he has suggested, don't exist independently of each other in human populations; instead, different microorganisms form a complex equilibrium. Grmek believes that pathocenoses have gone through four upheavals: the Neolithic shift to agriculture, the medieval migrations of peoples of the Asian steppes, the biological exchanges following the discovery of the New World, and the present combination of declining infectious disease and accelerating worldwide exchange of pathogens. Virulent strains of AIDS probably were at a competitive disadvantage in populations widely infected with diseases like tuberculosis and malaria, even though TB can spread more rapidly in populations already infected with the AIDS virus. The retreat of TB in Europe and the United States and of malaria in Africa may have changed the balance of pathogens to the advantage of virulent HIV. Grmek's hypothesis is still difficult to prove or disprove, but if it is correct, malaria, tuberculosis, and other infectious diseases actually helped keep virulent HIV in check.[30]

When medical researchers tardily recognized the existence of AIDS as a syndrome, it appeared at first that in the industrial countries, too, it was destined to become a catastrophic epidemic comparable to the bubonic plague, cholera, and typhoid. Its concentration in certain stigmatized groups—gay and bisexual men, intravenous drug users, and their partners—increased the sense of dread. Proposals circulated for quarantines and even compulsory tattooing of HIV-positive people. By the mid-1980s, Americans feared millions of deaths. In Africa and Asia, the spread of HIV is indeed confirming the worst fears of early estimates. The World Health Organization (WHO) has calculated an increase of adult HIV in Africa south of the Sahara from 2.5 million cases to 6.5 million between 1987 and 1992, with fully 18 million infections projected by the turn of the century. The WHO already recognizes a million cases in South and

Southeast Asia. In the United States, AIDS claimed 166,467 lives between 1982 and 1992; 45,472 new cases were reported in 1992.[31]

Less than five years after the most catastrophic predictions, AIDS began to appear in a new light, no less alarming but different. Daniel M. Fox and Elizabeth Fee began to point out in the late 1980s that as AIDS chemotherapy improved, as research burgeoned, and as hospitals established programs for long-term care, AIDS was behaving less like a plague and more like cancer. (Changes in behavior following public education campaigns may have also played a part.) By June 1989, the director of the U.S. National Cancer Institute, speaking at the international AIDS conference, was explicitly calling AIDS a chronic illness, one directly comparable to cancer. The analogy proved all too apt. As in the cancer crusade that began in the 1960s, early hope yielded to pessimism. The chairman and chief executive of Merck declared in 1988 that the company expected to find an anti-AIDS drug within five years. By 1992 he acknowledged to the *Wall Street Journal* his "enormous disappointment." The rapid mutations of HIV, as of other viruses, appear to make eradication almost impossible.[32]

The drug most widely used to arrest and reverse the multiplication of the AIDS virus, AZT, received approval in 1987. It has been thought to delay the onset of AIDS symptoms—though this effect is now disputed—and to retard the action of the virus in people with AIDS. But it is, at best, no cure and it has severe side effects. For those very reasons, it resembles many cancer treatments. As with cancer, hospital-based medicine can deal with acute episodes without halting the progression of the disease. But AIDS is also like other infectious diseases: AZT-resistant strains of the virus are a growing problem. AZT may have a social revenge effect. By prolonging life, as Anderson and May have pointed out, it may help spread the virus in the community if it does not reduce the infectiousness of the virus and if people being treated don't take precautions to avoid transmission. All in all, then, technology *can* hold back the spread of the AIDS virus, but not (yet) technology in the form of antiviral drugs or vaccines. It is rather the technology of vigilance upon which we must rely for now: autoclaves, disposable syringes, condoms, rubber gloves, and other safeguards.[33]

The Price of Living

So far we have seen revenge effects linked to efforts against acute and infectious illnesses. But can efforts against chronic illness have revenge effects of their own? Here there seems less room for technology. To the contrary, physicians and epidemiologists urge healthier living: quit smoking, reduce intake of saturated fat and cholesterol, eat more fresh fruits and vegetables, exercise regularly, sleep adequately, and so forth. Changed habits have helped cut the incidence of heart disease significantly over the last generation—once more, what matters is the cumulative effect of small acts of vigilance rather than some new technology. But attempts to avoid chronic illness are not all free of risk.

Using technology as a shield against environmental danger can actually increase exposure by multiplying the hazard: a repeating effect. A cigarette, for example, is a drug-delivery system for nicotine. Cigarette smokers depend so much on a steady level of nicotine that when they try filtered low-tar-and-nicotine brands, most unconsciously compensate for the reduced volume of nicotine by inhaling more often or more deeply, and even by blocking the tiny ventilation holes in filters that are supposed to dilute their intake with air. Some even break off filters, thinking the tobacco itself is safer. (The tobacco in these cigarettes has the same nicotine and tar content as that in other brands.) While brands with low tar and nicotine are in themselves a slightly smaller health risk than standard smokes, manufacturers market them as alternatives to quitting. An editorial in the *American Journal of Public Health* speculates "that the existence of low t[ar]/n[icotine] cigarettes has actually caused more smoking than would have occurred in their absence and thereby raised the morbidity and mortality associated with smoking." And it continues that technological solutions to smoking's health hazards "precisely fit the thinking emblematic of an addicted individual; the user tries to solve a problem in a manner that lets drug consumption continue without interruption." Just one alternative delivery system, the nicotine patch, offers a way to reduce intake gradually—and generally only if used as part of carefully supervised therapy. (Filter materials also need evaluation. The vaunted Micronite filter of Kent cigarettes in the mid-1950s contained crocidolite asbestos that may have compounded cancer deaths in smokers even after the formula was discontinued later in the decade.)[34]

Sunscreens protect conditionally at best against carcinogenic ultraviolet light. The most common types of the 1990s stop one form of solar radiation, ultraviolet-B (UVB), responsible for burning the outer layer of skin. But the absence of visible tanning conceals the penetration of lower levels of the skin by another form, ultraviolet-A (UVA). If a sunblocking product encourages people to feel safer while sunbathing without a visible tan, and if it does not block UVA effectively, it could be an encouragement to get more UVA exposure. The epidemiologists Frank and Cedric Garland have reported evidence that sunscreens may contribute indirectly to the increase of melanoma—a form of cancer deadlier than the squamous-cell cancers of the epidermis. They also believe sunscreens may interfere with the skin's natural synthesis of vitamin D. The Garlands acknowledge that statistics linking sunscreen use with melanoma are hard to assess, since those with the greatest UV exposure are also heavy sunscreen users. Most other epidemiologists and dermatologists do not accept the Garlands' conclusions, but sunscreen manufacturers are careful not to claim cancer prevention.[35]

Dieting

Even more widely used than cigarette filters and sunscreens—and equally subject to revenge effects—are the technologies of weight control. Since smoking does regulate eating to some extent, the decline of tobacco has helped shift attention to diets and nutrition. Dieting was uncommon in the 1950s and 1960s, when surveys showed that only one person in ten was trying to lose weight. By 1985, 55 percent of women and 41 percent of men responding to a *Psychology Today* survey reported dissatisfaction with their weight. In 1989 about a quarter of all men and 40 percent of women were reducing.

(Medical guidelines of the time suggested that 39 percent of men and 36 percent of women weighed too much for their height.) Fashion played a part, too. Top models and actresses are much thinner than average people and than their predecessors were in the 1960s; one estimate is that their body fat is slightly more than half the 22 to 26 percent of healthy women of normal weight. Diet organizations, books, and foods are a $30-billion-a-year industry.[36]

The health-conscious decade of the 1970s, with its ethic of personal responsibility for health, probably made the difference. Even critics of dieting and of excessive health-consciousness agree that many gains were real. Consciousness of exercise and nutrition and lower rates of smoking and drinking helped reduce deaths from heart disease by 40 percent, and of stroke by more than 50 percent, between 1970 and 1990. Yet the link between overweight and cardiovascular diseases, hypertension, diabetes, and other diseases may not be as strong as some influential insurance-industry studies once suggested. As the effects of smoking, hypertension, and elevated cholesterol are accounted for, fat people are not necessarily sicker than normal-weight or thin people.[37]

As long as overweight (and normal-weight) men and women consider fatness unhealthy, the sense of responsibility can have alarming consequences. The psychologist Kelly D. Brownell has pointed out that it can lead people to think they have more control than they do. Fat people face discrimination in jobs and services—and even unprovoked insults from strangers. The poor are more likely to be fat not only because they eat more fats and starches, but also because fat people are more likely to be poor. So oppressive is the stigma of self-indulgence and gluttony that of one group of patients who had surgery for weight reduction, all would rather have diabetes or heart disease than be fat again. Nine out of ten would rather be blind.[38]

Unfortunately for those who need or want to lose weight, the human body has plans of its own. We still understand very imperfectly how the body's weight is regulated. The hypothalamus now seems to be only one of several regions of the brain that control long- and short-term mechanisms. As physiologists and nutritionists point out, it is amazing that human weight fluctuates so little. Even a tiny long-term imbalance of diet—as little as 0.03 percent of food consumed—appears to account for adult weight gain between youth and old age.[39] Unfortunately for dieters, the body has evolved to cope far better with famine and shortage than with abundance. According to one widely held theory, it seeks an equilibrium or set-point weight that varies with each person's physiology. The body interprets any significant loss of weight as deprivation and compensates by slowing its metabolism. The dieter, compelled to consume fewer and fewer calories to lose each additional pound, understandably becomes depressed and preoccupied with food and eating. It becomes ever more likely that the dieter will break the regimen. Then the body—true to its evolutionary heritage—tries to regain weight as fast as possible. The psychologist C. Peter Herman believes the dieter pays "interest" on the "debt" of lost weight, gaining somewhat more after the diet because of "defensive adjustments that 'protect' the body against future dieting attempts by raising the regulated level of body weight/fat. As the old ad had it, 'It's not nice to fool Mother Nature.' "[40]

Compounding the body's own readjustments, lapsed dieters are often prodigious binge eaters. The increasing self-denial of a diet makes weight loss appear an all-or-nothing proposition. In one study, subjects were told that they were evaluating food tastes and then given ten minutes to judge ice cream. Some of them were also given milk shakes to evaluate before they were offered the ice cream. Nondieters who had drunk milk shakes ate less ice cream than those who had not, but the reverse was true of dieters. Dieters given milk shakes actually ate even more than a control group of dieters who had not eaten beforehand. Dieters, Herman and his colleague Janet Polivy believe, treat a single lapse as a release from their daily quota. Fear of breaking the diet paradoxically creates a sense of failure that becomes an excuse for bingeing. And dieting, by making control of food intake a more conscious, deliberate act, seems to impair satiety. Dieters are slower to receive that "enough already" message that protects most other people against bingeing. In a survey of the Harvard Class of 1982, alumni who had dieted during college were more likely than others to have gained ten pounds or more.[41]

The set point of human weight may not be fixed as firmly as

skeptics about dieting once thought. It appears possible to lose moderate amounts of weight slowly without the hunger that triggers the revenge effects that come from limiting calories. The nutritional psychologist David Levitsky and his colleagues, in a rare long-term controlled study of human eating patterns, discovered that women eating exclusively low-fat foods could lose weight steadily and persistently without consciously limiting calories—about half a pound a week, or 10 percent of body weight a year. This regimen allows low-fat desserts; it probably works mainly by substituting lower-calorie carbohydrate and protein for higher-calorie fat. People disciplined about avoiding fat can reduce their weight slowly and safely. Still, the technology of low-fat food can have revenge effects of its own. As the food writer Trish Hall pointed out, a piece of nonfat cake weighing as much as an average 65-calorie apple may have 560 calories.[42] Especially outside a laboratory setting, repeating effects can cancel the benefits of low-fat eating.

The obese have to take another approach. For them, weight control does not seem to be a one-time event but a continuing process. The alternatives are unpleasant: continued obesity, with great social disadvantages and possibly medical problems; constant and painful hunger; or lifetime medication. Physicians and patients are redefining obesity as a chronic disease like diabetes or hypertension, to be managed, rather than as a problem to be solved. The price of this last definition is a lifetime of dependence on prescription drugs. Once more, technology does not so much solve a problem as create a new need for vigilance and repetitive actions.[43]

Health Crisis and Medical Outlook

The same technology that was so effective in treating traumatic and acute infection also promoted chronic disability and pain. There are signs that as early as a hundred years ago, longer illnesses were a side effect of longer life. Pneumonia, bacterial infections, and other acute complications are now less likely to end the lives of chronically ill patients. Rapid evacuation, prompt treatment, and improved techniques have multiplied the survival rate of the military wounded and of civilian accident victims alike. We are living longer, but thanks in part to technological advances, we are also more likely to suffer from conditions that medicine can only manage, not cure.

Even our control of acute illness has turned out to be much more tenuous than we had imagined. We have made infectious agents into moving targets; the antibiotic triumphs of the 1940s and 1950s were not victories but only a respite. The generations that followed exploited and began to exhaust what turned out to be a finite reservoir of effectiveness. Antibiotic resistance is itself a chronic condition, a response so flexible that our best hope appears to be not to defeat it but only to manage and delay it with a combination of medical conservatism, limits on agricultural use, multiple-drug strategies, new vaccines, and the unceasing deployment of new agents.

Most ominously, the ravages of AIDS have shown how expanding continental and world transportation (and twentieth-century technologies of intravenous drug administration and of blood transfusion) can disperse fatal viruses as well as bacteria on a global scale. With other emerging viral and bacterial diseases, it has eroded the old colonial-era distinction between "endemic" tropical diseases and Western epidemics. AIDS and other rapidly mutating viral diseases have also shown the limits of the nineteenth-century goal of targeting specific causes. In extending the lives of people with HIV, AZT and other therapies are slowing the progression of AIDS without halting it, adding it to the managed chronic illnesses.

Environmental disruption and mass migration have led to fatal pandemics so often that it is hard to be optimistic about the short run. But there is also a positive paradox of research. The nineteenth century's own sense of crisis made possible the public health measures that succeeded in sharply reducing the impact of infection.

Industrial nations are beginning only now to realize how seriously their health may be threatened by the unchecked spread of new infections in the tropics. A worldwide network of regional research centers and early warning stations at the cost of hundreds of millions of dollars a year (out of a total U.S. medical bill of over $690 billion) would be a bargain if it could intercept a single new infection in time to develop strategies, vaccines, and therapies. Even within the United States, early recognition can save countless lives. As far back as 1976 there were hundreds of new cases of a rare respiratory infection otherwise occurring only in organ transplant patients taking immunosuppressant drugs. The infectious disease specialist Dr. Robert T. Schooley argues that with an advance warning system, AIDS could probably have been detected by 1977 or 1978.[44]

In the longer run, the evolutionary biologist Paul Ewald has suggested a new strategy in the campaign against AIDS and other viral diseases: taming viruses instead of trying to eradicate them. Viruses naturally become less virulent as they coevolve with their hosts, not only because killing all hosts would simultaneously destroy the last viruses, but because making toxins takes precious energy. Other things being equal, benign strains have a competitive advantage that will assert itself in time. Why has virulence, then, sometimes increased? Rapid changes of sexual partners and spread of the virus by intravenous needles make it unnecessary for the virus to spare the host in order to reproduce. In West Africa, where family structures are largely intact, HIV-2 kills less quickly than HIV-1 does in East Africa. Significantly, in monogamous Islamic Senegal, there is a low rate of progression to full-blown AIDS. The rapid rate of change of AIDS antigens creates daunting obstacles for both antiviral therapies and vaccines; altering behavior to lower the virus's transmission rate will not eliminate HIV, but it can help domesticate it. A less mobile virus can't afford to kill its host. Blocking transmission not only prevents disease but provides an evolutionary bonus: a shift to less deadly forms. If the conditions are right, mild and even harmless viruses may finish first.[45]

There are even some grounds for hope about chronic illness itself. Peptic ulcers, long thought to be the bodily expression of stressed executive living, may be the work of treatable bacteria. Real evidence for risk factors in any chronic disease must be epidemiological, not experimental. The negative side of relying on statistics is what Lewis Thomas called an "epidemic of apprehension," of alarms about nutritional and environmental risks. But skepticism is not nihilism. It is a call for more rigorous statistics, even if people can't be assigned at random to vary their occupations, diets, and smoking and drinking habits. The lesson from randomized trials can help scientists develop more effective ways to judge the quality of epidemiological data. An important item of good news is that radical changes in diet are not necessary to reduce the risk of cancer, arterial disease, and stroke significantly.[46]

The prevalence of chronic problems is the dark side of stunning improvements that nobody would want to reverse. But chronic illness contradicts at least one dream of advanced technology. Despite all of the technology available to help people with disabilities, many conditions demand large blocks of time for unautomated human care. Even advanced rehabilitation equipment needs skilled operators. Most of the burden of daily care falls on spouses, adult children, and parents.

Medicine stands accused of shifting human responsibility for health to chemicals and machines. Of course, it sometimes does just that, especially in the last weeks and months of life. But more often than not, it does just the opposite. Far from automating health, medicine for better or worse throws more and more responsibility on routine performance by states, families, and individuals. Tuberculosis is resurging in the United States only in part because drugs suddenly lost their effectiveness. Its return is also due to the reversal of a historic rise of urban living standards, and to the related neglect of the day-to-day work of detection and treatment. Health promotion also demands routines that can't be automated, from aerobic exercise to dental flossing to hand-washing by health workers. With no vaccine in sight for the near future, the prevention of AIDS and other sexually transmitted diseases also depends on repeated use of mechanical barriers. Even an innocent walk through the woods now seems to require self-examination for tick bites. Automobiles, typewriters, and wristwatches may need less scheduled maintenance. We need more.

4

Environmental Disasters: Natural and Human-Made

Our exceptional record for increasing life expectancy and managing acute episodes has focused our concern not only on the rise of chronic ailments but on the physical and chemical world around us, natural and human-made. Far from losing confidence in science and technology, citizens of the developed world have come to expect ever-higher standards of accuracy and protection. The psychiatrist Arthur J. Barsky has noted a decision (later reversed) against the U.S. Weather Service for failing to replace a damaged ocean databuoy before a storm killed several fishermen. It takes a big reduction in disease and famine to let people start worrying about physical hazards. But once everyday threats have been sufficiently reduced, extraordinary ones take on more importance. In fact, throughout the severe floods and droughts of the U.S. Depression era, "natural hazards" did not exist as a scientific or academic field. Nor, despite the influence of writers like Aldo Leopold and Rachel Carson, was there a field of "environmental studies." A certain sense of well-being was needed before men and women could advance to a new level of worrying.[1]

Concern was justified. The number of the world's natural disasters exceeding a hundred deaths did rise sharply in the 1960s. But while catastrophes happened more often, there were fewer of the upheavals that had taken far more lives in the early twentieth century:

the Indian drought of 1900, the Soviet drought of 1921, and three floods in China in the 1920s and 1930s, each flood producing over half a million casualties. Humanity was doing better but was more conscious of its insecurity. Television helped by personalizing losses that had once been little more than abstract newspaper and radio stories.[2]

The story of environmental hazards in North America and in other developed countries is a special case. As in medicine, our response to emergency conditions and to local problems has been superb. But this also has another side. By intensifying our protection against some forms of natural danger, we have sometimes only shifted greater liability to the future: a rearranging effect. We have traded acute problems for gradual but accumulating ones. This is especially true of the environmental disasters affecting energy.

The Erosion of Disaster

If healthier living has made us sicker, or, rather, has changed the nature of our sickness, safer living through technology has also had its price. Just as we have developed unprecedented means of recovering from an injury or infection, we have a superb network of structures, devices, and social institutions to shield us from hazards. But on a closer look, safety technologies and relief programs have broadened and shifted rather than eliminated our insecurity. We have defused problems by diffusing them. We have exchanged risk to human life for greater exposure to property damage, and then distributed the cost of that damage over space and over time. We have assumed an increasing burden of vigilance along with our protection. Technology is again taking its revenge by converting catastrophic events into chronic conditions—even as natural catastrophes persist.

With its high-rise buildings, complex food-distribution systems, and vulnerable railroads and pipelines, the industrial world *looks* more sensitive to calamity than the flatter, simpler, slower-paced Third World. Things would seem to be closer to collapsing at the first harsh blow. Even a casual reading of headlines suggests the opposite, however, and statistics confirm that technological living is safer living. One international comparison of the period 1960–80 revealed that below the middle level of national income, fatalities rose sharply with the poverty of countries. Low-income nations had over 3,000 deaths per disaster, high-income ones under 500. Japan lost an average of sixty-three people in forty-three disasters, Peru an average of 2,900 in thirty-one. Deforestation is promoting floods, landslides, and drought even as millions of people are pushing into new and more dangerous surroundings like mountain foothills.[3]

Rich countries lose fewer lives because they have built up systems, from the local fire department to the U.S. Army Corps of Engineers, that identify danger and that warn, reinforce, shield, and evacuate. Democratic politics has responded well to threats of catastrophe. Billions of dollars in public works have channeled and dammed rivers, constructed seawalls, and thrust up barriers to landslides. Not all these projects necessarily protect property in the long run; in fact, they may eventually lead, as we shall see, to even greater property damage than would have occurred without them. But they are at least initially very good at what most people would say is the main purpose of safety engineering: preventing loss of life.

Unfortunately, what makes life more secure can expose property, and to some extent life as well, to new risks. When people in the Third World occupy hazardous terrain, they usually have little choice. Population pressures and inequities of landownership may force them to live in danger. For some North Americans and Europeans, relocation also appears out of the question. But for others it is the feeling of safety inspired by technological solutions that lets them choose dangerous areas for natural beauty and other amenities. From the canyons of southern California to the foothills of Colorado and the shores of Florida, many beautiful places can be unstable. The same tectonic, atmospheric, and biological facts that have made landscapes interesting can make them hazardous. Suppressing catastrophe, and promoting an illusion of safety, can lead to a new generation of risks.

Storms and Floods

Flood myths are almost universal, for good reason. Catastrophe was one of the first revenge effects of agriculture. The soil is most productive in alluvial plains and deltas. Farming societies cannot exist without the fertility of these rich lands. As recently as the 1940s it was estimated that alluvial soils then fed a third of humanity. Because population has grown rapidly in river valleys in the last fifty years, this estimate is probably equally accurate today. Farming on a floodplain produced and still yields great benefits—for a risk that is still not easy to estimate. But the peril of flooding was not a revenge effect; it was a trade-off, and one that hundreds of millions of people have still not been able to reject.

Technology has changed the risks of living in a floodplain or tropical storm zone in three ways. First, it has rationalized the risk by letting us map the area most likely to be struck, limiting new development and giving warnings. Second, it has improved structures to control storm surges and flooding. And third, it has made (motorized) evacuation easier and faster than ever.

Especially in the Third World, the *partial* application of these technologies can have revenge effects. Flood-control systems encourage settlement in flood-prone areas, increasing the risk to life and property. One of the world's worst "natural" tragedies, the East Pakistan (Bangladesh) cyclone of 1970, took over 225,000 and possibly as many as 500,000 lives partly because engineering works designed to control high tides and salt had encouraged massive settlement on reclaimed land that appeared to be protected. As recently as 1985, another storm surge in Bangladesh killed 10,000. Yet great storms don't necessarily kill so many. Technology may aggravate, but it also mitigates. Two years after the Bangladesh cyclone, the floods brought by Tropical Storm Agnes in the United States inflicted $3.5 billion in material costs, at the time more than any other disaster in American history, including the Chicago fire of 1871 and the San Francisco earthquake of 1906. But because hundreds of thousands of people could be evacuated in North America, the death toll was only 118, less than 1 percent of the fatalities from the Bangladesh cyclone.[4]

When Hurricane Andrew struck south Florida and Louisiana in 1992, it exceeded even Agnes's record, leaving 150,000 people homeless and causing an estimated $15 to $20 billion in property damage. But the death rate was lower than it had been during Agnes: fifteen people in Florida, one in Louisiana, and four in the Bahamas. More people may have been injured in removing debris and repairing damage than in the actual storm—hurt by tree branches, glass, and gasoline and kerosene fumes. Andrew killed so few people at least partly because a new Doppler radar system could analyze the fine structure of distant wind patterns. National Weather Service meteorologists could predict the storm's path with unprecedented accuracy, to within thirty miles. Since 1900, deaths from tropical cyclones in the United States have declined from six thousand per year to only a few dozen; yet property damage has soared to over $1.5 billion in recent decades because of the amount of new construction in hurricane-prone areas. This in turn has probably been encouraged by better warning and evacuation systems as well as by insurance; casualty policies covered an estimated $7.3 billion of the damage from Andrew.[5]

Our success in dealing with floods and storms has brought three revenge effects. First, better predictions encourage a confidence that can be perilous. As the landscape historian John Stilgoe has pointed out, when trains were so punctual that people set their watches by them, any exception to the schedule could be lethal to people who "knew" that the tracks were safe. Likewise the better weather predictions become, the more dependent we become on them. And consider the logistics of evacuation. Anyone who has driven in a Florida rush hour can imagine what it would be like to try relocating large numbers of people in a short time. Yet by 1986 the American Meteorological Society (AMS) could warn that many hurricane-prone areas in Florida and on the Texas coast could require as long as twenty to thirty hours to evacuate, while sudden acceleration of a storm may not leave more than six hours' notice. Surveys suggest that coastal residents usually think that one or two hours should be enough.

Long lead time presents a dilemma. Doppler radar cuts false

warnings in half, but evacuation calls may still be wrong a third of the time. And nobody can predict how congested highways might be in a massive evacuation, especially one extending to a dense center like Miami Beach or New Orleans. The AMS pointed out a second risk. Severe hurricanes were rarer in the 1970s than they had ever been. When Andrew struck in 1992, many coastal residents had never lived through a major storm.[6]

Protection systems continue to improve as a result of past episodes. A fifteen-foot seawall built in Galveston after the six-thousand-fatality hurricane of 1900 held during Hurricane Alicia in 1983. And after five incidents in Pennsylvania were declared federal disasters in the mid-1980s, a computerized emergency information network was developed that could save lives in other states by tracking situations and resources for faster response. Despite better forecasting and tighter government measures, coastal states and communities are unlikely to choke off the growth that is so important for them. The cost of better coping with disaster is spread among insurance policyholders and taxpayers and thus becomes a charge against the entire economy.[7]

Protection encourages more development of the coasts. It also discourages house buyers from demanding stricter construction standards; available construction technology could reduce property damage sharply, but at a higher price to buyers or at a lower profit for developers. And it never really rules out the possibility of a new level of catastrophic revenge effect if even a single important assumption turns out to be overly optimistic. Floodplain zoning, for example, bars new housing within an area at risk for a major flood every hundred years; but new suburban housing can and does press up against this still risky border. The long-term risk, at least to property, remains.[8]

Drought

Droughts, like floods, show how technology linked to values and political institutions can make an acute, locally urgent problem into a national and chronic one. The most fatal natural disaster of the

Third World sometimes seems to be only an irritant in the developed world. In both north and south, cultivators push the land to its limits and beyond. In Africa, South America, and India, population growth and overgrazing have made the land vulnerable to the shortages of rainfall to which traditional agriculture was adjusted. In the United States, the booming wheat prices of the First World War and Henry Ford's gasoline-powered tractor in the 1920s encouraged the illusion that "dry farming" would make wheat-growing on the southern Great Plains both sustainable and profitable. But whether they realized it or not, farmers were cultivating semiarid lands subject to recurring drought.

Through the 1920s, tractors pulling new-style disk plows broke the soil with unprecedented speed as businesspeople bought and plowed up hundreds of thousands of acres of grassland. Expensive combines replaced gangs of threshers; small farmers went into debt to stay competitive. As the environmental historian Donald Worster has suggested, the realities of natural rainfall cycles and the limits of the soil were no match for the indomitable optimism of the plains farmers. The low wheat prices of the Depression (leading to a desperate need for increased planting) and a severe and prolonged drought from 1930 until 1936 made the plains synonymous with furious black clouds of topsoil—raised, contemporaries recognized, by the very efficiency of the disk plows in uprooting native grasses and pulverizing the earth.

And the ecologists of the 1930s saw an angry nature turning against its human controllers. The leading popular treatment of the Dust Bowl, Paul Sears's *Deserts on the March,* admonished that by "revers[ing] the slow work of nature that had been going on for millennia," plains farmers had caused "deserts, so long checked and held in restraint, to break their bonds." The mainstream of the New Deal, far from urging more humility in the face of environmental forces, presented instead an enlightened Prometheanism of relief, insurance, water projects, and shelterbelts. No wonder that when better times returned after the Second World War, ambitious farmers were again breaking the soil, protesting the cost of maintaining terraces and shelterbelts, and fighting efforts to restrict their acreage. Drought returned in the mid-1950s, bringing new dust storms, and the "filthy fifties" joined the "dirty thirties" in the region's vocabulary. And after another wave of expansion following the 1973 OPEC oil shock and the Soviet grain sale, there were still more dust storms in the mid-1970s, and a major drought in 1983.[9]

A combination of politics and technology turned an acute regional disaster into a chronic national problem. Crop insurance and credit have shifted much of the cost and risk of drought to the taxpayers and consumers. And the old-style catastrophes might still return. The geographer Richard Warrick and others have warned that if these social systems and technologies were stressed enough by a future drought, the results could disrupt national and even international economies far more than the Dust Bowl itself did.[10]

EARTHQUAKES

Earthquakes also show how technology exposes more people to "natural" risks yet can protect their lives from the same risks. Industrial-age building and transportation methods did not create earthquake hazards. Wherever people are densely settled, especially in port cities that are open to the massive tsunamis that may follow earthquakes, casualties could already be staggering before a single steamship or railroad appeared. The Lisbon earthquake of 1755 killed 50,000; there had been deaths in the hundreds of thousands in Chinese earthquakes centuries earlier, and 300,000 in Calcutta in 1737. The industrial society of the nineteenth century made earthquakes more hazardous mainly by gathering more people into cities. In a sparsely settled interior like America's, huge releases of energy may harm few. The New Madrid earthquakes of 1811–12 were strong enough to reshape the landscape, alter the course of the Mississippi, and topple chimneys as far away as Richmond. One was even felt in Boston. Yet only a few lives were lost.[11]

Twentieth-century technologies can multiply the havoc of major earthquakes. Because of our concentration in urban areas depending on long-distance transport and transmission lines for drinking water, electricity, and electronic data, a miscalculation of seismic risk might in the worst case have the impact of a Chernobyl or Bhopal. As it is, earthquakes still take an average of ten thousand lives and cost $400 million a year worldwide. Seismologists agree there is a strong chance of a major earthquake affecting northern or southern California by the year 2020; estimates have ranged between 10 percent and 60 percent probability. Yet according to Allan G. Lindh, chief seismologist of the U.S. Geological Survey in Menlo Park, the $100 million cost of a major California earthquake is modest compared to the $15 billion annual benefits of agriculture, energy, and shipping that the San Andreas Fault has made possible. Without California's mountain-building and faulting it would be hard to imagine the Central Valley, San Francisco Bay, the gold fields of the nineteenth century, the Peninsula. The same features that make California earthquake-prone make it resource-rich as well as visually beautiful. (Similarly, the towns of the American Midwest are built along rivers and in low-lying areas for transportation and water. Tornado vortices follow these features.)[12]

Global statistics confirm that lives are better protected against earthquakes now than they were earlier in the century. A growing proportion of earthquakes have taken human life, and world population has soared since 1900, yet the annual death rate from quakes was actually slightly less from 1950 to 1990 than it was in the first half of the century: down from about sixteen thousand to fourteen thousand a year. Most of this gain appears due to protection, not prediction. Earthquakes are still difficult to forecast, even in the few areas like California where exposed faults are closely studied. Chinese seismologists may have saved hundreds of thousands of lives by correctly predicting an earthquake near Haicheng and other northern cities in 1975, yet they could give no warning for another catastrophe the following year. In 1976, the city of Tangshan lost 250,000 lives—a sixth of its population—and most of its housing to an earthquake of magnitude 7.8.[13]

The technologies that reduce earthquake casualties are those of preparation. Seismology, soil mechanics, and engineering cannot prevent vast damage. But they can, in Allan Lindh's words, "turn earthquakes into events that, while terrifying, take very few lives."

We are going beyond national maps of seismic hazards to "micro-zoning" of sites for soil conditions that can multiply the effects of ground motion or lead to landslides or to the liquefaction that was so prominent in the 1989 Loma Prieta earthquake. Since the 1970s, new instruments have made it possible to measure surface waves and have revealed that earthquakes can subject buildings to acceleration equal to or exceeding the force of gravity—many times what older building codes had assumed. Techniques exist to make buildings far safer in shocks than they are. We know the hazards of unreinforced masonry. Special standards for schools and hospitals have been tested in actual earthquakes. Large buildings can probably be safe in severe earthquakes if correctly reinforced. Bolting houses and mobile homes to their foundations can prevent fires and explosions.[14]

Stronger construction in richer countries makes fatalities much lower than in the Third World—especially in the slums of cities like Cairo, where hundreds died in the earthquake of October 1992. During the twentieth century, China has suffered an earthquake death for every $1,000 in property damage, whereas the United States has experienced only one death for more than $1 million in property damage. Compare, too, the Tangshan casualties with those of an equally powerful earthquake that struck Valparaíso, Chile, in 1985: it killed only 150 of one million residents. Seismic design of Chilean urban buildings made the difference. In the Loma Prieta earthquake of 1989, the high-rises of the financial district, designed to meet seismic codes, survived. The low frame buildings of the Mission District, built before the enactment of stricter codes enforcing new engineering standards, collapsed and burned.[15]

The greatest risks, human and material, may not be where earthquakes have caused the greatest damage recently but where they are likely but less familiar. Lisbon, despite the fame of its Richter 9.0 earthquake of 1755, may be ready for another that could once more produce record fatalities. In the United States, the greatest risks to life may not be in California but in regions where the danger has received less attention: near St. Louis and in the Pacific Northwest, and even in the high-rise cities of the Northeast. California's faults release great energy when the earth's plates move relative to each other;

but fault lines also dampen shock waves and check their spread. Eastern rocks are older and denser than Western ones. A Midwestern or Northeastern quake would affect a far larger area than a quake in California. The Federal Emergency Management Agency has estimated that a daytime quake of 7.6 on the Richter scale could kill over 2,500 people including 600 schoolchildren in Memphis, Tennessee, alone, leaving over 230,000 homeless. Damages from a Midwestern quake could exceed $50 billion.[16]

Conspicuous earthquake risk has a positive side. California has learned to get ready, if imperfectly, because it has known disaster. Of course, events may once again show that safety measures are inadequate, as they did when the reinforced Nimitz Freeway in Oakland collapsed during the Loma Prieta earthquake in 1989. The Northridge/Los Angeles quake of 1994 and the Kobe event of 1995 destroyed and weakened buildings, bridges, and transportation lines once hailed for their advanced, robust design. Concentration of populations in densely built suburbs may increase the risk. But most engineers believe that improved education, warning systems, maps of fault lines and expected hazards, better building design, careful zoning, and stricter building codes can continue to reduce the risk to life. Indeed, the most serious problems of Northridge arose not from inadequate building standards but from failure to enforce them. And even losses, up to a point, have a bright side. Like tropical storms, earthquakes can't be prevented by technology, but their hazards can be reduced and the cost spread over time (in systematic reinforcement programs) and over taxpayers in regions that are less exposed to the hazard. It has been public policy to turn acute local economic risks into chronic long-term national debt. In fact, California economists credited federal earthquake relief in early 1994 with accelerating the state's financial recovery by a full fiscal quarter.[17]

The developed world's experience, for all of the alarming gaps and uncertainties, shows how well technological change can domesticate though not eliminate natural catastrophe. But this security has two high price tags: the long-term cost of relief to the national economy, and the local burden of remaining prepared for long disruptions of essential services.

FIRE: "SMOKEY'S REVENGE"

Unlike earthquakes and tropical storms, forest fires are as much human-made as they are natural. Huge fires have happened relatively recently without apparent human influence. In Siberia in 1915, for example, a record drought helped bring about two vast fires, one of which burned an area as large as Germany; the smoke has been compared to the volume that would result from a nuclear war. But most American conflagrations, like urban fires, were at their deadliest in the late nineteenth and early twentieth centuries. In the Great Lakes forests, the technology of concentrating people and intensifying production outpaced the technology of preventing and containing combustion. The logging industry piled dead branches and leaves on the forest floor, while sparks from steam locomotives created a ubiquitous fire hazard. In just one of the Great Smokes, the 1871 Pestigo fire, fifteen hundred lives were lost.[18]

Twentieth-century technology has made possible a precarious control of catastrophe. Aviation has extended early warning from the ten-mile horizon of the fire tower and made it possible to drop smoke jumpers (actually a Soviet innovation) and water in remote areas. Infrared cameras can identify firelines through smoke. New foams can hold back flames using small amounts of water. Communications can mobilize resources rapidly. While fire can still be deadly where development pushes to the edge of forest and wildland, casualties are now remarkably low. Not one person died in the Yellowstone cataclysm of 1988. In August 1992, there were 65,000 fires covering 1.1 million acres in the Northwest, but no death occurred even among the 13,000 firefighters at greatest risk. The largest recent tragedies among fire crews are well below historic records. Twelve firefighters perished when high winds fanned a conflagration out of control on Storm King Mountain near Glenwood Springs, Colorado, in July 1994, but this was still much fewer than the twenty-five killed in 1933 in Griffith Park in Los Angeles, let alone the seventy who died in the Idaho and Montana fire of 1910.[19]

The superb modern response to disaster, however, has had

revenge effects of its own. It has helped create a Western forest more dangerously flammable than ever. In the original forests of the West, grassland separated large trees—a landscape sustained by periodic natural fires that burned off small trees and brush. During the Second World War, the (real) threat of Japanese incendiary balloons targeted at the Pacific forests galvanized the U.S. Forest Service's Smokey the Bear campaign. As intended, Smokey sold Americans on avoiding the creation of fire hazards, and of course on supporting the firefighting budget of the Forest Service. But the talking bear could not expound fire ecology. Decades before *Silent Spring* and the new environmental consciousness, his Madison Avenue handlers drew no distinction between bad human-set fires and good natural ones. Fire was unhealthy for bears and people, invigorating though it was for the forest. (Only later would wildlife scientists discover that surprisingly few large mammals perish even in the biggest forest fires. They seldom even need to run and may feed peacefully with the blaze in view.)[20]

Ironically, nearly every time a fire was suppressed, the forest itself was found to change—a process the environmental writer Charles Little has called "Smokey's Revenge." Fire suppression accelerated the transformation that logging had begun. By 1968 an article in *American Forests* described much of the west slope as "a doghair thicket of young pines, white fir, incense cedar, and mature brush—a direct function of overprotection from natural ground fires." Accumulating vegetation in the forest allowed fire to climb from the understory to the crowns of mature trees, where it could spread more rapidly than ever. When drought turns it into an insect banquet, the new-style forest is ready to explode into flame. And letting loggers remove dry, insect-ravaged lumber is not necessarily a solution. If clear-cutting loggers leave slash (branches and needles), as they usually do, the forest may burn even hotter than it would have before.[21]

We can begin working at restoring the Western forest, but it probably has been changed forever. In the altered environment, reintroducing fire can have the revenge effect of killing precisely the old ponderosa pines that we want to protect. As debris ("duff") has mounted at the base of these trees, their roots have extended above the soil into this loosely packed layer. Trees that would once have withstood fire now succumb to it. Even allowing fire to smolder at low temperatures lets heat penetrate much deeper into these relatively exposed roots than fire scientists once believed.[22]

Just when destructive fires in Western forests and wildlands are becoming more likely and more intense, nature lovers are building closer to them. Growth-minded officials in areas like the Sierra foothills have obliged by easing restrictions on building in fire hazard zones. In California, 1.5 million dwellings are close enough to wildland to be at risk; in Oregon, 187,000. Their owners' suburban ways invite unwelcome surprises. House-shading trees, rustic shakes and shingles, and winding and leafy driveways all invite wildfire to make fatal moves—and obstruct firefighters. The fire historian Stephen J. Pyne found in the 1991 Oakland hills fire traumatic proof that an age of "exurban" fire "has brought wildland fire back to the people." (The people helped, too, by planting thousands of trees on formerly open land which had been swept for centuries by periodic fires, just as others built houses on floodplains.) In fact, the final twist of fire history, according to Pyne and other forest ecologists, is that the Western forest has now been so transformed by logging and firefighting that it can never go back to what it was before human intervention began. Most of the great conflagrations since the 1970s have been unplanned consequences of "prescribed burns" deliberately set to reduce the risk of major fires by controlled burning. One fire set to prepare habitat for the rare Kirkland's warbler ran out of control, killing firefighters and destroying an entire village. Yet keeping hands off natural fires also has its risks. When the managers of Yellowstone National Park let a series of fires take their course in 1988, the fires multiplied to seventy thousand, burning 4.2 million acres before a $350 million effort halted the flames. Fires on adjacent private lands, promptly suppressed, caused far less damage. Neither fighting every fire nor letting each one take its course is the answer. In Pyne's words: "There is too much combustion and not enough fire. The fire that does exist is maldistributed—too much of the wrong kind in the wrong places and at the wrong times, not enough of the right kind in the right places and at the right times."[23]

The Western fires of 1993 showed even more clearly the revenge effects of firefighting (and communications) technology. With more and more suburban refugees able to live and work in the "urban interface" of picturesque, isolated forest-edge housing, firefighters found unprecedented risk. The long-term gains in fire crew safety, achieved in remote forested areas, were threatened by new settlement patterns and new responsibilities. The need to protect lives and houses led to the deaths of dozens of firefighters and the destruction of added wildland: a rearranging effect. One Montana-based writer commented, "People have died this year to make valleys like mine safe for lawn mowers." Arson was the immediate cause of some of the worst fires, but these became conflagrations because of a fatal combination: flammable houses set down in wildlands where smaller fires had been routinely suppressed. Meanwhile the reluctance of managers to set controlled fires is slowly building up catastrophic accumulations of fuels, and big fires are growing bigger still. America has a fire deficit so severe that a single region, the Blue Mountains of Oregon, needs 100,000 acres of annual burning to restore its health.[24]

Looking back, firefighting technology and skills have had two revenge effects: helping make forests more flammable and encouraging people to move to their edge, where their property and sometimes their lives are in greatest danger. But even the Oakland and 1993 fires had far fewer casualties than the conflagrations of the late nineteenth and early twentieth centuries. What is new is the constant watch against threats to the forest's self-selected hostages. In one way or another the response has to be more vigilance: either in limiting settlement, mandating construction and clearance standards, deterring arson, or preparing to fight more and fiercer fires.

Beaches: The Ocean's Revenge

The seashore has its hostages, too. Along with the destruction by storms of life and property, there is a chronic consequence: erosion of beaches. Most coastal geologists now find the shoreline in constant natural motion, driven by waves and storms. Over three-quarters of it is in retreat. The rate of retreat has increased in the last several hundred years for reasons that geologists still do not understand. Rapid retreat occurs even in remote areas where landscape appears largely untouched by human intervention, as in the barrier islands on Colombia's Pacific coast. More severe storms and rising sea levels are just a few of the processes that scientists suspect of shrinking beaches. The world's oceans are more than a foot higher than they were a hundred years ago. Further global warming is likely to double the rate, and some projections call for a gain of between four and seven feet in the next century.[25]

Coastal engineering, stabilizing the shoreline against natural forces, is ancient. But until the nineteenth century, seashores were workplaces—sites of seaports and fishing villages, barriers guarding reclaimed farmland—not playgrounds. Romantic poets and artists taught aesthetic admiration of the shore. The fashion of seaside holidays, and later of year-round bourgeois seacoast living, followed. Cultural change converted a natural flux into a technological challenge.

Human activities many miles from the shoreline can affect the shape of the coast. Because waves do not strike beaches at perfect right angles, their energy creates currents parallel to the shore, according to one widely accepted theory of shore processes. If this theory is valid, the current is slowly but relentlessly moving particles toward submarine canyons, where it is lost to the beach. Where sand is carried off naturally it can also be replenished by particles transported by the rivers that empty into the sea, or by erosion of coastal rocks. The beaches of Long Island are really hundreds of billions of fragments of the Montauk Point cliffs and other sources, moving westward by millimeters toward their final resting place in New York Harbor. Meanwhile at Sandy Hook, New Jersey, 436,000 cubic yards of sand drift northward every year.

Where there are no sources of particles like the cliffs, dams trap the sand that would otherwise rebuild beaches. After millions moved to the Los Angeles area, attracted in part by access to natural beaches, new development required flood-control dams that have impounded millions of tons of sand needed for the same beaches. (At Santa Monica, a million cubic yards move southward annually.)

Inland dams creating new recreational shorelines also have the revenge effect of starving coastal shorelines.[26]

Nature takes its most perverse revenge when we try to preserve the shore itself. The nineteenth- and twentieth-century gentrification of the coast turned beachfront property into some of the nation's highest-priced land. Its owners and residents are accordingly some of the highest-priced and most politically influential professionals and corporate managers. Calls for federal, state, and local action have been hard to resist, even in the budget crises of the 1990s.

Coastal engineering starts offshore with breakwaters, built parallel to the coast, of boulders and concrete. By deflecting the energy of waves, they create safe havens for ships and let sand accumulate on the part of the shore that is spared the sea's natural pounding. But even though many breakwaters are modeled after natural features like sandbars and barrier islands, they usually have unnatural results. The rearranging effect is that this sand, which normally moves like a slow river, never drifts to the beaches farther down the shoreline. These beaches consequently erode as their updrift neighbors' beaches expand.

The next line of defense is the edge of the sea. Three main kinds of structures are supposed to protect the shoreline. Property owners build small barriers of wood or other materials to maintain a boundary between the sea and their lawns and terraces. These are bulkheads. Government agencies construct massive stone and concrete protective formations along the shore below the high tide line. These are seawalls. And they also build long walls of stone extending into the water at right angles to the shoreline to capture sand that would otherwise wash down the shore. These are groins. As beachgoers we are so used to these protective technologies that they almost seem part of the natural landscape of beaches. We are aware of them only at the extremes: where no more sand is left and only seawalls remain, as along many parts of the New Jersey coast.

Did armoring the coast actually promote the beaches' disappearance? Geologists, civil engineers, landowners, and developers debate the question vehemently. On some points there is agreement: groins build new beach behind them while starving the beach down-drift, resulting in a familiar scalloped pattern. On other points there is still too little evidence to make a firm judgment. Because sand flows in both directions between beach and upland, structures can work both ways. But we know enough to say that at least in some places, stabilizing technologies have destabilizing, revenge effects. By stopping the erosion at the bases of cliffs, seawalls can reduce the flow of sand. Seawall building is also contagious; erosion increases at the ends of a seawall, prodding neighbors into building their own. Beaches often appear to recover more slowly from storms when they are protected. (In fact, replenishing beaches by pumping offshore sand may also increase the rate of beach erosion by reducing the damping effect on waves that the sand had while underwater.)[27]

People concerned about the coasts are likely to continue to dispute when and where environmental revenge effects are happening. Whatever more rigorous research may show, it is clear that the shoreline is a zone of chronic technological difficulty. Just as logging and fire suppression alter the forest's composition and fire ecology, compelling more and more vigilance, so beach protection feeds on itself by establishing a new order that needs constant and ever more costly maintenance. And the order is as political as it is technological.[28]

Ironies of Energy

Just as technological change has made disaster more bearable but also more feared, it has made energy use less environmentally disruptive but also more worrisome. The need for fuel drives worldwide deforestation in the tropics, threatening far more species and degrading much more land than the equally profligate use of fossil fuels and nuclear energy in the industrial countries. Western technologies continue to use energy and almost all other inputs more efficiently from one decade to the next, and they have the knowledge to do even better. Revenge effects come not from using more advanced technology but from accepting deceptive solutions in place of costlier ones. And like other revenge effects, they transform a problem by spreading it in space and time.

In the nineteenth century, mine accidents and especially explosions happened on a scale hard to imagine. In some counties in Pennsylvania's anthracite country, between 1.5 and 3 percent of all miners died or were seriously injured each year in the period just after the Civil War. Today, firedamp explosions are so rare that coal mining's occupational injury and illness rate is actually below the average of manufacturing, and the death rate has dropped to 0.04 percent annually. Mining may still have tragic chronic consequences, especially black lung, but there is no doubt that its catastrophic problems have receded.[29]

Just as mine calamities were accepted as the price of industrial might, smokestacks were prized as its emblems. They flaunted sooty plumes in manufacturers' and utilities' advertisements, on their letterheads, even—especially—on their stock certificates. In yesterday's gritty cities, effluence signified affluence. Smokestacks removed some of it from the immediate surroundings of the source. They made the soot go up; and like Wernher von Braun in Tom Lehrer's song, their builders did not ask where it came down. Power plants, factories, and smokestacks were concentrated in and near cities, maintaining an alliance that had existed for centuries: urban residents in all social classes had access to more goods, services, and information in return for their sacrifice of personal space—and sometimes of health as well. Indeed, the smokestacks of industry probably were not nearly as harmful to urban air as the smoke from coal-burning home and business furnaces, the excreta of horses, and later the emissions of automobiles.[30]

As environmental consciousness grew in the 1960s, the belching smokestack became a negative symbol of the excessive price of prosperity. Oil and oil-powered electricity appeared cheap and clean; houses and industries often converted, sometimes choosing even cleaner natural gas. But many plants, especially electric power stations in the Ohio Valley between Pittsburgh and Cincinnati, were and are still fired by coal. America's vast coal reserves seemed, along with conservation, the answer to the Arab oil embargo and the energy crisis of the 1970s. Burning coal, however, especially high-sulfur Eastern coal, produces tens of millions of tons of sulfur dioxide and nitrogen oxides each year. Transported through the atmosphere, these chemicals acidify lakes and soils, reducing the productivity of forests, lakes, streams, and cropland. The National Acid Precipitation Assessment Program (NAPAP) report, released after a decade of research in 1990, discounted the most drastic predictions of environmental collapse from acid rain deposition. It questioned the part of acid rain in forest damage. It pointed to other sources of lake acidification, which it considered a localized problem of New England and the Adirondacks. But nitrogen and sulfur oxides and ozone unquestionably are unhealthy for fish, wildlife, and human beings.[31]

The NAPAP report also confirmed what critics of the 1970 Clean Air Act had been saying for years. When utilities built high chimneys in the early 1970s to meet local pollution standards set by state regulators, they were promoting the long-distance migration of sulfur and nitrogen oxides: a rearranging effect. An EPA report of 1976 concluded that using tall stacks to reduce ground-level concentrations had produced "the indirect-tendency [sic] . . . to increase the amounts of pollutant emitted to the atmosphere." Until that year, in fact, utilities routinely used what were called "intermittent control systems" that let them release more pollutants when atmospheric conditions were favorable.[32]

The EPA tightened its rules on tall stacks in 1976, but the face of power generating and smelting had already been changed, permanently. The average height of electric power station stacks along the Ohio more than doubled from 1950 to 1980, from 320 to 740 feet. In 1981, the United States had twenty smokestacks over a thousand feet high. A few of the largest reach twelve hundred feet, well above the Eiffel Tower and just short of the base of the Empire State Building's antenna. The tall stacks undoubtedly benefit their immediate surroundings. When the smelters of Sudbury, Ontario—for decades one of the unhealthiest places in North America—installed a 1,250-foot stack in the early 1970s, they complied with the Ontario government's mandate "to dilute and thus disperse the smelter's gases." Sudbury's air was cleaner, but the result was a massive streak of gases exiting at up to fifty-five miles per hour at nearly 700 degrees F. But in Canada as in the United States, the relief of the immediate region did not dissipate the gases harmlessly, as government and corporate officials had believed and promised they would. Instead, the tall stacks made the chemicals a still greater interregional problem. The gases can rise to twice the height of a thousand-plus-foot stack. At lower altitudes, sulfur dioxide tends to settle on trees, structures, and soil within sixty miles or so of its source. Propelled instead to much higher levels, it remains in the atmosphere and undergoes a series of chemical reactions in which pollution is converted to sulfuric and nitric acid suspensions called aerosols. These can join with other sources of pollution to form acid concentrations that ride prevailing winds toward the Northeast and Mid-Atlantic states, sometimes for hundreds of miles. The resulting pollution may cut the autumn view from Shenandoah Park's Skyline Drive from its former sixty to seventy miles to ten miles or less. In the Virginia mountains, with little or no soil, half the fish species of some rivers (and all the acid-sensitive invertebrates) have disappeared. Summer rainstorms bring the acids down to earth. Many lakes and rivers are naturally "buffered," alkaline enough to neutralize the transported acids. The Ohio Valley itself, with its limestone soils, is far less vulnerable to acid rain than the Adirondacks and Canada.[33]

Compounding the irony of the tall stacks is another revenge effect. Beginning in the 1950s, electronic precipitators in chimneys were trapping the soot particles that had once deposited films of grime on neighboring buildings. Only decades later did research show that the soot had actually protected its surroundings even as it had dirtied them. Sulfur and nitrogen oxides in smokestack emissions reacted with it to form environmentally benign compounds. This had a parallel in Washington, D.C. There, until recently, the nightly illumination of the Lincoln Memorial attracted so many midges that spiders feasted on them, and sparrows on the spiders and midge debris; attempts to scrub away bird droppings and spiderwebs only compounded the damage by infusing automobile exhaust particles into the marble, weakening it. (Changing the timing of the lights appears to have helped.) In each case, suppressing a visible and symbolic problem had the revenge effect of magnifying an even more pervasive and serious one. Cleaning up the soot probably was still a net gain, especially for the neighbors of the plants—particulates now appear even more hazardous than they did forty years ago—but not as large a benefit as it seemed. In fact, a cleanup can even *produce* pollution under special circumstances. The restoration of Michelangelo's Sistine Chapel frescoes, which a minority of scholars continue to denounce as vandalism, has helped increase tourist visits. Yet well before the project's completion, the heat of visitors' bodies, vapor of their breath, and sulfur in the dust were already creating an invisible acid rain within the chapel.[34]

Some of the revenge effects of acid rain in the 1970s and 1980s were more political than technological. As Bruce A. Ackerman and William T. Hassler showed in *Clean Coal: Dirty Air,* a "bizarre coalition" of Eastern high-sulfur coal producers and environmentalists influenced the 1977 amendments to the Clean Air Act to require all plants to use flue gas desulfurizing technology—"scrubbers"—instead of old-style, cheap water treatments or shifting to low-sulfur Western coal. Scrubbers, especially early models, are sensitive systems that need constant monitoring and skilled maintenance. They generate sludge by the ton. Because existing plants did not have to be retrofitted, industry had a powerful incentive to stretch out the lives of high-pollution facilities. If the Clean Air Act had not been amended in 1990 to permit polluters to sell their allowances, the result of the 1977 legislation might have been to add 170,000 more tons of sulfur dioxide reaching the East from the grandfathered Midwest plants.[35]

Some utilities and industries might have intended to dump their sulfur dioxide in other states or regions in whatever form their airborne travels and chemical encounters might give them. Some environmentalists were emphasizing this problem as the stacks were built. But most engineers, executives, and regulators honestly believed that transformation and dispersion would be largely harmless. Weren't their critics always proclaiming that the sky would fall? And some scientists did find evidence suggesting that human disruptions of the environment could cancel each other out. In a reverse revenge effect, large parts of the country may be protected from acid rain by the interaction of the acids with carbonates in the dust particles that farming and construction whip into the atmosphere. New England suffered as much as it did from acid rain just because its forests had grown back so vigorously—because it was so green. In the end it may have been not only the NAPAP report but the Rio Summit that devalued acid rain as an issue. Acid rain receded as concern over global warming advanced. The tall stacks movement shows how ironic the results can be of attacking a symbol like the dirty plume of a low smokestack.[36]

Oil Spills: Dispersing Pollution

When the *Exxon Valdez* hit Bligh Reef off the Alaska Coast in 1988, the murky discharge of 35,000 tons of crude oil was an ethical Rorschach test. To some it was simply another example of human failure—resulting from flaws of character and responsibility, and of course from drinking on the job. To others it expressed the heedlessness of corporate capitalism at its worst, the inevitable outcome of putting profits above safe operation. And to still others, the real fault was neither the captain's nor the corporation's but the consumer's: an inexorable price of the industrial world's insatiable hunger for energy. In fact, as the next chapter will argue, great oil spills threaten species diversity far less than some other consequences of shipping do. They may not even be the ugliest fruits of marine traffic. One report in 1986 estimated that ships and drilling rigs were dumping hundreds of thousands of tons of plastic debris into the world's oceans each year, the U.S. Navy alone accounting for sixty tons a day. Plastics strangle birds and seals, poison turtles, and fatally ensnarl whales. But litter, excepting medical waste, isn't newsworthy. Television spreads the more spectacular ugliness of spills electronically around the world, and it would take a very stony-hearted policy analyst to argue that shippers and governments are already spending too much money to prevent them.[37]

Decades of megaspills from the world's growing supertanker fleet before the *Valdez* affair suggest that the problem is indeed structural. The even larger wrecks of the *Torrey Canyon* (120,000 tons) in

1967 and the *Amoco Cadiz* (220,000 tons) in 1978 had already shown it was worldwide. Europe's coasts have suffered much more than America's. And the U.S. National Research Council and others have pointed to a potential technological revenge effect. Computerized design lets naval architects model the result of stresses on larger and larger ships; the *Seawise Giant* of 1980 has a deadweight of 565,000 tons, though today's very large crude carriers (VLCCs) have typical deadweights of about half of that. Far from making shipping safer, new design technology—like Humphrey Davy's mining lamp, which initially resulted in deeper shafts and more accidents—encouraged owners to push the limits of risk. They specified lighter, high-tensile steel which saved fuel but could rupture when repeated stresses began to produce small but potentially deadly cracks. Stronger steel is less ductile and more likely to break under some circumstances (in the early 1980s, wings fell off airplanes, storage tanks exploded, and hip implants cracked). It is harder to weld properly, and shipbuilders don't always provide a suitable internal framework.[38]

The search for safer designs shows that the conversion of catastrophic problems to chronic ones can sometimes be reversed. When treaties in the 1970s forbade ballast water in empty petroleum tanks, they slashed the steady pollution of seas and harbors from the flushed water. Unfortunately, oil in the higher-riding ships with separate ballast tanks has ever since been under greater pressure relative to the sea. If a hull is ruptured, more oil pours out. The most popular recommendation for tanker safety, adding a second hull about a meter or more from the ship's exterior, might let petroleum vapors seep into the space between hulls and explode. Safety inspections, already covering twelve hundred kilometers of welded seam on the vast single hulls, could be so daunting that more sources of leaks could be missed. Keeping tanks partly empty to reduce pressure relative to the sea might retain most oil within tanks in small accidents, but it could increase dangerous stresses in high waves. Pumps to maintain negative pressure after accidents could also promote explosions. And owners argue that larger numbers of smaller, safer ships might actually result in more accidents and more oil spillage than conventional supertankers do now. (They also warn of a social revenge effect: if liability under the U.S. Oil Pollution Act is too risky, serving the U.S. market will be left to doubtful operators.) Fortunately there are technological solutions that in time will reduce the risks of marine accidents. Intermediate decks can limit spills from the largest tankers. Automatic ocean sounding and Global Positioning System satellites can cut the time and expense of producing more accurate and comprehensive nautical charts. And offshore petroleum receiving stations like the Louisiana Offshore Oil Port, linked to the mainland with underwater pipelines, appear to reduce the risk of collision.[39]

Once a spill does happen, it is ugly. There is a grandeur in natural hazards, even where they devastate natural habitats, as the eruption of Mount St. Helens did in Washington State and Hurricane Andrew did in the Southeast. The horrific images of an oil slick, the struggles of oil-coated seabirds and mammals, the contaminated shorelines—all assault cameras and consciences. They call for technologies to repair what technology has wrought. Unfortunately the record of cleanup technology has so far been filled with revenge effects of its own. England experienced some of these problems as early as 1967, when napalm failed to burn off the oil from the *Torrey Canyon* and shores and harbors were treated to ten thousand tons of chemical dispersants. These turned out to kill many of the remaining crustaceans and other animals and plants that the oil itself had spared. Even in the 1990s, dispersants have potential revenge effects. By breaking down petroleum into minute globules that will mix with water and sink below the surface, they relieve some of the unsightly signs of a leak. They also keep blobs of oil from washing up onto the shoreline and from contaminating sediments. But as the petroleum is free to sink below the water's surface, it is also more likely to harm the reproduction of organisms on the seafloor, from fish eggs to lobsters.[40]

Tidying up an oil spill mechanically can have even more serious revenge effects. The $2 billion cleanup of the *Exxon Valdez* disaster relied heavily on hot water applied to the shoreline through high-velocity pumps—Exxon's response to outrage. A later independent report for the Hazardous Materials Response unit of the National Oceanic and Atmospheric Administration (NOAA) showed just how unexpected the consequences of purification could be. David Kennedy of NOAA's Seattle office explained: "The treatment scalded the beach, killing many organisms that had survived the oil, including some that were little affected by it. It also blasted off barnacles and limpets. And it drove a mixture of sediment and oil down the beach face, depositing them in a subtidal area richer in many forms of marine life—one where there hadn't been much oil." A report commissioned by NOAA suggested that the high-pressure cleanup had disrupted rock-surface ecology by destroying mussel and rockweed populations, "relatively tolerant" to petroleum, making surfaces more vulnerable to waves and predators. Fewer mollusks also encouraged opportunistic algae to preempt surfaces from rockweed and red algae. The oil flushed from the surface killed hard-shelled clams and crustaceans in intertidal and subtidal zones. It apparently also reduced the productivity of the eelgrass that shelters young fish and shrimp. The water pressure itself was even more damaging. At up to one hundred pounds per square inch it disrupted the natural sediments of beaches, both gravel and sand, smothering clams and worms.[41]

Rescuing animals from spills may also result in revenge effects. Of the 357 sea otters saved from the spill and treated by veterinarians and volunteers, 200 could be returned to the sea. A number of biologists now believe, though, that these spread a herpes virus to otters in eastern Prince William Sound that had avoided the spill itself. The transplanted otters also died in unusual numbers. While some form of the virus appears to be endemic in the waters off Alaska, the treated otters had lesions that could have transmitted the disease, or a more virulent strain of it. Stressed animals are potentially dangerous, and some biologists and veterinarians now favor keeping them in captivity. They also point out that efforts to save the most seriously injured otters may only have made them suffer longer.[42]

The *Torrey Canyon* and especially the *Exxon Valdez* disasters show the perils of purification. Contamination anywhere in the world can become so unbearably visible that it seems to cry out for equally televisable remedies. Exxon officials still defend cleanup methods even though most scientists now believe a less costly strategy—for Exxon, too—would have been more effective. This does not mean that cleanup never works. It should not discourage us from reducing and correcting our degradation of nature—whether chronic or acute. But the big marine spills underscore how complex natural systems are, and how creative and flexible human management of them has to be. It is fortunate for us that crude petroleum seeps into the ocean on its own, since natural selection has already engineered bacteria that thrive on it. In fact, the existence of natural pollution of Prince William Sound sped its recovery; local spruce trees produce hydrocarbons related to those in the Prudhoe Bay crude spilled by the *Exxon Valdez.* (Refined products are less likely to find preadapted bacteria.) Recovery rates have varied from one site to another, but a study by the Congressional Research Service specialist James E. Mielke underscored the ability of marine ecosystems to recover from even severe impacts. Fishing and hunting have a far greater impact than oil spills do on those species that are harvested; most species recolonize polluted areas quickly.[43]

The closer we look at marine oil pollution, the less catastrophic and the more chronic it turns out to be. During the 1980s, global oil pollution from sea and land disasters like spills, shipwrecks, and fires declined from 328 million gallons to between 8 and 16 percent of that figure annually. In 1985, tanker accidents accounted for only 12.5 percent of oil pollution—not much more than natural marine seeps and sediment erosion. In spite of the ban on ballast water in tanks, routine bilge and fuel oil pollution from tankers was almost as bad a problem as tanker accidents, and other "normal" tanker operations caused as much pollution as the last two combined. Municipal and industrial sources put nearly three times as much oil pollution in the sea as all tanker accidents combined.[44]

On land, too, it is smaller leaks, seepage, and waste-oil storage—not catastrophe—that pose the most dangerous threats to both wildlife and human health. The U.S. Fish and Wildlife Service has estimated that more than twice as many migratory birds died after landing in open ponds and containers of waste oil in five Southwestern states alone in one year as were lost in the *Exxon Valdez* spill. Tank farms and pipelines on a Brooklyn site have been slowly leaking over one and a half times the spill of the *Exxon Valdez*. Another tank farm in Indiana is being forced to remedy leaks that could have been three times as large. Rusting pipes, bad welding, leaking valves, and sloppy maintenance account for most of the loss. Leak detectors are so unreliable that in January 1990, 567,000 gallons of heating oil were discharged from an Exxon facility in New Jersey where warnings had been ignored after twelve years of false alarms. And the problem extends to the retail level. Richard Golob, publisher of an oil pollution newsletter, has calculated that at any time, 100,000 of America's 1.5 million underground fuel storage tanks are leaking or starting to leak; the safer tanks that local service stations are required to put in their place may leak anyway after careless, cut-rate installation. In fact, the chronic leakage problem can turn into a catastrophic risk of explosion if the electrical conduits needed by the new systems are not sealed expertly.[45]

Back to Nature? The Revenge of the Stove

While there are many low-intensity alternatives to conventional power, they are no more immune from revenge effects than the massive technologies that we worry so much about. Early-nineteenth-century Americans heated their houses and water with wood-burning fireplaces and stoves. Coal, especially cleaner-burning anthracite, became the preferred fuel by the turn of the century, but the open wood-burning hearth retained powerful symbolic ties to the family life of a more wholesome past. Even now, a new working fireplace is one of the few residential improvements that (in contrast to a swimming pool, for example) is likely to increase the market value of a house more than it costs. But there are revenge effects in striving to be natural.

One is the extraordinary demand on land. Michael Allaby and James Lovelock have estimated that heating a three-bedroom house entirely with wood needs three hectares (about 7.4 acres) of woodland for a sustained yield of fuel. Widespread use of wood in populous areas of the United States—and few Americans now live outside cities and suburban agglomerations—would mean diverting cropland or second-growth forest to firewood plantation use. The energy cost of cutting, air-drying, and transporting firewood long distances is not trivial.[46]

Pollution is an even more serious revenge effect of the urge to live naturally. Wood is innocent of contributing to acid rain; it has negligible sulfur. Burning wood produces carbon dioxide, as other organic fuels do, but growing trees absorb CO_2, evening the score. But as Allaby and Lovelock point out, wood smoke contains chemicals like benzo(*a*)pyrene, dibenz(*a,h*)anthracene, benzo(*b*)fluoranthene, benzo(*j*)fluoranthene, dibenzo(*a,l*)pyrene, benz(*a*)anthracene, chrysene, benzo(*e*)pyrene, and indeno(*1,2,3-cd*)pyrene—known carcinogens shared with cigarette smoke, which after all is just another product of burning vegetation. The more slowly and efficiently stoves burn, the more of these chemicals they produce. In addition, a stove operating at low temperature through the night may emit thirty-five to seventy grams of particles an hour; some of these particles are now thought to suppress the immune system, increasing the likelihood of respiratory illness. Many householders magnify the problem with stoves that are too large, producing more emissions (and hazardous creosote in the chimney) when operated at the necessary lower temperatures. And atmospheric emissions from thousands of small units are much more difficult to trap or neutralize than those from large, central ones. Towns in the valleys of New England, the mountain states, Oregon, and Washington were so susceptible to woodstove pollution that many began drastic limits on woodstove use. Telluride, Colorado, even made a homeowner wanting to install a new stove pay two other people to remove theirs. Missoula, Montana, installed a light on a water tower to warn residents to extinguish wood fires during air pollution warnings.[47]

By 1987 the U.S. Environmental Protection Agency (EPA) had issued guidelines to cut the particulate emissions to five grams per hour. Prices of some models jumped by hundreds of dollars as manufacturers added catalytic converters. Production of new woodburning stoves declined by 80 percent in the late 1980s.

Today a well-designed stove or enclosed hearth can be an efficient way to use wood that would otherwise be wasted. But the failure of wood burning as an alternative energy movement proved that simpler, smaller, and more natural doesn't necessarily mean healthier. It also points once again to the difference between nineteenth- and twentieth-century environmental problems. The wood or coal smoke that often descended over our ancestors' towns and cities was a health menace but a localized one. Today's vast central generators cause far less environmental damage per unit of output. That's why people are working to replace millions of internal combustion engines with battery-stored electricity from central generators.

Catastrophe is hardly a thing of the past. Natural hazards are damaging property at rates well beyond inflation and the growth of the national income. In spite of political innovations like floodplain zoning, we are pushing toward the very areas where nature puts us most at risk from tropical storms, mudslides, and forest fires. Our political and economic institutions offer powerful incentives to ignore danger. In Virginia, three thousand dwelling units and businesses have been built on a sandy strip in Hampton Roads Harbor originally formed by a hurricane nearly 250 years ago; an equally powerful storm could wipe it out. A direct hit on Miami could have tripled Hurricane Andrew's cost of $25 billion. If Hurricane Hugo had come ashore in Charleston, South Carolina, rather than at a nearby park, a twenty-foot wave of water would have devastated the city. A big storm could leave twenty feet of water in downtown New Orleans and flood evacuation routes. Around the world, weather can still break up supertankers with far more petroleum on board than the *Exxon Valdez* or even the *Torrey Canyon.*[48]

Federal disaster aid and private insurance, plus a cyclical lull in major storms, have helped reduce perceived risk and have encouraged a 50 percent increase in shore residents on the Gulf Coast and East Coast in only twenty years. With each generation, part of the collective memory of the last terrible events is lost. When the cycle returns, its fury falsely seems unprecedented. Meanwhile technological and social devices for protection—whether physical barriers, evacuation procedures, or disaster relief—are overwhelmed. Once more, our cleverness may catch up with us.

When we think of advanced technology, we usually imagine structures and devices that work with less and less human intervention. New automobile engine and fuel injection systems have added months to routine maintenance intervals. Computers need professional servicing less often than mechanical typewriters did. But the technological "solutions" for catastrophic risks show another, unintended side of technology. Safety takes *more* human attention. Sometimes the technology is traditional: maintaining levees and dikes. But new technology requires regular work, too: from testing residential smoke detectors to inspecting double-hulled tankers and maintaining earthquake-reinforced bridges and elevated roads. Our past success in suppressing forest fires and in deferring the natural migration of beaches has tied us to the long-term task of keeping up the artificial regimes we have created. The maintenance compulsion that Albert O. Hirschman identified—the necessity for vigilance that technology imposes—applies as much to natural hazards as to human systems like roads and aviation.

5

Promoting Pests

As terrifying as environmental catastrophes may be, they have a redeeming virtue: they are self-limiting. After an earthquake has released its energy, a fault may move little for tens, hundreds, or even thousands of years. A storm dissipates at sea. Even without human suppression, a forest fire runs its course. A power plant spews sulfur and nitrogen compounds and particulates only so long as it is in operation. A tanker or pipeline has only so much fuel to leak.

The pest, usually a less visible hazard than the physical or chemical one, is a more persistent, open-ended one. A pest is any plant or animal that flourishes by taking advantage of human-made environment change—in a way that injures human interests. The alteration can be the result of either a modified or a disrupted habitat. It can also be a means for migration to new territory. Whatever is changed, the pest competes successfully for resources with existing organisms, either domesticated or preferred wild ones, and may soon displace them entirely. Pests may also be directly or indirectly harmful to human health. Natural hazards are feared, but pests more often than not are hated as well.

Although infrequent but dramatic oil spills are more visibly destructive, the daily routine of world shipping has killed far more wildlife and endangered more species by spreading pests than by fouling seas and shores. John Balzar, writing in the *Los Angeles Times*, acknowledges that the *Exxon Valdez* spill killed hundreds of thousands of birds, but notes that by spreading rats to over 80 percent of the world's islands, shipping is condemning millions more. In the last four hundred years, rat predation has eradicated more species of land and freshwater birds—mostly on oceanic islands—than all other causes combined. Once established, rats are so difficult to eliminate that when they were suppressed on a New Zealand island with a surface of less than a square mile, a documentary film commemorated the feat. Environmentalists now recognize that the greatest threat to the abundant wildlife of the Aleutians and the Pribilof Islands is not from supertankers but from tiny creatures stowing away on smaller ships and barges. Pamela Brodie of the Alaska chapter of the Sierra Club told the *Times*: "It bothers me personally how we set our priorities. We tend to ignore the chronic problems—which can be much more serious—in favor of the occasional accidents."[1]

Pests are a problem in environmental ethics, including rats—our fellow omnivores, camp followers, and laboratory surrogates. We despise them. Few people would begin to weigh a rat's right to eat birds against the preservation of native bird colonies—even though similar birds were routinely exterminated in a designated wildlife sanctuary when they were found to endanger planes at New York City's Kennedy Airport. Where should we draw the line in correcting the effects of our own disruptions? Thanks to our garbage, rats now mature more rapidly in cities, and grow larger, than they did in their original grassland habitat; house cats, once their ancient enemies, now are as likely to feed alongside them as to devour them. The cowbird is a native bird, but now it threatens songbirds as a nest parasite because our patterns of agriculture, settlement, and road building have broken up woodland habitats.[2]

It is one thing to build nesting boxes for Eastern bluebirds that our agriculture has displaced and endangered. It is quite another to trap and drown the house sparrows that occupy these boxes, wring their necks, or put them in a sack tied to an automobile exhaust pipe—all techniques recommended in a pamphlet distributed by the North American Bluebird Society. Is it the house sparrows' fault that they too fit the holes designed for the bluebirds? More to the point, are they to blame if development has dangerously reduced the number of the bluebirds' preferred nesting places, decaying trees?

To many environmentalists, any intervention, even in favor of "struggling" species against "aggressive" ones to correct the results of human intervention, can amount to biological fascism. One distinguished environmental historian wonders whether a campaign to eradicate invasive plants in the Everglades might not be Nazi in spirit. The garden writer Michael Pollan and others have noted that Heinrich Himmler supported a movement to promote native German plants and garden designs to the exclusion of foreign organisms and landscape ideas. Other gardeners and nurserymen deplore a prejudice against new and useful plants, regarding as futile the search for an authentically native landscape. Even the smallpox virus has its advocates. Many microbiologists, ecologists, and philosophers question the ethics of destroying the last remaining laboratory stocks of it fifteen years after the last known naturally transmitted case of the disease. Some believe pragmatically that we can never be sure we will not need the organisms in the future. Others object that we have no right to extinguish any other life-form.[3]

Pests reflect humanity's stunning success in modifying its environment by intensifying production. Farming land means replacing a natural plant community and, usually, growing a single crop. When the original cover is removed, long-dormant seeds of pioneer succession plants germinate, and others invade the disturbed land. Even a low-yield, preindustrial field of crops is an artificially uniform habitat. Its homogeneity rewards organisms, notably insects, that specialize in eating it. Natural selection assures that these organisms appear and flourish. Recently this process has accelerated. As farmers mechanized in the late nineteenth century, they also bred and selected their crops for easier machine processing. These uniform cereals, vegetables, and fruit also invited pest specialists. Moving a crop to a new habitat may turn a previously unimportant species into a serious pest, as happened when potatoes were introduced to the American Southwest from South America. A local insect that had limited its diet to the wild sandbur took on a new menace, and identity, as the Colorado potato beetle. And the twentieth century's Green Revolution of higher-yielding cereals has produced not only more food for people but greater losses to insects, weeds, and microorganisms. Farmers traditionally had selected for adaptation to natural conditions—including characteristics that resisted local pests. By the 1970s, pathogens, weeds, and pests were shrinking the world's food production by about half, if losses both before and after harvest are included.[4]

The Hazards of Improvement

It's a truism that human life is impossible without some pollution. Because, as we have seen, a prosperous industrial society seems to be good for people's health, a decline of economic output threatens it. Morbidity and mortality rates from a number of causes jump during depressions. This doesn't mean we can't make large gains. Most consumer goods could still be made to be more energy-efficient. Americans could shift (back) to the denser land-use patterns typical of Europeans and Japanese if they had the political will. Products could be redesigned for easier recycling. New control devices are already reducing industrial emissions. Many or most of these matters involve not revenge effects but trade-offs: price-competitive consumption goods against environmental values.

Consider polluted harbors again. Despite the continuing problems of oil seepage and chemical releases, most of America's and Europe's rivers and harbors are cleaner than they have been in years. Fish species have returned that were absent for decades.

Too bad cleanliness is part of the problem. We have already seen that soot from smokestacks helped neutralize acid rain before it was precipitated out, and that dust that farmers and builders pour into the atmosphere can do the same. The clouds of mud and toxic substances in America's harbors, we are learning, had surprising benefits. While killing game fish or making them inedible, they also poisoned the animals that attacked wooden harbor structures. Harbors became dirtier than ever but, for human purposes, more stable.[5]

Reducing pollution changed all this. It helped not only sea bass but the marine animals that feed on the wood in piers and bulkheads. These crustaceans and mollusks, like many other pests, have been associated with humanity for centuries, plaguing Christopher Columbus himself. The tar and oil of the mid-nineteenth century killed off the creatures. These substances were concentrated enough in parts of the harbor that merely sailing into port could treat a ship's hull free. Later generations of petrochemicals also suppressed the populations of boring animals and even deposited a protective film on steel structures.

By the 1980s, wooden piers in the cleaned-up harbors were showing signs of renewed attack. Siphon-equipped shipworms (*Teredo navalis*), mollusks growing to nearly an inch in diameter and two feet long, were burrowing random channels in the wood with their spiny shells. Tiny gribbles (*Limnoria lignorum*), finishing the job, chewed the hollowed-out pilings to their core. The efficient intestinal bacteria of these organisms helped them to reduce timber at rates that amazed New York Harbor officials by the early 1990s—the diameters of pilings were reduced from up to a foot to only a few inches in just two years. For unexplained reasons, whether harbor structures contained different wood or the creatures have improved their performance, 150 years earlier, in the 1840s, it took not two but up to seventeen years to destroy a piling. Although coating structures with creosote and other protective chemicals can slow the attack, in another resurging effect, creosote-resistant strains of boring animals soon appear, while the chemicals eventually pollute the harbor themselves. Plastic coatings work but need laborious wrapping.

Marine borers do have surprising positive uses. The shipworm's method of lining the tube it digs with a chemical from its body reportedly inspired Marc Isambard Brunel's movable shield for constructing tunnels—a method still used after many refinements over a century later. And the harbor crisis is helping create a new market for millions of recycled plastic milk cartons in the form of borer-proof piers. Improving the environment may improve the environment after all. But until present structures are phased out, the borers are yet another chronic problem, and one estimated to cost hundreds of millions of dollars in New York Harbor alone.

What shipworms and gribbles do to wood and harbors, the shipworms' cousins the zebra mussels (*Dreissina polymorpha*) are doing to waterworks, rivers, and lakes—but on a scale of billions rather than hundreds of millions of lost dollars. Over more than two hundred years, these two-inch-long, gray-striped mollusks extended from their habitat in the Aral, Black, and Caspian seas and the rivers feeding them to the waters of western Europe. In German-speaking Europe their rapid spread earned them the romantic popular name *Wandermuschel.* By the early nineteenth century they had turned up in England. Not until the 1980s were they able to join the tide of immigration and trade to North America. Already in the late nineteenth century, the practice of taking on harbor water as ballast (as opposed to dry bulk materials like bricks, cobblestones, sand, and lead) accelerated the worldwide dispersal of marine organisms. The discharge of ballast water in harbors exchanges plankton and hordes of small organisms. A ship's tanks may carry dozens of species, and a harbor in Oregon has become home to fully 367 nonnative taxa. But the zebra mussels have been unique in their visibility, range of dispersion, and immediate destructiveness.[6]

Within Europe, zebra mussel eggs and free-swimming larvae (veligers) traveled locally in ballast water; but transatlantic voyages were long enough to kill any young mussels that had managed to survive the filth of the harbors. Nor could adult zebra mussels endure the salinity of the open ocean. Europe built its public waterworks just as the mussel was spreading in its rivers and lakes. European systems were engineered to minimize the impact of obstructions by the mussels. Today, some have dual intake systems that are shut down alternately for cleaning. European engineers also specify short pipes and shun the right-angled sections where mussels flourish.[7]

In the late twentieth century, mussels crossed the Atlantic barrier as they had traversed the English Channel 150 years earlier. Marine scientists are not sure why the mussels spread so much more rapidly in North America than in nineteenth-century Europe. There, they were present in Germany by the 1830s but did not reach Swiss lakes until the 1960s. On this continent, they were first found in Lake St. Clair only in 1986, yet were established throughout the Great Lakes and major Midwest river systems by the mid-1990s. Part of the reason may be the disturbance of the Great Lakes ecosystem by logging, but a series of otherwise positive innovations prepared their way. After the Second World War, European harbors (like American ones) became cleaner, and ships became far faster, taking as little as a week to cross the Atlantic and thus retaining more oxygen in their ballast water. With the opening of the St. Lawrence Seaway in 1959, ships from Europe with their bivalve stowaways moved directly into the Great Lakes, and the mussels could invade the rivers feeding them. Unaware of the potential disaster, U.S. and Canadian governments initially allowed unlimited release of water ballast in the Great Lakes. The size of the ships and the volume of water released were massive by nineteenth-century standards, possibly accelerating the spread of the mussels. Despite restrictions in the 1990s, zebra mussels now circulate throughout the fresh waters of North America. The 1993 Mississippi floods heightened their circulation in the Midwest. Now they use not only ballast water of commercial ships on the Great Lakes, but the bilgewater of pleasure boats—not to mention the hulls, hull openings, outboard and inboard motors, pumps, anchors, propellers, shafts, and other ship parts. Worse yet, fishery employees may be unwittingly helping to destroy their future catch by moving the destructive mussels among natural waters and hatcheries as they stock rivers and lakes with game fish.[8]

Zebra mussels retain in adulthood a mass of threads called a byssus that lets them attach themselves to many surfaces. They use it. In North America they have alarmed manufacturers and public works authorities by proliferating in the intake pipes of factories, power plants, and water systems, blocking the flow of water. They also clog the water channels of locks and dams along the Mississippi, accumulating on the interior walls of pump valves and threatening to block the flow of water needed to cool air compressors operating the locks. One female can produce from forty thousand to one million larvae a year. Tens of millions of adults have been found in two-foot-diameter pipes. In the intake pipes of a large Midwestern electric plant, mussels have reached a density of 700,000 per square yard.

The mussels threaten not only public works but other aquatic organisms. They attach themselves to native mussels, smothering them and raising concerns about extinction. (Illinois has suspended the harvesting of native freshwater mussels until more is known about the prospects for their survival.) Their numbers and efficient filtering reduce the supply of nutrients to fish and to other shellfish. Although zebra mussels, like other exotic organisms, often are cited as "natural pollution," they ironically clarify lakes and streams even while they impoverish them. Where excessive algae are a problem, the mussels can work apparent wonders. Russians use them to purify canals; Netherlands biologists are studying their value in managing lakes. (In some Dutch lakes, mussels filter the entire body of water in a month or less.) Wisconsin scientists have found that they eliminate 95 percent of the parasitic cryptosporidium protozoa that infected hundreds of thousands of Milwaukee residents in 1993. But the mussels' positive side has a negative side of its own; in absorbing contaminants, the mussels transfer them to lake beds and shorelines, poisoning some species of ducks.[9] Conversely, the most popular and effective treatment, chlorine, is toxic to other organisms. Some biologists hope for deliverance by natural predators. Divers in Lake Michigan have found that native freshwater sponges of the Spongilla family share the zebra mussels' attraction to waters moving around breakwaters and piers. These sponges grow around the mussels, which are immobilized by their own filaments.

Parasitic worms and microorganisms can be as opportunistic as mussels and crustaceans in taking advantage of environmentalist measures and healthier eating habits. In the late 1980s, physicians on the West Coast and in Hawaii began to notice dozens of cases of painful infection by a nematode larva, *Anisakidae*, one of the revenge effects of environmental protection. Marine mammals have been flourishing under legal protection for over twenty years; sea lions alone have multiplied sixfold, reaching an estimated population of 177,000 by 1992 and decimating the steelhead population of the West Coast. Even more serious for people is what happens to the resurging mammals' parasites. Tiny crustaceans ingest from their feces and spread to fish—some of which are eventually eaten uncooked as sushi and sashimi, creating a small epidemic.[10] Residents and vacationers on Italy's Adriatic coast have a related problem: the chemicals that have replaced phosphates in "environmentally friendly" European detergents have eliminated algal blooms, but in interacting with other minerals and chemicals found naturally in waters, they have also formed floating mats nourishing vast colonies of bacteria. These microorganisms in turn secrete a slimy, malodorous foam to attach themselves to the mats. Boaters are nostalgic for the algae, and phosphate lobbyists are campaigning to make their material legal for detergents again.[11]

Of course, it was not foolish to clean up harbors. And water ballast costs much less and is far easier to control than other ballast materials. There are effective ways to slow the inadvertent spread of marine organisms; exchanging ballast water on the open seas will kill the stowaways. Redesigned marine structures and a variety of chemical and physical weapons are reducing some of the expected damage. But zebra mussels and other shipboard migrants will never be eradicated, only controlled. Species like the Asian clam *Potamocorbula amurensis* in San Francisco Bay are transforming their surroundings radically. They are another case of a chronic problem for which all solutions demand an added degree of vigilance.[12]

HAZARDS OF IMPROVEMENT: THE HOUSEHOLD

The quest for comfort can be as hazardous as the pursuit of environmental purity. Living standards in North America and Europe have increased substantially over the last fifty years. As Ruth Schwartz Cowan has shown, twentieth-century domestic technology did not mean the end of the household as a productive unit: "Households are the locales in which our society produces healthy people, and housewives are the workers who are responsible for almost all of the stages of the production process." Many household technologies, including those that lead to repeating effects (like more frequent washing), really do make people healthier. Central heating, for example, has caused chilblains—skin inflammations following prolonged exposure to damp cold—to become relatively rare.[13]

But even home comforts can be hazardous to our health. Some have nothing to do with pests or bacteria. Vacuum cleaners and shampooing machines helped promote the wall-to-wall carpet as an emblem of middle-class comfort and health for generations. After the Second World War, manufacturers introduced a new method of gluing tufts of synthetic fiber to backing, which was originally jute and then polypropylene. As the near-universal choice for houses and offices alike from the 1950s, the new technology seemed to promise cleanliness; nobody can sweep dirt under a nailed-down rug. But environmentally it was (and is) an unknown quantity, especially in the "tighter" houses and offices of the last twenty-five years, in which insulation and weatherproofing have reduced the exchange of indoor and outdoor air. The fibers, latex glue, and backing are fused in an oven in a complex process that can leave two hundred known substances in the carpet. Many people believe chemical emissions from their carpets are making them sick, and a few have filed consumer complaints and lawsuits. Carpet emissions, in some experiments, appear to kill mice, although scientists at the Environmental Protection Agency have not been able to replicate the best-known study, by Anderson Laboratories. Still, years before the Anderson tests, when nearly a hundred EPA employees became ill in its redecorated office in 1987 and 1988, the carpeting was blamed. The agency subsequently removed nearly 27,000 square yards of carpet and introduced a carpet- and chemical-free zone for sensitive staff.[14]

In fact there are proved health hazards of wall-to-wall carpeting. One is occupational: installers risk arthritis and other joint problems when they kick material into place with knee-mounted stretchers. But the problem affecting the largest number of people is that wall-to-wall carpets seem to have helped spread a pest—dust mites—that has in turn favored one of the most chronic diseases of childhood and adulthood: asthma.

We have already seen how hygiene can foster sickness: how young upper-middle-class adults were at greater risk of polio than age-mates with a dirtier upbringing. Hay fever is another disease associated with higher living standards. When the English physician Michael Bostock wrote the first description in 1819, he was one of the few known sufferers. In fact, as hay fever and other allergies multiplied in the nineteenth century, it was not working-class children growing up amid industrial haze but instead the scions of the best households that were affected. Epidemiologists are beginning to believe that large families, messy play, and early infections could have helped condition children's immune systems not to gear up against a common substance like pollen when they first encountered it. The protein that mediates hay fever, IgE, appears designed to defend the body against worm infestation. The allergist and historian Michael Emanuel has speculated that hay fever results from IgE deprived of its original target, noting that "man evolved with his parasites and there may be a price to pay for their removal." (Other medical historians believe nineteenth-century industrial emissions and the rise in smoking were largely responsible. Bostock himself grew up in the industrializing North and worked with harsh laboratory chemicals.)[15]

Comfortable middle-class households turned out to be technological and social systems that produced not only healthy people but chronically sick ones as well. The warm, humidified, well-insulated Western home is as comfortable for pests as it is for human beings. The medical entomologist John W. Maunder has evidence of "a vast flea epidemic" throughout Western Europe and large parts of the United States; taken together, the world's fleas probably weigh more than its people. From 1991 to 1992 alone, the number of requests for flea extermination in England increased by over 70 percent. Fleas are starting to appear even in middle-class households, and pet owners have yet to admit the need for disinfecting the whole house—not just the dogs and cats. The cat flea, *Ctenocephalides felis*, spends nearly all its time in carpeting or on draperies waiting for a cat or other warm-blooded host; ten thousand may be lingering, with only two dozen on a host at any time. People in flea-infested quarters, shunned by visitors, are said to cope with their loneliness by acquiring more cats.[16]

With the virtual end of bubonic plague, fleas are more of an annoyance than a menace. But other arthropods are a more serious matter. The carpet, larger than ever, is a country club for dust mites as well as for fleas, just as the mats produced by those "environmentally friendly" detergents in Italy are floating resorts for bacteria. Cousins of the spiders, and less than a fiftieth of an inch long, the mites live on tiny flakes of dead skin in common dust. They thrive in the warmth and humidity of well-insulated, centrally heated housing. Cleaning may not work, since cold-water detergents apparently help save dust mites as well as energy. The fecal pellets of dust mites contain a powerful airborne allergen, *Der p* 1, which stimulates the immune system to inflame the airways. A study of children growing up in England showed that those from houses with high levels of dust-mite allergens were up to five times as likely as others to become asthmatic by their teens. Researchers estimated that children were being exposed to as many as 500,000 fecal particles per gram of house dust. Some of the worst risk factors are luxury goods like down comforters and pillows, and finely woven oriental rugs. But even simpler improvements in living standards may inadvertently spread mites and promote disease; the adult asthma rate increased 5,000 percent in part of Papua New Guinea after people began wrapping their heads in their recently introduced blankets at night.[17]

Vacuum cleaners, long promoted for healthy living, actually make this problem worse, according to another English study. They do suck up dirt, but they also bounce mite pellets into the air, where they may stay suspended for days on end as they sink back slowly into the carpet—a classic rearranging effect. Vacuuming can triple the density of suspended droppings. Only a few expensive vacuum cleaner models use high-efficiency particulate air (HEPA) filters that trap such micro-debris.[18]

Of course, carpets, drapes, and vacuum cleaners aren't the only promoters of asthma. Good construction and airtight insulation also contribute to the problem. Studies by U.S. Department of Agriculture entomologists suggest that cockroaches thrive in tightly built housing, and cockroach allergens are more prevalent. Materials shed by dogs, cats, fleas, mites, and cockroaches, plus secondhand smoke and industrial air pollution, might play a part, though asthma rates have sometimes risen even where air quality has appeared to improve. Poverty also increases the likelihood and severity of asthma. But since none of these risk factors is new, there is reason to think that the rise in severe cases of asthma comes at least in part from more home comforts and better insulation.[19]

Acute episodes of asthma can be fatal; about 4,600 Americans died of attacks in 1990, double the number ten years earlier. But for most of the more than ten million Americans who suffer, the disease is a chronic one. Inhalants can control symptoms and open airways. Corticosteroids produce no significant side effects when inhaled, but another common family of asthma control drugs, the beta agonists, may increase the risk of a fatal attack. If one theory is correct, pocket inhalers with metered doses of beta agonists may suppress the symptoms of an asthma attack while leaving the patient continually exposed to dangerous antigens—another medical revenge effect. The search for comfort, then, helps further a situation of prolonged discomfort in a significant minority of the population. And while the discomfort can be managed and controlled, it is only with constant vigilance, monitoring, and adjustment—the hallmarks of a chronic problem.[20]

Like the dust mites and more visible insects of the household, the most annoying organisms of agriculture are themselves revenge effects—animals that have flourished by seizing on resources we have assembled for them, or by accompanying us into territories where their natural predators are absent.

Transportation and air conditioning helped open the American Southwest for year-round living after the Second World War. A rising number of allergies sent hundreds of thousands of people from the Northeast and Midwest to seek a healthier working and retired life in the deserts of Arizona. "Send your sinuses to Arizona" became a legendary television pitch for antihistamines. And for a number of years, the arid Southwest delivered on its promise. It was never perfectly pollen-free; recent research has shown that pollen from at least one favorite native plant, the paloverde tree, may become windborne and irritate allergy sufferers. At first there probably were not enough landscaped specimens of these trees to be a noticeable problem. In any case, most native Arizona desert plants are pollinated by birds, bees, butterflies, and other insects, not by the wind.[21]

If the new residents of the Southwest had followed traditional housing patterns, building adobe houses right up to their property lines, Arizona might have stayed largely allergy-free. But the migrants did not really want to go completely native. Like settlers everywhere, they grew nostalgic for the plants they had left behind. In the 1950s and 1960s, Arizonans started to build "ranch" houses with lawns. They brought Bermuda grass with them and showered it with the water that federal projects were diverting from Western rivers for their benefit, in the best traditions of self-reliant American individualism. Golf courses, country clubs, and resorts followed. The grass became a major seasonal producer of pollen. Lawns also provide food and moisture for allergenic molds, which increased by nearly tenfold in Tucson after the migration began.[22]

Grass and spores are a relatively small problem compared with the wind-pollinated trees and plants that newcomers brought: olive, mulberry, cottonwood, sycamore, pecan, ash, and elm. Most of these produce large quantities of pollen over intervals of several months each spring. Warm temperatures sometimes extend the hay fever season to as many as ten months. Now Arizona has a relatively high concentration of people with pollen and mold allergies, about 230,000 in the Phoenix area alone, according to one estimate. Since 1985, Tucson has banned new olive and mulberry trees, with a corresponding reduction of 35 to 70 percent in tree pollen. Still, the city has to send out hundreds of letters each year reminding people to cut their grass before it grows to seed.[23]

Fortunately for Arizona's allergy sufferers, an almost pollen-free variety of olive tree was discovered not long ago in Swan Hill, Australia. Its blossom drops without opening, and the little oily pollen that it does produce is too heavy to become windborne. It is now widely available. So too are sprays to stop older trees from pollinating. Plant biologists are beginning to talk about transplanting an engineered bacterial olive gene originally developed for pollenless corn. It could, they say, produce a nearly pollenless landscape within a decade. It is always possible that the new varieties may have revenge effects of their own, but these will not be known until the new trees are widely planted. So far, none has appeared. In the U.S. Southwest, at least, technology may indeed have the last laugh.[24]

The Frustrations of Extermination

Revenge effects come not only from the quest to make our own surroundings more comfortable. They also arise from the attempt to extirpate the pests that surround us. The impulse to slaughter whatever appears to threaten livestock and crops is almost certainly as old as agriculture. In the nineteenth century, cultivators and ranchers nearly wiped out many of the larger predators of North America and Europe with the not terribly high technology of firearms, poison, traps, and habitat destruction—with regrettably little knowledge of what might truly support agriculture. Not until the turn of the century did some states eliminate the bounties they offered for dead hawks and owls. Later-twentieth-century opinion, reflecting presumably enlightened urban and suburban environmentalism, substituted protection for persecution. There is new respect if not affection for predators as capstones of ecosystems, mirrored in books and films like Farley Mowat's *Never Cry Wolf*. It has generally benefited not only the animals themselves but their surroundings.

When respect shades into sentimentality, however, revenge effects are sure to follow. Wolf hybrids, bred with sled dogs and German shepherds, can forget their lowly place in the chain of being and turn into New Age pit bulls without provocation. Rangers must warn visitors to U.S. national parks to keep their distance from bears—perhaps another unintended consequence of the Smokey campaign—and protected alligators are again a menace in Florida. Paradoxically, the one group of carnivores that has retained its ancient dread, the shark family, may be the one most threatened by humanity.[25]

Internationally, rare and protected or otherwise popular animals can become pests with the disruption of their habitats. New Zealand parrots—scavengers and omnivores before the introduction of sheep to the islands—somehow acquired a taste for the fat around the animals' kidneys, and began attacking live sheep. Pandas in China, under intense stress from human intrusion in their habitats, have been known to raid penned-up sheep. (Of course, they are too slow to threaten free-range flocks.) Gray squirrels, harmless to forests in their native North

America, have become woodland pests in the United Kingdom. The thicker phloem layer of beech and sycamore trees grown widely spaced in British tree farms ("plantations") is filled with delicious sap, which appears to encourage squirrels to strip the bark for access. In growing numbers, the white-tailed deer and Canada geese of the United States have adapted all too well to a combination of resurging forests, encroaching suburbs, clipped lawns, and corporate ponds.[26]

The most notable revenge effects come not from our efforts to control the larger pests—some of which, like the geese and deer, retain human admirers—but from our attempts to crush the omnipresent smaller ones.

Nineteenth-century farming, as its penchant for slaughtering resident wildlife suggests, was no Arcadia of stone-ground wholesomeness. The historian of science James Whorton, in *Before Silent Spring*, reminds us that American farmers used a fearsome array of toxic copper- and arsenic-based chemicals against molds, fungi, and insects. Preparations with deceptively colorful names like Paris green and London purple joined arsenate of lead and other substances as persistent poisons that endangered not only workers applying them but livestock, consumers, and, of course, the plants themselves. The burgeoning profession of economic entomology boosted the use of arsenical pesticides, dismissing the reservations of old-fashioned farmers about the chemicals' health effects and high cost. Consumers would have to eat hundreds of pounds of fruit to get sick, the entomologists countered. The head of the Department of Agriculture's Bureau of Entomology declared fears of injury from spraying to be "utterly groundless," and his professional colleagues deplored the warnings of "a few ignorant alarmists." Doggerel exhorted:

> *Spray, farmers, spray with care,*
> *Spray the apple, peach and pear;*
> *Spray for scab, and spray for blight,*
> *Spray, O spray, and do it right.*[27]

In fairness to the entomologists, manufacturers were also using arsenic liberally, even in children's toys and coloring paper. It was only after the First World War that a few entomologists and physiologists began to study how cumulative doses of lead and arsenic compounds could cause neuritis, stomach disorders, skin disease, and cancer—all chronic side effects of a more intensive style of agriculture.

The paradox of pesticides—the chemicals that were supposed to replace discredited metallic compounds—became an even greater public issue. On the eve of the Second World War, a Swiss chemist named Paul Müller found what he and an entire generation thought was a miracle insecticide, DDT. Marketed in Switzerland, it was applied to over a million residents of Naples in 1944, saving them and Allied troops from an incipient typhus epidemic. It crushed insect-borne epidemics on the islands of the Pacific theater. Remarkably, it seemed perfectly safe. As Rachel Carson later noted, the body absorbs little DDT when the chemical is externally applied as a powder, though it can accumulate small amounts of DDT in its liquid form. Thus DDT appeared to offer almost miraculous protection from acute insect-borne illnesses without raising any of the risks to human health that the older generation of inorganic insecticides had posed. Even workers with heavy, prolonged exposure seemed to suffer no ill effects. Only one case of a directly lethal exposure is known: some DDT powder, confused with flour, was cooked in pancakes. (It was actually much less toxic than the potentially lethal organophosphate chemicals like parathion deployed after it was banned. Hundreds died working with them.) Postwar entomologists, armed with DDT, prepared to crush what one popular book of the period—ironically by a future satirist of science, Anthony Standen—assailed as the Insect Invaders. The magazine *Popular Science* foresaw "total victory on the insect front."[28]

DDT came to menace the future *because* it seemed so safe in the present. Wartime entomologists discovered that aircraft could spray it economically diluted in an oil-based solution requiring a quarter pound or less of DDT per acre. After 1945, former military pilots continued the air war on bugs with surplus planes, including big transports retrofitted for spraying and dusting. Direct contact did not seem to hurt people, whereas the inorganic pesticides that preceded DDT had chronic as well as acute effects.

Slow arsenic poisoning has long been a cliché of mystery writing. Similarly, the most alarming damage done by DDT was not the immediate deaths of birds and fish from massive spraying but the effects of its invisible accumulation in their tissues and especially in their reproductive systems, including its thinning of eggshells. Eventually, the public began to worry more about the slow buildup of the compound in body fat and the risk of cancer, not to mention the hazards to those who actually worked with insecticides. Banned in 1972, DDT remains a suspected human carcinogen, but only on the basis of limited animal tests. Like other revenge effect technologies, DDT defused one problem by diffusing another. Yet abandoning DDT has had revenge effects as well, especially the contamination of aquifers by new generations of water-soluble pesticides.[29]

DDT also showed for the first time the power of the resurging effect. In the teens of this century, entomologists began to report that some orchard pests were starting to resist the inorganic chemicals used against them. Cases remained relatively uncommon, though, because these chemicals attacked multiple sites in the target animals; they acted in a way that offered little scope for metabolic defenses. DDT and other synthetic organic substances changed all that. They opened the door to a new class of natural defenses: metabolic enzymes that could dexotify the new poisons.[30]

There is a perverse logic to the spread of resistance genes. We saw it in antibiotic-resistant bacteria. The more effective a pesticide, and the more widely and intensively farmers apply it, the greater the potential reward for genes that confer immunity to it. In Sweden and elsewhere in Europe and North America, DDT-resistant flies appeared as early as 1947. By the mid-fifties, only ten or fifteen years after the Naples campaign, body lice in many parts of the world were already unaffected by DDT treatment. So were many farm, orchard, and forest insects in the United States.

The resurging effect was not limited to the unnatural selection of chemically self-protected strains. DDT actually fostered the reproduction of some insects by killing their natural predators. In the rubber and oil palm plantations of Malaysia, where there had never been a serious insect problem despite the hot and rainy climate, the application of DDT for a relatively small infestation of cockchafers led to a new plague of caterpillars, followed by still heavier insecticide doses and widespread defoliation. An entomologist called to investigate the case discovered that the poisons had wiped out the wasp parasites that had kept the most troublesome caterpillars in check. Unlike the wasps, the caterpillars could shield themselves from the poison by curling up while it was applied. Even after the poisonings were stopped and the parasites returned, the caterpillars were still on the rampage. The poisoning apparently had synchronized their life cycles so that vast numbers of caterpillars appeared at once, overwhelming the wasps. Stomach poisons harmless to the wasps finally kept caterpillars in check. In North America and Britain, use of DDT (beginning in 1949) against another insect, the codling moth, promoted the red spider mite to a major pest of fruit trees, again by killing its natural insect predators. In fact, in small doses DDT even appeared to make the mites breed faster, by mechanisms that still are not understood. Fortunately gardeners of the 1990s now can order other, harmless, mites that prey on red spider mites and die off when their hosts are gone.[31]

Deployment of DDT revealed yet another revenge effect. Like antibiotic-resistant bacteria, strains of insects surviving DDT (or any other widely used chemical) tend to resist other compounds as well. Insecticides select not only for defenses against themselves, but for genes that make the animal better adapted to other aspects of its environment. Once this selection has taken place, it seems to make a population of insects more resilient in general. Even after DDT or another insecticide is withdrawn and immunity is gradually lost, it is reacquired more quickly. Intensification of the battle against insects seems to harden the enemy's defenses through natural selection. Megadoses build superbugs. While DDT has been banned in the United States for over twenty years, worldwide resistance to pesticides of all kinds continues to grow steadily. There were over five hundred resistant insect and mite species alone in 1990, and overcoming resistance requires higher doses and less cost-effective alternatives. Despite the vast improvement of agricultural yields since the Middle Ages, the proportion lost to insects, diseases, and weeds combined has not changed in the last fifty or five hundred years: it remains about a third. Internationally, the return of malaria has been due less to local bans on DDT than to increasing resistance of both mosquitoes and malarial parasites to pesticides and drugs of all kinds.[32]

Insects learn to resist not only poisons but environmentally friendly chemicals like those that disrupt their development by mimicking juvenile hormones. In the 1960s, insect physiologists considered hormone mimics resistance-proof, but an experiment using a mutagen showed that a single genetic change could increase resistance by a factor of one hundred. Used intensively enough, even natural control agents like hormones, parasites, and predators can select for resistant pests. After heavy applications of the bacterium *Bacillus thuringensis* (Bt)—hailed as an effective, nontoxic agent—some pests have become resistant to it as well. The quantities present in nature, where Bt outbreaks are rare, had been too small to exert selective pressure for resistance genes. Bt's popularity changed that. Meanwhile another bacterial weapon, milky spore, has lost credibility against the Japanese beetle.[33]

Revenge effects don't make either chemical pesticides or "natural" agents useless. Resistance has changed the strategy and tactics of combat from roundhouse slugging to biological judo. Subtle, time-winning gambits mean more than the search for a devastating first strike. Growers have been limiting doses, delaying them until a threshold of economic damage is reached, applying pesticides less often to smaller areas at more carefully calculated times. They have been alternating and coordinating them with introduction of natural enemies. (Unfortunately, some pet owners still combine insecticides, a practice that helped make multiple-resistant fleas among the most difficult insects to control in the early 1990s.) And they have even been reintroducing nonresistant strains of pests to mate with small populations that are beginning to develop resistance. These ploys don't eliminate the underlying resurging effects, but they do buy time in delaying them.[34]

Regrettably, agribusiness is still killing weeds in the good old way: dousing them with more chemicals. In ten years, the number of herbicide-resistant weed species has grown from a dozen to more than a hundred worldwide. To permit use of longer-lasting herbicides in the U.S. Midwest, seed producers now offer corn that tolerates one of the most popular but persistent chemical groups—chemicals that, after having been sprayed in soybean fields, would otherwise damage the young shoots of the next corn crop. (Farmers plant corn and soybeans in rotation.) But by using the same herbicides year after year instead of in rotation, farmers growing herbicide-resistant crops will inadvertently promote herbicide-resistant weeds. Still worse, there is growing evidence that genes from crops can find their way into weeds, which may not only be their next-door neighbors but their unruly cousins. If this happens, the resistance genes can spread so rapidly that superweeds could join the ranks of multiple-resistant problem organisms.[35]

Fire Ant Follies: The Vietnam of Entomology

Disastrous as it has been in the long run, DDT at least had the virtue of working for a few years—and even now, in limited, carefully defined circumstances, it can still work, according to its defenders. But the pesticide campaign mounted by the Department of Agriculture's crusade against fire ants, a struggle that continues, was not only damaging to wildlife but counterproductive. Natives of the upper Paraguay River floodplain where Brazil, Argentina, and Paraguay meet, red fire ants (*Solenopsis invicta*) first landed in Mobile in the 1930s. They soon proved to be among the insect world's fastest and fiercest colonizers. Not only did they overwhelm the native species of fire ant by building over forty mounds per acre against the natives' four or five, but they came to dominate the related and previously introduced black fire ant (*S. richteri*) as well. Like other pest species, fire ants specialize in changing and disrupted landscapes, including flooded riverbanks in South America, nurseries and sod farms here. They take to rich suburban lawns just as fleas and dust mites do to household carpeting. In fact, their colonies are five to ten times as dense in the United States as they are in their native territory, probably because South American parasitic flies (phorids) disrupt their foraging and reduce their ability to compete for food with other insect species. The phorids are still not present in the United States.[36]

S. invicta ants still produce more bloodcurdling anecdotes than provable economic damage. Because they kill just about everything that comes too near their mounds, they fight boll weevils as well as attack birds, reptiles, and small mammals. (Ironically, the boll weevil is one of the few major pests that are losing the war against pesticides; it is vanishing from the Cotton Belt.) A mound can be up to a yard high and five feet in diameter, linked to a wider tunnel system and housing hundreds of thousands of ants. Although a number of children have died from multiple bites, fire ants are not a life-threatening menace to people but a chronic nuisance. They hold on with their jaws as they inject venom, and while the results are not as intense as a bee sting, the itch is worse than a mosquito bite. In infested areas, between 30 and 60 percent of the population are stung each year, and tens of thousands of Southerners seek medical treatment for fire ant bites. Children playing in the grass are at highest risk.[37]

S. invicta meets any definition of a pest. It devours seedling trees. It kills young calves and fawns. Its mounds obstruct and damage farm equipment. With its boundless appetite for insulation, it even disables traffic signals and other electrical equipment. But until recently it was not a fearsome economic threat. It does eat germinating crop seeds, but its favorite dish is the grubs and larvae of other insects, and it generally doesn't attack mature crops. It is not in the same league as the corn borer, the Colorado potato beetle, and other crop and livestock pests. What probably helped raise its status in the world was the resurging effect that an eradication campaign made possible.

The agricultural historian Pete Daniel has revealed the background of the chemical fire ant campaign in the power struggles of the farm bureaucracy. A network of agencies, including the land grant universities and the Agricultural Research Service (ARS) of the U.S. Department of Agriculture, emerged from the Second

World War believing that biological control was outmoded. In the Cold War climate, the future appeared to belong to new chlorinated hydrocarbon insecticides like DDT, which could defeat the fire ant as the atomic bomb had beaten Japan and was containing the Soviet Union.

The first pesticide sprayed massively by the ARS with congressional authorization, dieldrin, was twenty times more toxic than DDT and drew unsuccessful protests from the Fish and Wildlife Service of the Interior Department—all the more after studies showed that the dosage of active ingredient per acre was 60 percent higher than necessary. Heptachlor was substituted, then discontinued in the early 1960s in favor of another pesticide, mirex, which in turn was found to harm wildlife and marine life, and possibly to cause cancer in human beings. Meanwhile biologists were discovering that the ants were killing sicker and weaker animals, as good predators should, while pesticides were wiping out quail and other wildlife indiscriminately. Not until 1978 did the spraying program stop. By then, the USDA had sprayed millions of acres, spent $200 million, and left more fire ants than ever.

As some of the early DDT sprayers discovered, heptachlor, mirex, and the rest killed not only fire ants but their natural insect enemies—for example, species that eat fire ant queens. Armed with a genetic heritage of explosive reproduction and colonization, *invicta* not only recovered swiftly but moved into the niches its insect enemies and competitors left behind. One study at the University of Florida has shown that a broad-spectrum insecticide helped fire ants increase their share of the resident ant population from 1 percent to 99 percent in only four years. By 1990 they occupied 400 million acres in the South and Southwest. And the campaign may have promoted an even more ominous trend that already dominates Florida: densely spaced supercolonies, as many as five hundred per acre, each with a hundred queens or more, resulting in average densities of over 175 ants per square foot and peak densities of over 500 per square foot.[38]

Single-queen colonies compete with one another and attack individual ants that stray into them from neighboring colonies. But multiple-queen colonies, linked by tunnels, seem to form an extended fighting organization capable of wiping out almost all other forms of insect, reptile, bird, and rodent life in its path. This is all the more enigmatic because such behavior was unknown in the ants' original South American habitat. Since multiple queens were first observed in 1972, after the heptachlor and mirex campaigns had been under way for fifteen years, pesticide spraying may have promoted the change by inadvertently selecting for genes previously expressed only rarely.

Whatever their origin, the new colonies now appear to have a foothold in Southern California and are said to be poised to move up the West Coast and to extend their domain in the Southeast at least as far as the Mason-Dixon line. They are so resilient that they migrate not just in potted plants but as part of insecticide shipments. Workers can defend and relocate queens, thereby reestablishing colonies so fast that fire, boiling water, and most poisons are ineffective in combating them. While there are baits that will slowly kill the ants, scientists and politicians alike have abandoned eradication for control. Control in this case means exercising constant vigilance to live warily with a chronic nuisance.

Excursus: An Electrical Alternative?

The problems of suppressing insects chemically make it natural to wonder whether there might be revenge effects in attacking them electrically. In fact, insect electrocution is a venerable theme in America's technological history. We have seen in the first chapter that "bug" was telegraphers' jargon for a hidden fault in the circuits, but it had a related and literal meaning for the operators. Western Union's city offices were notoriously dirty and insect-infested. Thomas Edison himself, as a young telegrapher desperate to "debug" his desktop in 1868, invented a pioneer electrical roach trap rigged to the office battery.[39]

While Edison chose not to pursue this line of work, insect-electrocuting devices were big business a century later. U.S. sales of products with trademarked names like Zappers, Bugwackers, and

Bug Blasters reached a peak of nearly $100 million in 1984. Sales have declined since then, with Asian imports battering domestic producers, but the industry is still large enough to have its own trade association. The products work similarly: intensified ultraviolet light lures insects to an electrically charged grid of alternating high-voltage current. Fluid in their bodies closes the circuit—the current arcs even if they don't touch the grid—killing the insects by dehydration and action on the nervous system. Suburbanites supposedly enjoy hearing the zapping sounds.[40]

The relief may be only illusory. In the early 1980s a team of scientists at the University of Notre Dame did a field test using a range of backyards adjoining a drainage ditch or other desirable insect habitats. Notre Dame graduate students (acting as bait) sat in yards with and without the bug zappers; they collected the mosquitoes as they began to bite. When the investigators later studied the insects killed, they found that the overwhelming majority were gnats. Female mosquitoes, the only ones that draw blood, accounted for a paltry 3.3 percent of insect remains. More important, the presence of a zapper did not reduce the number of attempted bites. Mosquitoes prefer warmth and carbon dioxide to ultraviolet light. And the insects breed so rapidly that even if zappers were more efficient, they would be unlikely to reduce their numbers significantly.[41]

Even if successful, insect electrocution devices might have the revenge effect of selecting for mosquitoes and other insects that avoid ultraviolet light, just as electric fly killers in some Mississippi barns are said to have selected for a tendency not to land on walls. But the real revenge effect of zappers may be yet another chronic one: promoting allergies. If used indoors or close to food preparation areas, electrocuting traps are known to produce allergenic debris, especially since some moths arc as long as thirty seconds. High voltage tends to fragment insect bodies and disperse particles. Over four hundred papers have been published on rhinitis, conjunctivitis, and asthma among people working with insects or inhaling insect parts. Researchers point to the physical characteristics of moth and butterfly scales that keep them suspended in air. To make things worse, pathogens carried by flies and other insects may be aerosolized and spread, negating the most important reason for killing them. When intensification of the battle goes too far, insects can bite back even postmortem.[42]

It may in the end seem that all intensification of agriculture, and of the struggle against household dirt and insects, has been some terrible mistake, that all we need is to return to nature. And indeed we are doing some things overintensively. Even after abandoning mirex and DDT, we are nevertheless multiplying some pests by killing their natural enemies or by disorienting the behavior of insect predators with sublethal doses. David Pimentel and his colleagues list over a dozen pests helped by pesticides that destroy predators. In cotton fields, these include cotton bollworms, tobacco budworms, cotton aphids, spider mites, and cotton loopers; in apple orchards, three species of mites, three of aphids, and two of scales persist. Pimentel has placed the price of chemical interference with natural enemies in the United States at over $520 million, about half due to extra control costs and half to lost production.[43]

Many farmers are discovering that "low-impact" agriculture can actually be more profitable than intensive chemical applications. Farmers and gardeners have cut back insecticide use significantly since the 1970s. They can get many of the benefits of chemical treatment and avoid many of the costs and revenge effects by using much smaller quantities, timed and applied more accurately. Some are making more money without chemicals than they did with them, partly because their prices jumped sharply in the early 1960s. (The market has been at least as powerful a damper on pesticide use as any environmentalist group.) Like antibiotic producers, agricultural chemical manufacturers continue to find ways to keep ahead of the insects; but they are finding that even in Asian countries where there is an overwhelming dependence on heavily treated high-yield rice crops, cutting back on spraying is actually increasing harvests.[44]

On balance, pesticides can increase crop production safely, if used in the right places, at the right times, and in the right quantities. Even most of their scientific critics acknowledge their value, at least for the near future. Taking into account all their adverse environmental and health consequences, and their cost, our nutrition and health would suffer if we suddenly tried to do without them. Just as California earthquakes taught lessons about collapsing buildings and bridges that were written into each new generation of building codes, biological revenge effects are not futile. The question is what lessons we should draw from failures like mirex and equivocal successes like DDT. If we insist upon the search for a new wonder substance that will effectively eliminate a given problem, the result will probably be the same: the adaptability of more resilient creatures will inevitably win. If we learn from revenge effects we will not be led to renounce technology, but we will instead refine it: watching for unforeseen problems, managing what we know are limited strengths, applying no less but also no more than is really needed.

6

Acclimatizing Pests: Animal

Some pests made their fortune through the optimism of nineteenth- and twentieth-century professionals—scientists, naturalists, and horticulturists—who believed the earth could be improved simply and relatively cheaply by the optimal rearrangement of its creatures. Exchanges of plants and animals between Europe, Asia, and the New World went on for centuries. Horses went west over the Atlantic, where they helped transform the culture of many Native American peoples; potatoes went east and became the staples of an industrializing Europe. Much of this happened without significant scientific study, thought, interest, or support.

There were already stirrings of a transplanting movement in the late eighteenth century. Sir Joseph Banks, First Lord of the Admiralty and a botanist of world distinction, sent the *Bounty* on its fateful voyage to obtain breadfruit, a plausible miracle crop for the New World tropics. This was not exactly a philanthropic mission; Banks's main purpose was to help make West Indian plantations more profitable by finding a new and cheaper crop to feed the slaves. (After another ship succeeded, the slaves found breadfruit so unpalatable that planters and authorities dropped the project.)

The Acclimatizers

Warwick Anderson, a historian of science, has traced the rise and decline of the acclimatization movement in the nineteenth century. Eighteenth-century scientists, doubting that species could thrive removed from the soil and climate of their origin, believed Arabian horses would degenerate into donkeys unless breeders took great care to slow—they could not stop—the decline of the animals. Naturalists of the nineteenth century advanced a new account of nature as a process of change and adaptation, in which the geographic origin of a plant, animal, or human group no longer necessarily limited its dispersion. The expansion of European empires opened new frontiers for the collection of wildlife and vegetation and inspired new dreams of adapting plants and animals to thrive in new surroundings. Overseas territories could yield beneficial new crops and livestock; they could also grow familiar ones transplanted from the metropolis or from other colonies.[1]

Nineteenth-century transportation encouraged acclimatization. A passenger voyage between one of the Atlantic ports and New York, which would have taken four to six weeks at mid-century, took only two weeks with the coming of steamships in the 1880s. By the early twentieth century, South Asia was no more than two weeks from Europe, and East Asia four weeks. Breeding stock of animals and plants could travel as fast freight. Each day gained increased the chances that enough individuals of a species would survive to propagate in North America. Even on long voyages, new equipment improved the chances for survival. Nathaniel Ward, a London physician, invented a sealed, moisture-retaining glass container that let plant collectors carry their own miniature greenhouses. The contents survived salt spray, extreme temperatures, and other dangers to which the crudely made glass-lidded boxes of earlier ships had exposed them. Plants in transit no longer demanded the constant attention of untrained sailors preoccupied with other duties. By the early 1840s the survival rate had changed from one in twenty to nineteen in twenty cases. Nevertheless, according to the garden historian

Kenneth Lemmon, intrepid explorers had successfully introduced a vast number of exotic plants to England even before Ward's cases appeared. In fact, their immediate consequence was a revenge effect as far as plant life of the tropics and the South Pacific was concerned. The ability to preserve specimens in heated greenhouses and transport them in sealed cases actually devastated many species in the wild. In Lemmon's words, "Orchidomania broke out, and for most of the remainder of the century orchid hunters were everywhere tearing these fantastic beauties from their natural haunts. . . . Whole tracts of forests were mown to the ground to get at the epiphytal treasure."[2]

It was in France that acclimatization blossomed as a theoretical and practical movement, with notable consequences for the United States. Isidore Geoffroy Saint-Hilaire proclaimed a new discipline of "zoötechnia," the applied science of manipulating animal and plant life, in an 1861 treatise, *Acclimatization and Domestication of Useful Animals.* To acclimatize a species meant "to impress upon its organization those modifications that will enable it to live and to perpetuate its species under new conditions of existence." Geoffroy was a campaigner as well as a theorist. He founded an Acclimatization Society that grew to eighteen hundred members by the late 1850s and an internationally famous Jardin d'Acclimatation in the Bois de Boulogne, a zoological garden that exists to this day, though in much-altered form. The society established eucalyptus and bamboo, silkworms and Chinese sacred pheasants—experiments that did not always survive the occasional hard winters of the South of France. It even introduced Burchell's zebra as a draft animal in Paris. Not that Geoffroy disdained *Equus caballus.* On the contrary, he was convinced that the indigenous horses of France could be a superior source of nourishment for the people. Accordingly, he believed that human diet, like the rest of nature, should adapt itself to the findings of science.

Geoffroy Saint-Hilaire established another society: of *hippophages* (horsemeat eaters), philanthropic gourmands whose periodic, well-publicized *banquets hippophagiques* encouraged the masses to explore a marvelous new source of cheap, lean protein. The *boucheries chevalines* of twentieth-century Paris and the horse steak until recently on the menu of Harvard University's faculty club show Geoffroy's enduring influence. His official report to the Minister of Agriculture and Commerce urging introduction of llamas and alpacas was ahead of its time, and its endorsement of kangaroo farming appears to have found no takers. (The animals, he advised, "grow rapidly, attaining a great height and producing excellent meat in abundance," besides having "a wool-like hair suitable for diverse uses.")[3]

Geoffroy's influence extended throughout Europe, from London to Palermo and even Moscow—surely a challenge to the acclimatization potential of the kangaroo. In acclimatizing itself, the movement changed markedly from country to country. In Algeria and elsewhere in the French empire, local acclimatization societies pursued bold schemes for transforming their surroundings. (The director of the experimental gardens of Algiers once declared that "the whole of colonization is a vast deed of acclimatization.") Warwick Anderson and Christopher Lever have written about England, for example, where the patronage of sporting aristocrats and venturesome gentleman breeders took the place of French state sponsorship and institute connections. The London Acclimatization Society's founder, the eccentric surgeon Frank Buckland, was an omnivore who did not stop at the horse or even the kangaroo, but essayed elephant trunk soup and roast giraffe. More seriously, Charles Darwin and Alfred Russel Wallace wrote on acclimatization, studying the experience of breeders and naturalists. Both Darwin and Wallace believed plants and animals adapted to new surroundings mainly by natural selection, but they also believed that the new environment could sometimes produce an inheritable change in an individual organism.[4]

In the United States, organized acclimatizing remained small, even when measured against the scale of the English movement, which never exceeded three hundred members (even after a merger with the Ornithological Society) and was dissolved in 1866. Five years later, the American Acclimatization Society was founded, but only one document, its charter and bylaws of 1871, is known to survive and in a unique copy. By the time it was formed, acclimatization was already starting to lose scientific favor in Europe, though colonialism supported a lively medical interest in acclimatizing European people to the tropics. (A lecture by Andrew Balfour, head of the Wellcome Bureau of Scientific Research in London, published in the *Lancet* as late as 1923, reviewed decades of research on the effects of climate on the physiology of transplanted Europeans and Asians.) But even though very few Americans joined acclimatizing societies, interest in naturalizing new species continued to grow. Without extensive overseas possessions for settlement, but with a vast territory, America appeared a country ready for biological transplantation.[5]

In this century, generations have struggled to preserve the native species of the United States, but earlier it appeared that North America lacked biological diversity. The founders of the American Acclimatization Society did not elaborate in their charter. They said only that the society's aims were "the introduction and acclimatization of such foreign varieties of the animal and vegetable kingdoms as may be useful or interesting" and "the discovery and development of valuable properties in species not hitherto brought into the service of man." In fact, not much is known about most of the incorporators themselves, but one name stands out: Eugene Schieffelin.[6]

Sparrows and Starlings

Schieffelin was born in New York City in 1827, the seventh and youngest child of Henry Hamilton and Maria Theresa Bradhurst Schieffelin. On his death in Newport in 1906, his *New York Times* obituary mentioned no profession or activity save an extensive list of clubs and societies to which he belonged. Uninterested in business, Schieffelin left the family pharmaceutical firm while still a young man. A biographical yearbook for 1907 praises his "rare intellectual qualities, the result of inherited tastes and talents, as well as of careful study and cultivation in literature, the fine arts, and the sciences," and points to his "accomplishment of manners and address, and . . . unusual conversational gifts." He was a church historian, an "expert and learned genealogist," and the founder of an organization called the Colonial Order. Schieffelin was a portrait painter as well, noted for "exceptionally sympathetic execution." His life-sized portrait of General Philip Schuyler hung in the august St. Nicholas Society, of which he was a member. Among his gifts was a deep knowledge of ornithology, according to the biographer, who then points without irony to what would become Schieffelin's true source of fame: "To him we are indebted for establishing in this country the English starling."[7]

Generations of writers have mocked Schieffelin as a sentimental fool, in part because he allegedly wished to introduce all the birds mentioned in the works of Shakespeare to Central Park. But neither the biography, the bylaws of the Acclimatization Society, nor other contemporary sources mention any such Shakespearean project. The story is probably a later speculation. Schieffelin began his acclimatizing career in 1860 as an early sponsor of introduction of the house sparrow to North America, in hopes, apparently, that the species would exterminate caterpillars infesting the trees of Madison Square, where he lived. He may or may not have been aware of a similar attempt made nearly a decade earlier by the staff of one of America's leading scientific institutions, the Brooklyn Institute, forerunner of today's Brooklyn Museum. Probably seeking, like Schieffelin, a natural predator for caterpillars, it unsuccessfully released eight pairs of birds in 1851 and imported another and larger group in 1852. Some of the birds were placed in the tower of the chapel of Brooklyn's Green-Wood Cemetery, where a keeper supplied them with grain and nesting boxes. Over the next twenty years there were other mass releases, including one of a thousand birds in Philadelphia in 1869. Apparently Schieffelin, the Brooklyn Institute staff, and the other acclimatizers were unaware that European farmers considered the sparrow a pest. Although the birds were promoted in America, societies for the destruction of house sparrows had existed in England since the middle of the eighteenth century.[8]

The English house sparrows turned out to prefer grain, fruit, vegetables, plants, and trees to insects. Above all, they had no stomach for caterpillar hair. But they did have technological change on their side. The nineteenth-century proliferation of horse-drawn vehicles—itself an unexpected consequence of the volume of railroad traffic—gave the sparrows a special niche as recyclers of the grain in horse droppings. They drove native swallows, finches, and bluebirds from their nests and competed with the caterpillars' real predators among the birds: robins, orioles, and cuckoos. The hardy and adaptable bird became as much a nuisance in urban parks and buildings as in the countryside, and was intelligent enough to avoid traps and poison. Yet acclimatizing was such a strong impulse that introductions continued as late as 1883, when sparrows from San Francisco were released in Stockton, California. Many sparrow enthusiasts were actually European immigrants homesick for their native fauna; others believed, like Schieffelin, that the birds could control insect larvae. Fortunately there was one unintended but effective control on the number of sparrows: the automobile's rise and the horse's decline cut off an intermediate source of their livelihood.[9]

In the thirty years following his modest part in bringing the house sparrow to America, Schieffelin (no doubt helped by his fellow acclimatizers) tried to introduce song thrushes, chaffinches, bullfinches, skylarks, and nightingales, all without success. Finally, he released a hundred pairs of starlings in New York in 1890 and 1891, at least one pair of which nested under the eaves of the American Museum of Natural History. Years later, a curator at the museum wrote that Schieffelin had visited the bird department regularly to ask whether any of the starlings had been reported and was delighted to hear a nest was so close at hand. If Schieffelin and the museum staff had known what we have since learned about starling biology, the successful spread of the birds would have been no surprise.

Starlings not only are highly intelligent birds and renowned mimics, but are also prolific, able to rear two broods every season. Experiments in the 1980s also found starlings to be among the most aggressive birds known. They compete fiercely for nesting sites, and the losers fight back by sneaking their eggs into the winners' nests—or simply by shoving the eggs out of a nest and taking it over. Nestless birds make their raids in gangs. As many as 10 percent of starlings die by the beak of another starling. Their tactics against other species occupying a choice tree cavity are easy to imagine.[10]

They are also extremely mobile, finding new territories up to fifty miles from the nests where they were hatched. Some birds establish year-round residence, while others migrate in patterns that vary from season to season. Biologists believe that this apparent randomness has helped the birds increase their range and number year after year. When starlings move, they do so on a large scale. Flocks numbering up to 200,000 move in formation with amazing precision at up to fifty miles per hour, and they combine in far larger numbers. They are infamous for their attraction to urban trees and buildings, including the Corinthian columns of the U.S. Capitol and the grounds of the White House. Chemical and electrical repellents can displace them, but usually only to nearby trees and buildings that have not been starling-proofed. One roost in California had an estimated population of five million before an extermination campaign began in 1966, only twenty-five years after the birds were established in the state. No wonder starlings have spread to all forty-nine continental U.S. states and throughout southern Canada.[11]

GYPSY MOTH

Schieffelin was already in his sixties when he introduced his delightful starlings. But the days were ending when gentlemen amateur acclimatizers could undertake to enrich the wildlife of North America, Australia, and New Zealand (to name just those areas where their influence was greatest). "Acclimatization" still rated over six pages of the eleventh edition of the *Encyclopaedia Britannica* in 1910 by Alfred Russel Wallace himself, but the eighty-seven-year-old biologist underlined at the outset that the plants and animals that thrive in new environments (like European weeds in the United States) don't need gradual adaptation to their new surroundings. On the other hand, organisms that have been acclimatized to new surroundings, even those as hardy and widespread as the potato, usually can't survive competitors and predators without human intervention.[12]

We might have expected the rise of scientific professionalism in the nineteenth century to reduce significantly the perils of animal and plant introductions. Unfortunately, it didn't. The professionals did introduce a much higher level of security and more sophisticated precautions. But because they were so careful, and knew themselves to be so careful, they felt free to experiment with organisms much more dangerous than the exotic creatures preferred by older generations of naturalists like Isidore Geoffroy Saint-Hilaire and the Eurocentric fauna of enthusiasts like Eugene Schieffelin. The first great representative of the new peril was the scientific illustrator and naturalist Léopold Trouvelot.

Where writers have scorned Schieffelin as a Shakespearean dilettante—slighting his connections to an important movement of his time—others have disparaged Trouvelot as a greedy and heedless entrepreneur. And this label, too, is unjust or at least incomplete. Trouvelot was a respected scientist who showed little recorded interest in business ventures. He arrived in America in the 1830s as a political refugee from the France of Louis Philippe and was soon associated with Louis Agassiz and other biologists and natural historians of the Boston area. His main specialty became astronomical illustrations. Research libraries continue to treasure his albums of plates, among the very best work produced just before astronomical photography began to make them obsolete. A century later, scholars still thought them important enough to qualify Trouvelot for the *Dictionary of Scientific Biography (DSB),* the retrospective Who's Who of world research. But even his *DSB* entry began by acknowledging that Trouvelot is now best known for his accidental release of the gypsy moth.

We remain ignorant of the details of Trouvelot's experiments. His motives are not the mysterious part. They were pinned on a commodity that has helped shape history: silk. For centuries, even millennia, silk was profitable enough to support a four-thousand-mile network of caravan trails across Central Asia between China and the Middle East. Silk had been a princely commodity; it is found in the wrappings of pharaohs' mummies. The budding consumer economy of the nineteenth century appeared to offer limitless possibilities to families willing to plant mulberry trees and tend *Bombyx* moth eggs and larvae. Because tending the silkworms, unwinding the cocoons, and reeling the fiber to make raw silk needs skilled but low-paid workers (traditionally women and children), silk production is a tempting but potentially ruinous business. Industrial organization usually failed in the nineteenth century; it worked best as a seasonal family enterprise in Asia and the Middle East.[13]

Beginning in 1849, a silkworm plague ravaged the silk industry of the Mediterranean, especially in southern France. Like the almost-contemporary Irish potato blight, this was a revenge effect in its own right, a serious problem that a lack of genetic diversity allowed to become a devastating one. France lost five-sixths of its silk output. Production had soared from about 350,000 kilograms in 1805 to over 2.1 million in the early 1850s, only to drop again to its 1805 level by 1865. Louis Pasteur's investigations of the two diseases ravaging the silkworms were models of the analysis of parasitism, but even his program for recovery through systematic selection of healthy silkworms left French output at less than half its high point. There seemed to be a boundless opportunity for creating new silk industries, with healthier insects, elsewhere.[14]

By the time Trouvelot began his experiments with moths, every attempt to start producing raw silk in the United States had failed in the long run. James I in the early seventeenth century and the Paris-based promoters of the ill-fated Mississippi Company a hundred years later imagined in vain that sericulture could flourish in the American Southeast. Benjamin Franklin and Yale president Ezra Stiles were among other early enthusiasts, but their experiments in Pennsylvania and Connecticut fared no better. In the 1830s, speculators bid up the price of a touted variety of mulberry tree, *Morus multicaulis,* from $5 for a hundred seedlings to $500, only to be devastated by the financial crisis of 1837, followed by a blight of all mulberry trees in 1844. Nevertheless, none of this stopped America from becoming, for reasons fashion historians have never explained, one of the world's largest per capita *consumers* of silk. (Interestingly, Japanese cottage industry was America's largest source of raw product.) Trouvelot was only one of the Europeans—Frenchmen, Germans, Italians, and no doubt others—who saw gold in New World sericulture. When the California legislature established a bounty for new mulberry trees in 1865, speculators planted a million before the legislature revoked the reward.[15]

It was amid this biological and commercial crisis that Léopold

Trouvelot sought to apply his scientific interests. In the first two issues of the journal *American Naturalist* in 1867, Trouvelot reviewed nearly seven years of research into domesticating the American cousins of the Old World silk moths, members of the family Bombycidae. It had taken him five years to learn how to cultivate the only promising candidate, the Polyphemus moth, and he reported he had fully a million caterpillars under cultivation on five acres of woodland surrounded by an eight-foot fence. Netting supported by poles covered the whole area—to keep birds out rather than caterpillars in. Even so, Trouvelot's papers, lucid and precise in their account of the moth's life history, contain a prophetic observation. A single silkworm caterpillar eats 86,000 times its original weight in less than two months: "What a destruction of leaves this single species of insect could make if only a one-hundredth part of the eggs laid came to maturity! A few years would be sufficient for the propagation of a number large enough to devour all the leaves of our forests."[16]

Trouvelot's interest in the European gypsy moth (*Porthetrea dispar,* actually a native of Japan) remains a mystery. He seems to have been satisfied with the promise the Polyphemus moth (*Telea polyphemus*) had shown, while at the same time another naturalist was promoting the large Cynthia moth (*Samia cynthia*) as an alternative to *Bombyx mori,* the most common species in Europe and Asia but difficult to acclimatize. Why import a species that appears never to have been cultivated for silk in its homeland? Trouvelot may have hoped to breed the gypsy moth with another moth species, and he even delivered a paper on how to breed insects of different species, but the gypsy moth was of a different genus and Trouvelot had found by February 1867 that unions between the genera were sterile.

Moreover, Europeans had known the gypsy moth as a pest for at least 150 years. It had periodically devastated forests in Brandenburg, Saxony, Bohemia, the Crimea, and Belgium—not to mention Trouvelot's native France. When a storm blew away protective netting in 1869, Trouvelot realized the potential for harm to trees, tried to kill the caterpillars, and made a public announcement of what had happened. Not long thereafter, and probably to take advantage of the more liberal political climate of the Third Republic, Trouvelot returned to France.[17]

Despite the best efforts of their natural enemies in the New World, notably birds and parasitic insects, gypsy moth caterpillars remained a local pest in Medford for nearly twenty years. Elsewhere, Trouvelot's accidental naturalization of the gypsy moth was almost forgotten, although the moths continued to thrive, mainly in public woodland that formed a natural reservoir. European gypsy moth females cannot fly. The larvae do not travel far, either, even when disturbed, but they share with fire ants, house sparrows, and starlings a ravenous appetite for a greater variety of food than their local competitors. They overwhelm native tree pests, including cankerworms and tent caterpillars, by outfeeding them. The house sparrows introduced by earlier acclimatizers actually made the problem far worse because so many of the birds they displaced were natural predators of the gypsy moth larvae.

Sparrows and gypsy moths became partners in pesthood. "The caterpillars used the bird boxes occupied by the sparrows as a place of retreat," a Massachusetts report on the gypsy moth found, "and the female moths deposited their eggs in these boxes. Sparrows and caterpillars formed a sort of happy family in the bird houses, which swarmed with both birds and insects." What really made their population explode in the 1890s, though, were other trends in society and technology: suburban growth.[18]

Railroads and trolleys helped turn the villages surrounding Boston into suburbs well before the coming of the automobile. But the caterpillars traveled mainly by the growing number of horses and wagons plying the dirt roads that connected Boston with its hinterland, bringing produce, firewood, plants, and flowers to the city and returning with manure. Caterpillars dislodged from trees tended to hang from branches by their silk threads, landing not only on pedestrians' hats but on wagons and carriages. Cut wood could harbor egg clusters large enough to start new colonies. As growing numbers of wagons repeated their deliveries day after day, they helped spread self-sustaining populations of moths.[19]

By the early 1890s, federal and Massachusetts authorities had begun a campaign against the gypsy moth. Before the rise of organic pesticides, as we have seen, most pesticides were arsenic-based and far more toxic to people and animals than later synthetics like DDT. A tenth of the sprayers became ill from arsenic poisoning. The press reported that children were dying after eating contaminated food, and one editorial recommended that spraying crews be shot on sight. Without careful application, the insecticides could and did damage plants and fruit trees. Mechanical defenses worked—creosote destroyed egg masses, sticky bands and burlap wrapped around tree trunks trapped the caterpillars—but like the hand application of pesticides, these were costly remedies even at the low wage rates of the time.[20]

Could the gypsy moth have been eradicated before it established itself permanently? After it had spread to much of New England by the first decade of the twentieth century, the state of Connecticut began one of the broadest-based campaigns ever mounted against the insect. Authorities distributed thousands of pictures, hired trained scouts, and sent out crews of young men to band trees, pick caterpillars with forceps, and drown them in alcohol. Yet no matter how thorough they were in any locality, officials discovered new infestations in others. Motorized crews could cover larger territories, but it soon became apparent in the 1920s that the caterpillars were so widely dispersed in the state's forests that local victories were only temporary. Even in the 1890s, there probably were more clusters of gypsy moths in eastern Massachusetts than its residents realized.[21]

In the twentieth century, automobile commuters and truck drivers unwittingly continued what the market gardeners of nineteenth-century Massachusetts had begun. The gypsy moth continued to spread at a rate of about fifteen miles a year and reached Canada, Michigan, and Virginia. Air travel was probably establishing the gypsy moth population on the West Coast. Environmental disturbance also helped the moths extend their range; the forests that grew back after decades of relentless logging had a far higher proportion of oak trees, the caterpillars' favorites, than before. Nor was the problem limited to the descendants of Trouvelot's escaped colony. In the early 1990s, air and ship transportation introduced the Asian gypsy moth, with more ravenous caterpillars and flying females, on the East and West Coasts of North America.[22]

As New Englanders found a hundred years ago, gypsy moth caterpillars can be scarce for years, only to flare up spectacularly for a few years and then drop as sharply as they had risen. A peak infestation can reach 13 million acres, as in 1981; at other times, as in 1993, it may affect only 1.8 million. Even at this rate, gypsy moth infestation is yet another form of chronic environmental degradation. And as with other chronic conditions, crises alternate with uneventful intervals. Starlings are now among the natural predators controlling the gypsy moth population, though none of the caterpillar's bird predators can cope with the millions of caterpillars that emerge during explosive years. Botanists and entomologists of the late twentieth century have discovered that many species of trees emit chemicals called phenols that brake gypsy moth population growth by keeping caterpillars smaller, limiting the egg production of female moths, and reducing the size of the eggs themselves, producing a generation of scrawnier larvae. On the other hand, the phenols have a natural revenge effect of their own. They defend the caterpillars against their most serious disease, wilt virus, which is inactive in normal years but is able to reproduce as explosively as its insect host during gypsy moth outbreaks, infecting virtually all caterpillars. Tree phenols like oak tannin may help caterpillars increase their density while resisting wilt virus. (When the caterpillars chew the leaves, they oxidize the tannin, making it more lethal to the virus, some entomologists believe.) Oaks may even be a natural sanctuary for the caterpillars during outbreaks.[23]

Fortunately there are other agents for controlling gypsy moths. A Japanese fungus, *Entomophaga maimaiga,* first brought to the United States and tried against gypsy moths without apparent success in 1910–11, was reintroduced more recently, and its spores were found in caterpillar cadavers in England. Researchers now think the fungus may have persisted from its original introduction as well and may be responsible for some of the die-offs attributed to wilt virus. It could be a powerful control agent, though it is also likely that resistant strains of gypsy moth caterpillar will eventually appear if it is used widely and heavily.

Foresters can also deploy other viruses and chemicals. Many threaten biological diversity, especially the survival of beautiful species like the regal fritillary butterfly, more than the gypsy moth itself does. Some of them may risk the revenge effects of mirex against the fire ants: killing spiders and other predators of harmful insects. Others are harmless to other species but require messy production methods (culturing in caterpillars) and costly application. Paradoxically, spraying even with relatively benign insecticides like Bt may actually help the gypsy moth population prepare for resurgence. By keeping down early infestations, sprays keep gypsy moth populations from reaching the densities that promote the rapid spread of fungi and viruses and the subsequent total crash of gypsy moths. At Pennsylvania's untreated Hawk Mountain sanctuary, gypsy moths appeared to be completely absent after the disastrous infestation of 1990, while some insects survived in the neighboring sprayed forest. And a resident ecology graduate student has argued that by thinning oaks and promoting the growth of other trees in their place, the gypsy moth may actually be helping to restore the more diverse Appalachian forest that prevailed before logging and disease tipped the balance in favor of oaks.[24]

Carp

Carp also show how problematic new introductions can be. Although Eugene Schieffelin was a serious amateur attached to an international acclimatization movement with distinguished origins and Léopold Trouvelot was a respected if somewhat eccentric scientist, they were obscure compared to Spencer Fullerton Baird, who was one of the leading zoologists and administrators of his time. Named assistant secretary of the Smithsonian Institution before he was twenty-eight, he wrote outstanding systematic studies of North American birds and mammals, built the Smithsonian's and other research programs, and became secretary of the Smithsonian in 1878 after having established and headed the U.S. Commission of Fish and Fisheries.[25]

Baird was an energetic commissioner who began the scientific study of fisheries in America, struggled with problems like overfishing by fixed-net fishermen, promoted hatcheries, revived shad and salmon runs on the East Coast, stocked the Great Lakes with whitefish to halt a declining catch, and fought for stricter state regulation of fishing and control of pollution in the interest of long-term production. Like many contemporaries, Baird feared for the future food supply of the rapidly expanding American population and doubted that available grazing land would satisfy the need for "animal food." After decades of pollution and overfishing, the nation's fisheries urgently needed protection, restoration, and conservation to feed the country. Baird also resisted, unsuccessfully, the efforts of the American Fish Culture Association, a group of affluent hobbyists and private hatchery operators, to import fish and eggs freely and introduce them without government interference into any lake, river, or pond members might choose. He believed that fish should be introduced only after careful study and as part of a balanced program for maintaining local populations. Yet, Baird's personal favorite among the fish culture programs, his prime candidate for the American table, was the one that even in his lifetime became the most questionable: the German carp (*Cyprinus carpio*).[26]

Carp was an acclimatizer's dream: a hardy, fast-growing organism with an ancient history of cultivation, it was nutritious and was easy to care for, apparently satisfied with almost any vegetable matter as food. Unlike horsemeat, Geoffroy Saint-Hilaire's popular-priced protein, carp had an international heritage. The Japanese treasure their pampered carp as exquisite and lucky animals, emblems of masculine vigor. Prize specimens living a hundred years or more pass as heirlooms from one eldest son to the next. Introduced in the West two millennia ago, they have long remained an aristocratic ornamental fish. By the nineteenth century, carp were a staple Continental food; after Atlantic salmon and rainbow trout, they are still among

Europe's favorite sporting fish. The preferred basis of Jewish gefilte fish, among other favorites, carp was already a delicacy to Central and Eastern European immigrant families.

Technology gave carp its transatlantic opportunity. A resident of Newburgh, New York, may have introduced carp into the Hudson as early as the 1830s, but large-scale introductions of carp and other fish became possible only after the development of artificial propagation in France in the 1840s made fish hatcheries possible. Carp were raised in Sonoma County, California, in the early 1870s. At almost the same time, fish breeding was potentially transcontinental in scope, as a railroad car brought ten species and hundreds of thousands of fish from the East to the West Coast in 1873.[27]

Baird especially wanted to give the South a substitute for the northern trout. In reports from the mid-1870s he noted that carp could be raised on a vast scale—like the twenty thousand acres of ponds on the Austrian estates of Prince Schwarzenberg—or in nothing more than washtubs supplied with vegetable scraps, as in China. He observed that in Germany carp ponds were regularly alternated with grain and other crops. While Baird was aware that a number of native American fishes could be grown under similar conditions, since carp was such a known success, "there is no reason why time should be lost with the less proved species." Carp populations in America, however, were not of the desirable domesticated variety, so Baird engaged the German fish culturist Dr. Rudolph Hessel to import the German variety.[28]

Hessel had already contributed a thirty-two-page appendix on the carp to Baird's report for 1875–76, with detailed instructions for its cultivation. The fish, despite its historic association with some of Europe's noblest houses and most ancient monastic foundations, "is very easily satisfied, and will not refuse the offal of the kitchen, slaughter-houses, and breweries, or even the excrement of cattle and pigs." It is hardy and productive, sheltering itself through winters in its pond hideaways and producing hundreds of thousands of eggs. Hessel portrayed the carp as a delicacy commanding up to three times the price of other fish (except trout and salmon) in the markets of Paris.[29]

By 1877, Hessel had succeeded in transferring carp across the Atlantic, an achievement in its own right, to judge from the failure of Baird's attempts. With Baird's enthusiastic sponsorship and a speedy congressional appropriation, Hessel established National Carp Ponds at the foot of the Washington Monument. These were soon expanded into an acclimatization garden covering twenty acres of ponds, with plantings and at times a mini-zoo and winter skating to encourage visitors. When Fourth of July fireworks killed a number of young carp, Baird personally intervened and even briefed the park police to be sure his baby fish would not be endangered again. President Grover Cleveland and other notable visitors toured the ponds with enthusiasm.[30]

Hessel's diligence and Baird's promotion soon sold America on the German carp. Official publications of the Fish Commission claimed that pound for pound, carp cost only half as much as chicken to produce and could be raised on land unsuited for crops. American farmers rose to the bait. Tens of thousands of citizens requested carp through their congressmen—the commission preferred this system of distribution—and 298 of 301 congressional districts participated. As many as 350,000 carp went to over 6,200 individuals in a single year, not to mention gifts to Texas and other states that maintained their own ponds. The commission's fish, raised to lengths of two or three inches and shipped by rail in the autumn to assure mild weather, arrived with minimal losses. Carp had become the commission's most popular project. "Almost every farmer wants a carp pond in his front yard, back yard, or barnyard," observed one fish expert.[31]

Unfortunately, once distributed across the country, the fish did not behave as expected. Unless raised in clear water they often had a foul taste, though Baird insisted that they could be delicious if properly cleaned and cooked. He even published recipes. In fairness to Baird and Hessel, the instructions and cautions in their writings were clear. Carp need careful management. They can grow lazy, for example, and lose firmness when fed too routinely in one place. Hessel recommended adding a few pike to scare the carp into swimming away from them, thus staying vigorous. (Recent research shows that carp can bulk up defensively in the presence of pike; the effect of this maneuver on their eventual taste, however, is not known.) Carp also wander and can easily establish themselves in whatever bodies of water they can reach.[32]

Because the expansion of *C. carpio* coincided with growing environmental degradation of all kinds, the fish lost favor. In 1895, one of Baird's successors as Commissioner of Fish and Fisheries, George T. Mills, declared: "Time has now established their worthlessness, and our waters are suffering from their presence. As a food fish they are regarded as inferior to the native chub and sucker, while their tenacity to life and everlasting hunger give them a reputation for 'stayers and feeders' unheard of in any fish reports I have seen up to date." Overrunning the waters of the Great Lakes and the Upper Midwest, carp devoured the wild celery, wild rice, and pond weeds that had once sustained countless waterfowl. They filtered lake beds for food and roiled waters with bottom mud, robbing plants of light (thereby depriving young game fish of plant cover), disturbing sight-feeding fish, and nourishing surface algae that grew to block light further and rob game fish of oxygen. They became a special nuisance in reservoirs as America's urban populations and water systems grew. They have been accused of eating the young of other fish, though they are poorly equipped for it and probably don't often try it. (Eggs are another matter.) More recently, some specialists have argued that they have intensified the pollution of lakes from lawn fertilizer runoffs by stirring phosphorus from lake sediments.[33]

Carp, like house sparrows and starlings, multiply furiously and displace native organisms aggressively. A female may lay two million eggs in a season. Like other organisms branded as pests, carp resist environmental insult. They thrive not only on aquatic plants and insects but on almost anything people leave behind as garbage. They share the longevity of their Asian ancestors and relations. They rescue themselves from pesticide spills by manufacturing ethyl alcohol in what has been called self-ferment. In the absence of food they can survive for weeks or months by converting the alcohol to energy. Fat and lazy though they may sometimes appear, carp can move when they want to, jumping three feet or more in the air and escaping from ponds. They negotiate mazes as well as rats and migrate up to a thousand miles. They thrive in almost any kind of fresh and not-so-fresh water, in nearly any weather. Nuclear-plant-cooling systems in Connecticut kill most flounder larvae sucked into them, but carp love to linger around the discharge pipes in winter. No wonder they are now said to be the most common American freshwater fish.[34]

Where they are unwanted, they are difficult to remove permanently. In southern Wisconsin in the 1970s, state authorities dispatched crews of motorized carp removers with seine nets four thousand feet long that yielded fifty tons of carp and other rough fish a day to make room for species considered game fish. The carp had brains enough to huddle together on the bottom of lakes and avoided the seines. The fecundity of even a few remaining fish could quickly restore the carp population. And since the nets tended to trap older fish, the dredging made the carp population of lakes and ponds younger and thus more likely to compete directly with game fish and birds for insects.

The failure of netting did not change the conviction of many wildlife people that, as one Wisconsin state official put it, "as far as overall lake ecology is concerned, the only good carp is a dead carp." One Wisconsin county has sponsored a bow-and-arrow carp shoot with prizes. A number of states permit unlimited catches by any means at any season. By the 1980s and 1990s, fishing and environmental authorities had turned to poison to kill carp and other rough fish. One of the most popular, rotenone, is made from tropical plants and interferes chemically with fishes' use of oxygen. Despite the natural origin of rotenone, some samples have contained levels of toxic chemicals—including benzene, xylene, and trichloroethylene—that exceed Environmental Protection Agency guidelines. Chemical kills also destroy game fish and other aquatic life, despite rescue programs, without guaranteeing that a few surviving carp will not repopulate the lake or river. Sometimes rotenone fails to break down in a target reservoir, and the waters kill thousands of trout when released into rivers. Still, state officials insist the normal fish population, minus the carp, returns in several months to a year.[35]

With 32 million pounds of carp sold each year in the United States as recently as 1967 (up from under 4.6 million pounds in 1900), the fish has been a resource rather than, or as well as, a pest. It even has some admirers in the sport fishing world that once condemned it as a "rough fish" and pest. While anglers in the Upper Midwest and California campaign for purging the fish, others in Ohio and Missouri use new lures and tackle to bring out the fighter in carp. Still, it is a minor component in the American diet and not the economical alternative to chicken that Baird and Hessel imagined. Like most other naturalized animals and plants, it did not need to be acclimatized. The carp imported from Germany seem to have flourished without temperature controls or human selection. With careful supervision and regulations, they might have been cultivated in ponds without upsetting the balance of other marine life. Although the U.S. Fish Commission cautioned farmers many times against introducing carp in waters with existing fish populations, decades of intensive distribution and deteriorating conditions for native fish (not to mention the weak powers of state fishery authorities in the late nineteenth century) made the victory of carp almost inevitable.[36]

The carp story ends with a twist. In Asia, the indigenous silver carp—a close relation of the German carp imported by Baird—is not a pest but a victim. The ruthless invaders are North American fish. This time the agents are not headstrong acclimatizers or philanthropic scientist-bureaucrats, but pious (if ambitious) believers and fund-raising clergy. Korean Buddhists traditionally release fish in ceremonies celebrating respect for life. Once, monks would liberate a symbolic fish, bought from a fisherman, for the sake of all the faithful. Recently the observance has taken a new direction. More and more Buddhists are releasing fish to bring favor on their careers, despite criticism from Buddhist theologians. (They are even more upset at the fishermen who net the released fish, then resell them to a new group of worshippers, often stressing the animals to death after several rounds.)

Some of the two million released are imported species from fish farms. Among these are bluegills, descendants of fish introduced from the United States twenty years ago. Bluegill males protect fertilized eggs zealously, while both males and females devour the eggs of silver carp and other native species. Similarly, along the upper

Han River (Seoul is near the mouth of the Han), the introduction of another American fish, the largemouth bass, has destroyed populations of twenty-five native species. These fishes, the bass and the bluegill, are, ironically, two of those most often mentioned as casualties of the carp explosion in the United States. And just as carp were blamed for North American extinctions caused by pollution and other disruptions, waste and dam construction may have harmed the other fishes as much as the bass did. In any case, the practice is not easily changed. Korean Buddhist temples have few steady financial sources, and believers are willing to pay the equivalent of nearly $40 to release a fish.[37]

The Killer Bee

While it has been over a hundred years since the most controversial animal introductions, the risk of revenge effects is not over by any means. A reminder has been making its way toward the United States since the late 1960s—the Africanized honeybee. If the importation of insects had been forbidden in the Americas, we would have had no honeybees at all, but now the killer bee may displace many of its less aggressive naturalized cousins.[38]

American bee dealers may have imported queens from Africa as early as the nineteenth century; beekeepers, too, were acclimatizers. The Department of Agriculture imported African bee semen as early as the 1960s and kept hybridized colonies for several years. But none of these bees are known to have established themselves. The aggressive (or defensive, to their advocates) insects come not directly from Africa but from Brazil, imported experimentally by a prominent Brazilian geneticist of American descent, Professor Warwick E. Kerr. For a number of years, Kerr had been trying to improve the disappointing production of Brazilian bees, descended like North American bees from European stock adapted to temperate weather. Impressed by reports of the superlative honey yields of African species, he collected mated queens and bred them in an enclosed eucalyptus forest, much as Léopold Trouvelot had experimented in woods with netting. Kerr hoped to cross the African bees with gentler European-descended varieties to obtain a strain that was both high-yielding and manageable.

What Kerr and other entomologists did not realize was a crucial difference in the ways European and African bees establish new colonies. Living close to human settlements and predictable nectar and pollen sources, European bees are homebodies. They build large nests with extensive stores of honey. The African bees Kerr imported, *Apis mellifera scutellata,* are adapted to the unpredictable rainfall of East and South African highlands; one specialist has even argued for the term "highland bees." They not only swarm—sending part of the colony to establish a new one—far more readily than European bees, but unlike their European counterparts, they also abscond—abandon the colony to found another. And they move farther: as far as sixty miles from their old hives, as opposed to the European range of under ten miles. This behavior serves them well in Africa, where more energy has to be spent in finding new food. In the New World, absconding helped the African bees expand their range rapidly; thereafter, swarming multiplied the number of colonies and overwhelmed European bees both genetically and physically.

Traditionally, beekeepers try their best to prevent swarming, and absconding means a total loss to them. Furthermore, African bees, at least the descendants of Kerr's queens, have mixed reputations as producers. They are energetic even by bee standards, and especially in terrain like the highlands of Central America they have been top producers. Brazil's honey output multiplied after the Africanized bees prevailed. They are also robust. Old World mites and infections that have been ravaging New World colonies—inadvertently imported by jet aircraft and devastating to strains of bees that have lost their resistance over the centuries—do not seem to affect them. (Africanization is a drastic way to improve bee health, though, since there are disease-resistant European breeds, like the celebrated queens bred by Brother Adam of England's Buckfast Abbey.)[39]

Neither Kerr nor, apparently, anybody else knew these things about the African bees in the 1950s. He also did not understand that the explosive spread of the African bees lets them dominate new areas so thoroughly that virgin queens from European hives mate overwhelmingly with African drones, producing generations of hybrid queens and workers with increasingly Africanized traits. The African and hybrid bees defend their hives with massive retaliation against human beings and livestock that appear to be a threat. When swarms of the African bees, including queens, escaped from Kerr's experimental enclosure in 1957, hybridization with local bee strains began. The result was a new variety of honeybee that overwhelmed all colonies in its path.

By the mid-1960s the hazards of African bees were apparent in Brazil, and the USDA killed its hybrid colonies. While these bees had since been Europeanized and none seem to have escaped, isolated Africanized bee swarms have hitched rides on ships from South America and come ashore in Florida and California. In the Lost Hills, California, episode beginning in June 1985, more than 22,000 hives in a zone of over three thousand square kilometers were searched, and a dozen hives invaded by Africanized bees were destroyed. But after futile attempts to create Central American barrier zones, more African bees were found near Hidalgo, Texas, in October 1990. In July three years later, an eighty-two-year-old farmer, the first U.S. casualty of the invaders, died from dozens of stings after poking the wrong beehive with a stick in an abandoned house in the border town of Rio Grande City, Texas.[40]

Few if any bee specialists doubt that killer bees are here to stay, at least in the South and Southwest. Harder to estimate is how far north they can survive year-round. They probably will not make it all the way to New York City and Seattle, as pessimists predict. Argentina's experience suggests that the most likely range will probably be the southern third of the United States—appropriately enough, since Warwick Kerr's grandfather emigrated from Tennessee with his family in 1865 after the fall of the Confederacy. Latin American experience suggests that while few people will be stung to death, the risk of serious injury will be high enough that recreational and perhaps even commercial beekeeping will decline. Regulations on beekeeping will tighten. As in Canada today, shipments of queens from possibly Africanized zones will be embargoed. Where the Africanized bees have penetrated, beekeepers will have to introduce new, certified queens, preferably shipped from islands and regions around the world known to be free from the Africanized and hybrid strains. Beekeepers will also need stronger protective equipment and supplies.[41]

Once more, what appeared to be a catastrophic event turns out to be a chronic but manageable nuisance. Improvements in transportation after Kerr brought the first African bees to the Western Hemisphere by ship can also introduce or reintroduce the genes of gentler bees. Deaths from bee stings will be a minor public health problem that will probably be noticed mainly when the victim is a celebrity or the setting is an expensive resort. The economic damage from loss of honey production—and loss of crops where bees can't be used for pollination—will be significant but not disastrous. Indeed, Brazil has soared from forty-seventh to seventh place on world rankings of honey production since Africanization.

Still, it is doubtful that the Africanized bees will maintain their pace in temperate zones. Even in parts of South America, where their honey production is at its highest level, yields have often declined because absconding and stinging discourage beekeeping. The main uncertainty is cultural, not biological. In Brazil in the 1970s, a new and enthusiastic cohort of beekeepers, trained and willing to work with more defensive insects, replaced those who dropped out after Africanization. But that took at least a decade, and it remains to be seen whether insect-fearing and insect-spraying North Americans will be as adaptable. The naturalist and writer Sue Hubbell hopes so and prefers the phrase "bravo bees," arguing that right-wing Brazilian authorities used the "killer" label to retaliate against Kerr's socialist politics. But Warwick Kerr himself had turned to studying stingless bees by 1991 and told the writer Wallace White that if he had it to do over again, "I would leave those African bees where I found them."[42]

Today's Acclimatizing

Over the past hundred years, animal acclimatization has lost its standing as a movement but retained interest as a practice. The movement last flowered in the Soviet Union of the 1930s. Scientists at the Askania nature reserve acclimatized creatures from all over the world as part of the Stalin Plan for the Great Transformation of Nature. They bred zebroids (crosses between zebras and horses) and introduced muskrats and other fur-bearing species in the Soviet campaign for the subjugation—in practice the degradation and destruction—of the environment. Eventually Soviet acclimatization was discredited along with its senior partner, Lysenkoism, but as late as the 1960s the *Great Soviet Encyclopedia* still ran a generally positive article on the subject. In the 1960s Fidel Castro was still pursuing genetically dubious schemes like breeding a zebu with a Holstein to produce a new, high-yield supercow. (An English consultant was thrown out of the country for predicting—correctly, it turned out—that after the first generation the animals would have the worst qualities of both Holstein and zebu.) Outside the former Soviet sphere, the few surviving "acclimatization" societies have abandoned most of their former campaigns for the introduction of new species in favor of zookeeping, as in Australia, or habitat conservation, as in New Zealand.[43]

The end of the acclimatizing movement has not killed animal introductions. It has instead shifted the initiative to coalitions of bureaucrats, scientists, agri- and aquaculturists, and sports enthusiasts. In turn, their efforts have mobilized other bureaucrats, other scientists, and environmental activists against such introductions. And some people are torn between intervening and protecting native organisms. "As an ecologist and ichthyologist," one wrote, "I am opposed to wholesale introductions of exotics and transplants into waters outside their native ranges. However, as a fisheries biologist and a member of the sport fishing public, I can support this activity under certain conditions." Transplanting rarely can restore disturbed habitats, but it can sometimes promote recreation or food production. In the 1960s the Michigan Department of Natural Resources began to stock coho and chinook salmon in Lake Michigan. Commercial overexploitation, pollution, and predators had all but ruined fishing in the Great Lakes. While Pacific salmonids must be restocked each year, they were attracting millions of anglers spending billions of dollars to the Great Lakes by the 1980s. The salmonids fed on alewives that had previously washed up on beaches and blocked municipal water works. On the other hand, the salmonids' presence has also led to a reduction in the populations of the remaining native lake trout.[44]

Intervention does not always work, and systems do not always recover from mistakes. In Montana, where minute opossum shrimp were introduced in a number of small lakes in the late 1960s and early 1970s, kokanee salmon flourished as expected. But a strange thing happened when the shrimp migrated downstream to the much larger and deeper Flathead Lake. There, and in some other North American lakes, they stayed near the bottom of the lake during the day, when the salmon were near the top. They rose only at night for feeding—when the salmon were unable to get them. And they devoured much of the plankton that otherwise would have fed the salmon. The result was an exploding shrimp population, a virtual collapse of the salmon catch, the disappearance of eagles and other wildlife that depended on the salmon run, and the absence of tens of thousands of tourists and birdwatchers who had once thronged each autumn to see as many as a hundred bald eagles at a time at the salmon's spawning run in Glacier National Park.[45]

Consumer shifts away from meat, controversy over fur, concern about poultry diseases and tropical bird preservation, and the fear of insects have all helped intensify debate over animal introductions to lakes, rivers, and estuaries. Over half of all fish species designated as endangered in the early 1990s were under pressure from nonnative species. Most of these, in turn, had originally been introduced for game fishing. Not all of these problems are revenge effects. Some of the threatened native species are of no direct fishing interest. On the other hand, bass introduced in Maine have been competing with the remaining Atlantic salmon. Fishing can lose as well as gain from such experimentation.[46]

Although the Asian walking catfish was introduced by pet breeders, not fisheries, it shows how difficult it can be to predict the environmental consequences of an aquatic animal. A fish farmer told the Florida Atlantic University zoologist Walter R. Courtenay, Jr., a leading specialist on fish introductions, that most of a load of four hundred walking catfish had escaped from Styrofoam boxes in the rear of his old van in the 1960s during a thirty-five-mile drive north of Miami. This could have been the ultimate fish story in reverse, but somehow, like other invasive organisms, the catfish were literally able to walk away from captivity. Entering South Florida's extensive canal system, they soon were spreading over a large area of the state, where their hearty appetite and amphibian ways have doomed many native species. Breeders may have believed that the albino coloring of the first imported catfish would draw predators to any escapees, and so it did at first. But there was enough variation in their skin to make natural selection for darker shades possible, and the catfish now appear impossible to eradicate from Florida waters. It now climbs out of the canals to feast on other waterbound fish in the breeders' tanks, a revenge effect indeed.[47]

If the record of animal introductions were only a series of comic disasters, attempts would have stopped long ago. But new mammals, birds, and fish are brought to new climes all the time, partly for the sheer love of novelty—a novelty that is often harmless. Of the dozens of exotic animals exhibited at Isidore Geoffroy Saint-Hilaire's Jardin d'Acclimatation, none appear to have become established in France, as pests or otherwise. The same aquarium trade that occasionally leads to the release of aggressive species, whether walking catfish in Florida or guppies all over, also promotes love of the sea and its life. While some historians have argued that an interest in acclimatization is simply an upper-class demonstration of mastery, it is more plausible to see it as a love of variety that touches all segments of society. The revenge effect has always been that attempts to increase diversity may actually reduce it, as one hardy, invasive species crowds many native ones from their niches. For over a century, the people who misjudged the effects of introductions included scientists of high distinction.

Animals called pests have some traits in common: fecundity, intelligence, defensiveness, and especially adaptability to changes brought about by human activity. As the fates of carp and largemouth bass in South Korea illustrate, the identity of pest and victim may depend as much on value as on biology. Looks count. Even Canada geese and red deer are beloved animals compared with the equally ubiquitous but awkward Muscovy ducks, persecuted throughout suburbia. Bears, prized by the environmentalist, are pests to the beekeeper, whose bees may in turn (especially if Africanized) be pests to the unwary hiker. But if the hiker's presence leads to political demands for removing the hives, the hiker is also a pest to the beekeeper, as indeed the seeker of trophy fish may be to the commercial carp fisher. Why destroy a valuable food resource for the sake of the recreation of a few?[48]

Pesthood is not only social but seasonal. Carp undoubtedly created many problems for some native fishes, and took advantage of the degradation of others' waters. But the ichthyologist Peter B. Moyle has underscored an American distaste for all nongame fishes. It is also possible, though hard to document, that growing anti-immigration sentiment in the late nineteenth century helped make carp, a favorite of Asian and Jewish communities, unfashionable. (Perhaps, too, Spencer Baird faced the dilemma that anything cheap enough will be undervalued; even salmon has been unpopular when abundant.) Occasionally it even ascribes the spread of pests to political conspiracy. Propagandists in Stalin's Soviet Union blamed the spread of the Colorado potato beetle on the CIA, and dubbed it the "six-legged ambassador of Wall Street."[49]

When people are the invaders, pests may even be allies. William H. McNeill has argued that "looked at from the point of view of other organisms, humankind . . . resembles an acute epidemic disease, whose occasional lapses into less virulent forms of behavior have never yet sufficed to permit any really stable, chronic relationship to establish itself." Thus the tsetse fly and the sleeping sickness it carries until recently preserved the wildlife of the African savanna from human encroachment. Similarly, in the Adirondack Park in our own time, foes of development and the spraying of pesticides almost welcome the annual ravages of flesh-slashing blackflies as protection against the invasion of the woods by developers and second-home builders. The blackfly as "seasonal scourge," one wrote, "guards the sacred bough and preserves the natural mystery deep within the Forever Wild Adirondacks."[50]

Still, cultural judgments aside, it is a biological fact that some animal species are much better at invasion than others—better even than close relatives. Can we prevent unpleasant surprises in the future by predicting which species are most likely to become pests? The answer is probably not, or at least not very precisely. Usually only one or two of a number of closely related organisms establish themselves. While the house sparrow spread not only in North America but in South America, Southern Africa, Australia, and New Zealand, the tree sparrow (*Passer montanus*) of the same family took over ninety years to increase its range to central Illinois after its successful release in St. Louis in 1870. Yet *P. montanus* is found as extensively in its native Eurasia as *P. domesticus*. We probably would need to know much more than we do about the genetics of animals, and their parasites, to improve our ability to predict their fate. We don't know why, for example, some natural predators can be released successfully to control imported pests, while others fail even after careful study. The population biologist Paul R. Ehrlich has identified a number of possible invader traits: abundance in their original range, ability to eat diverse foods and survive in different physical conditions, short reproductive cycles, great genetic variability, ability of a fertilized female to colonize alone, larger size than related species, and links with people.[51]

Even with these guidelines, predicting invasions is difficult enough that it is hard to exclude animals as potential pests—unless they directly threaten agriculture as well as the rest of the environment. The Fish and Wildlife Service attempted in the 1970s to exclude thirty or so fishes that Walter Courtenay and other ichthyologists identified as potential invaders. When the aquarium fish industry threatened legal action, Fish and Wildlife withdrew the list. Since then, at least two of the warnings have been confirmed: the blue tilapia has come to dominate the lower Rio Grande and the peacock cichlid has established itself in southern Florida. More recently,

Courtenay has campaigned for more secure containment of imported fish in Florida, the center of the aquarium trade, and for warning fish buyers against the false humanitarianism of releasing unwanted aquarium fish in natural waters.[52]

In the looking-glass world of exotic introductions, kindness to one species can be deadly cruelty to another. Are the romantic wild equids of the American West "an integral part of the natural system of public lands," as decreed by Public Law 92–195 of 1971, or are they the horses (and burros) of the apocalypse for hundreds of other Western plant and animal species? Weren't boll weevils also part of a regional culture with an equally legitimate interest against eradication? In refraining from "fascist" campaigns against imports, are we really affirming benevolent neutrality in the interplay of the world's life forms, or are we tacitly supporting global homogenization and banalization? Are "pests" really to blame, or is it human arrogance in disrupting, dividing, and degrading habitats? These are complex and thorny questions for environmental ethics. The likelihood that disease-resistant genes from genetically engineered plants and animals will spread to their wild relatives makes them urgent questions as well.[53]

The main lessons of the technologies of expert animal acclimatizing are three: how the latent properties of natural systems defy the imaginations of even the scientific elite, how the search for more intensive production can backfire against productivity, and in general how undoing animal introduction can be as impossible as unscrambling an egg.

7

Acclimatizing Pests: Vegetable

"Plants can do everything animals can do, only more slowly," a biologist friend once observed. When it comes to invasion, plants can do even more than animals, sometimes faster. That is why the unintended consequences of new plants can be even more drastic than those of introduced animals. Plants and animals alike accompanied Europeans in the global extension of their settlements, described memorably by Alfred W. Crosby as a "grunting, lowing, neighing, crowing, chirping, snarling, buzzing, self-replicating and world-altering avalanche." Domestic cattle, swine, and horses brought by the Europeans to the Americas not only tended to run wild themselves; they depended on plants unintentionally introduced with crop seeds, plants that the settlers condemned as weeds. These plants have many of the qualities we have seen in "pest" animals. They are prolific. Their seeds disperse quickly and far; they germinate early. Some also form superplants, connected by underground rhizomes, that choke out other plant life. They displace and overshadow. They profit from the absence of the natural predators of their homelands. They thrive in harsh conditions and germinate after decades. Just as carp can proliferate in polluted reservoirs and starlings roost in city trees, weeds survive in hostile habitats. After all, they evolved to cope with the conditions left by the end of the Ice Age in Europe. In fact, the only thing that appears to stop weeds is their own ability to stabilize the soil. The longer an area has been left untouched, the fewer weeds it has. The European occupiers of North America and elsewhere have had to deal with so many pests because they disrupted the land so profoundly and continue to modify it so intensively.[1]

The botanist W. Holzner has defined weeds as "plants adapted to manmade habitats and interfering there with human activities." One definition of a true weed might be a plant with a built-in revenge effect, i.e., a plant that actually profits from attempts to eradicate it. Cutting plantain triggers new shoots from its hidden rhizomes. Burning wetland vegetation to destroy purple loosestrife exposes the soil to sunlight, thereby hastening the germination of still more loosestrife. Ripping out other species, as the garden writer Michael Pollan discovered of burdock, leaves a taproot that will send up more. The root of bindweed, a destructive vine that stifles its plant hosts, easily falls into pieces from which new plants sprout. "It is as though bindweed's evolution took the hoe into account," Pollan observes. "By attacking it at the root I played right into its insidious strategy for world domination." (The same thing happened in the 1930s when a Chicago physician led a campaign to eradicate ragweed. His corps of volunteers killed countless plants but inevitably dispersed pollen that increased the range and intensity of the city's ragweed crop.)[2]

Like the problem animals we have seen, some of the most serious plant pests have entered North America on the recommendations of experts—scientists, government officials, seed merchants, and nursery owners. David Fairchild, former head of the United States Office of Plant Introduction, mentioned casually in a travel book his organization's "introduction of nearly 200,000 named species and varieties of plants from all over the world." Novelty and perceived beauty have counted at least as much as utility. Some experts did not understand how dramatically new plants could transform the landscape. Others did, but failed to foresee all of the consequences that might ensue. And like controversial introduced animals, many weeds have their admirers and defenders: for example, growers and buyers of costly dandelion greens, or admirers of cheeky urban vegetation like the ailanthus, the tree that grows in Brooklyn and other cities where more sensitive species often succumb to fumes and vandals. (Ailanthus, too, tends to send out new shoots and reappear where it has been cut.) There are enough hardy urban plants that one botanist has found over four hundred in inner-city Cleveland; and others encountered rare orchids growing wild on the shores of a Schenectady, New York, lake and on southern Arizona rangeland. Still other plants endangered in their original habitats may ultimately be saved from extinction by their persistence as wild vegetation in abandoned English gardens. Sometimes weeds may be promoted, and the Netherlands has picking gardens, from which children and adults can take colorful weed flowers home. The motto of weed appreciators everywhere must be Ralph Waldo Emerson's definition—a plant whose uses are not yet known.[3]

There are also important differences between animal and plant pests that must not be overlooked. Weeds seize open, disturbed territory; they usually do not compete aggressively with native wild plants. Development divides habitats into smaller zones separated by roads, housing, and industry. As more regions are thus broken into islands and stressed by human activities, plant invaders can transform their habitats far more radically than even the hungriest and most prolific insect, bird, or mammalian newcomers. Reshaping the landscape, they can exclude other vegetation on which valuable and rare animal species depend. Weeds may only be taking advantage of more fragile habitats. But a small number have turned from squatters into environmental storm troops. Whether suppressed or allowed to run their course, they unleash natural warfare by their very presence.[4]

Invasive plants depend upon technology for their dispersal even more than their animal counterparts. We have seen how the gypsy moth achieved its foothold in the Boston area through premotorized suburban traffic; railroad cars, followed by automobiles and trucks, did the rest. Railroads also helped establish carp across the country within twenty years. Human transportation was just as essential for plant invaders. The earliest railroads changed the vegetation of North America, not only spreading weeds but even preserving weeds that had become rare plants where farmers had all but eradicated them. The fencing that railroads and farmers alike erected to protect the trains from cattle—cowcatchers did not really do the job—also saved the plants from beasts that would otherwise have munched them. Passing trains spread seeds throughout the network. They helped introduce the tumbleweed, *Salsola kali,* to the intermountain West. (It is not an indigenous American plant at all. It is a European weed that when dead and dry easily detaches from its roots to form one of the rare rolling objects in nature, the better to distribute its seeds.) As some of the early weeds grew rare in the agrarian landscape, the railroad right-of-way became both their protective reserve and their means of propagation.[5]

As North American settlement spread westward, the plants of the Old World invaded with it. Weed seeds traveled in the wagons of the pioneers and on their horses' hooves. Vendors used them to adulterate commercial seed shipments until the United States and Canada enacted pure-seed legislation in the early twentieth century. Government agencies and other national organizations spread weeds, too. U.S. Navy officers and diplomatic personnel routinely brought back plants and seeds they thought might be of economic or aesthetic interest. In 1839, the U.S. Congress gave the Patent Office funds for obtaining new plants from around the world. Crabgrass—actually various grasses of the genus *Digitaria*—was among its early discoveries. In fairness to the bureaucracy, these grasses are good forage crops in Europe, and immigrant farmers brought and prized them, too. It was not until the 1860s that Frederick Law Olmsted and other landscape architects popularized the idea of detached houses planted in a parklike expanse of grass. The introduction of the first compact lawn mowers about this same time helped make the suburbs literally, in the urban historian Kenneth Jackson's phrase, a crabgrass frontier. Two kinds of intensiveness, each reasonable in its place, combined with unfortunate results: crabgrass was hardy, tenacious, and easy to propagate. The lawn was a carefully managed zone rich in nutrients.[6]

The federal government also helped generously with another weedy plant that it later disavowed: hemp, or *Cannabis sativa.* Like crabgrass, hemp can be valuable. Thomas Jefferson grew it for fiber; so did countless nineteenth-century Americans. And it was not only legal but patriotic. For decades before the War of 1812, the U.S. Navy did its best to promote a domestic hemp industry. In 1877–78 alone, helped by the national railroad and postal network, the Department of Agriculture distributed 339 samples of hemp (and 343 of opium poppies). As late as the 1890s, Special Agent Charles Richards Dodge "in Charge of Fiber Investigations" for the USDA was promoting cannabis in the department's yearbook as "so generally cultivated the world over as a cordage fiber that the value of all other fibers as to strength and durability is estimated by it." (Birds relish hemp seeds, and breeders fought anti-marijuana legislation in the twentieth century.) Nevertheless, neither the Navy nor the USDA had much success with cannabis. As a fiber, plebeian hemp, curiously like aristocratic silk, demands skilled hand processing. Raising it has always been a cottage industry. Even before the ban on cannabis, none of the estimated hundreds of patented hemp-cleaning machines was effective. And yet hemp was grown widely enough—thanks to commercial seed catalogs as well as to government encouragement—that the wild form of cannabis has survived into the late twentieth century as ditchweed. Curiously, Americans believed until this century that only South Asian hemp (*C. indica*) was suitable for drug production. And ironically, the application of severe antidrug laws to marijuana in the 1980s and 1990s encouraged illicit growers to hybridize *C. sativa* with *C. indica*, breeding compact plants ideal for clandestine, often automated indoor cultivation. Production soared. In 1995, cannabis was the biggest American cash crop in dollar volume, exceeding corn and soybeans combined.[7]

In fact the botanist Richard N. Mack has argued that the seed trade not only introduced future pests but made them more serious by its methods. An invasive plant once had to make one journey at a time to extend its range. Catalog sales, effected by railroad transportation of seeds, could establish plants at the same time in locations hundreds or thousands of miles apart. Each of these in turn could become a secondary point of distribution. (Today, of course, air transportation is an even faster way to do the same thing internationally.) A woody plant distributed as a seed has already shown its ability to reproduce sexually in a way that one sold as a seedling, or as a plant grown from a cutting, has not. Nineteenth-century seedsmen did not breed today's sterile hybrids or select for domestication; their seeds were much more likely than today's to survive in the wild. And of course commercial seeds get more attention and grow more densely than plants that spread accidentally, and are that much more likely to reproduce naturally. The seed trade, especially before twentieth-century regulation and hybridization, diffused plants less as a wave and more as a rain of widely spaced missiles with living payloads. In at least one case in Nevada, a well-meaning state agent for agricultural experiments included actual dried weeds, seeds and all, in bulletins he circulated warning farmers to beware of them. These samples, in fact, may well have introduced the weed, Mack believes.[8]

Automobiles encouraged their own revenge plant, the puncture vine or *Tribulus terrestris.* This had been a rare species in California until the motor boom of the 1920s, when better roads and heavier traffic helped spread it throughout the state. The sharp spines of *Tribulus* seedpods deflated and weakened tires. The California Highway Commission recommended repeated (and costly) spraying of roadside plants with diesel fuel. It also acknowledged that unless neighboring private lands were also treated, "our work will avail nothing, as these areas will be reseeded from outside faster than we can eradicate." Modern tires have ended the puncture vine menace for motorists, but it remains a hazard for bicyclists in the American Southwest.[9]

Motor vehicles are not the only innovation that has changed a manageable plant into a runaway menace. Along the rivers of the Southwest, the salt cedar or tamarisk, a Eurasian tree, was introduced in the late eighteenth century for shade, wood, and flood control. It spread slowly during the nineteenth century. Then, in the early 1900s, it seemed to sweep through the Southwest. Like other invaders, it had fecundity and swift growth on its side; one tree sends forth 500,000 minute airborne seeds annually, and those that sprout can grow to ten feet in their first year. The plant ecologist Duncan Patten and others believe that dams have given the tamarisk a crucial edge by suppressing the natural flooding that favored the seed-release schedules of its native competitors, the cottonwoods and willows. Patten also points to the rise of grazing and the greater appeal of the native plants to cattle. These changes transformed a useful homesteader's plant into a rampant invader covering a million acres, displacing native trees, and dropping leaves that release so much salt that the soil may not support other vegetation. Declining productivity in turn depletes wildlife. Their spread also creates revenge effects for dam builders; thriving along the edges of reservoirs, they pump priceless Western water into the atmosphere. The tamarisks of the American Southwest are said to soak up twice as much water every year as all the cities of southern California use. Restoring natural river conditions and restricting grazing can limit their expansion, but control usually demands labor-intensive burning, poisoning, plowing, and even covering with plastic bags.[10]

KUDZU

Salt cedar is not the only beneficial plant to turn Frankenstein monster, and hemp is not the only government introduction that has resulted in a drastic reversal of policy. Kudzu (*Pueraria lobata* or *P. montana*), a semiwild leguminous vine native to Asia, thrives in the most neglected and abused soils. It diffuses by rooted nodes. The woody kudzu root is virtually a mini-tree, up to seven feet long and weighing as much as four hundred pounds. In Japan its powdered root is a prized starchy ingredient—delicate enough to be served as an appetizer on handmade porcelain in the nation's most refined Zen temple restaurants. The kudzu leaf adorns family crests. Introduced to North America at the Philadelphia Centennial Exposition in 1876, kudzu became a popular ornamental plant for shading porches. There it stayed; rarely can it propagate by disseminating its seeds.[11]

It was government agricultural experts who turned kudzu into a pest. (Hydrilla, kudzu's waterborne counterpart, entered the United States as an aquarium plant without any sponsorship by experts and thus cannot be counted as a revenge effect.) David Fairchild, then already a celebrated plant hunter for the USDA Bureau of Plant Industry, brought kudzu to America again around the turn of the century just as private farmers were starting to use it experimentally on worn-out pastureland. Before this reintroduction, Fairchild had to spend more than two hundred pre–First World War dollars removing kudzu vines that were climbing pine trees on his own property and bending them to the ground, yet he seems to have found no reason for cautioning others about the plant. Some farmers needed no encouragement. One Florida man, a former school superintendent, called for planting it on half the cultivable land in his county: "[I]t would bring us $15,000,000 to $35,000,000 instead of the measly little pittance we now get from King (?) Cotton."[12]

During the New Deal the U.S. Soil Conservation Service advocated kudzu as a plant that could restore Southern cotton lands devastated by insects, erosion, and the Depression. They were right to admire its vigor. Like weeds, kudzu flourishes in disturbed and marginal settings, resists insects and drought, needs no preparation of the soil or fertilizer, and above all grows rapidly. An acre of kudzu left unchecked, one enthusiast calculated, would expand to thirteen thousand acres in the course of a century. Growing at up to a foot a day in the spring, kudzu tendrils can extend sixty feet within a season and overgrow objects as high as forty feet. Its lush growth keeps the soil cool and moist. As a ground cover, kudzu cuts water runoff by 80 percent. Land planted with kudzu loses 99 percent less soil than land planted with cotton. It restores soil nutrients and makes good pasturage and hay, though milk from kudzu-fed cows is not always palatable. No wonder the Soil Conservation Service sent 73 million kudzu seedlings to farmers and Civilian Conservation Corps groups, and paid $6 to $8 per acre for planting it.[13]

By the end of the Second World War, a half million acres of the American Southeast were planted in kudzu. Private boosters of the "miracle vine" took over where the New Deal had left off. Channing Cope, the man who made the 13,000:1 estimate, founded the Kudzu Club of America, aiming to raise the national acreage to eight million. Yet only ten years later, farmers and federal officials alike were condemning kudzu as a nuisance plant. The vine had not changed, but agriculture had. Southern landowners were letting forests grow back; kudzu climbs, overshades, and kills saplings and even hundred-foot-tall trees. (In 1988 a U.S. Forestry Service official estimated timber losses caused by kudzu in Alabama, Georgia, and Mississippi at up to $175 million.) It hogs potential forest land. Kudzu was also less attractive to cattle and milk producers, who wanted more productive feed crops and were willing to spend more fertilizer and attention on them. Even for erosion control, new and more manageable plants had appeared. Partly because of its own Depression-era success, kudzu was becoming technologically obsolete. The Soil Conservation Service, which once promoted it, has officially been calling it a weed ever since 1970.[14]

Kudzu may have fallen from federal grace (not entirely, as we will see), but it no longer depends on it. Without diseases or North American insect enemies, it has continued to thrive throughout the Southeast. Unwelcome down on the farm, kudzu has invaded the technological infrastructure of the flourishing eastern Sunbelt with a vengeance. It pulls down telephone poles, blacks out neighborhoods by warming local power transformers until they trip, and shorts high-voltage lines on long-distance electric transmission towers. Once planted by highway departments to stabilize roadsides, it now obliterates traffic signs and spreads over bridges. On railroad grades, kudzu grows over the rails and slickens tracks as trains crush it. Locomotives eventually lose traction. In fact, kudzu can overwhelm and envelop nearly any stationary object—unmoved automobiles, sidetracked railroad cars, abandoned houses, even (so says Southern folklore) unconscious drunks.[15]

Kudzu is not as vigorous as other problem plants. It can die from accidental or deliberate overgrazing. But a remnant will always be out of reach; cattle don't climb trees, as the authors of one book on the subject point out. Killing the kudzu means finding and destroying all of its root crowns; just one can regenerate all that was uprooted. Farmers have to slash vines and trees around borders. The arsenal of herbicides used against kudzu continues to grow, but many products can harm desirable plants. And there is a special revenge effect in poisoning kudzu: exposure of the soil for up to a year after all vegetation has been killed can renew the very erosion that kudzu was introduced to halt. It is not entirely surprising, then, that soon after it stigmatized kudzu, the Soil Conservation Service was distributing samples again—cautiously. Perhaps Japan's refusal to provide more seeds was a sign that the plant is more a trade secret than a menace. In the 1970s and 1980s, its Zen connections, ability to grow without and in spite of chemicals, and value as "natural" food and fiber brought kudzu from New Deal to New Age.[16]

Still, it is not certain whether there can be lasting détente with kudzu. Scientists at Florida's Department of Natural Resources took notice in August 1992 when visiting herbicide specialists from Alabama and Georgia pointed out a few specimens growing along several miles of levee in South Florida's canal system. They had thought it would never move south of Ocala. They saw immediately that it could overwhelm not only the remaining native vegetation but even the previous aggressive invaders (which we will see) of the Everglades, turning the park into little more than a huge kudzu patch. Isolated as the plants turned out to be, they reminded biologists of how much vigilance a chronic problem organism demands.[17]

Multiflora Rose

If kudzu has settled into an uncertain middle age as a naturalized plant, even an emblem of a whole region, multiflora rose (*Rosa multiflora*) is still an unruly adolescent. Like kudzu, multiflora is an Asian plant that gardeners and homeowners adopted in the nineteenth century. It is an excellent rootstock and was hybridized with other rose varieties. It is a healthy plant with few insect pests. Birds and other wildlife love the fruit. Added to the hundreds of other rose varieties in common cultivation, it probably had a negligible environmental impact in the nineteenth century. Then, beginning with the Depression, the Soil Conservation Service and other federal and state agencies discovered and promoted it. Like kudzu, multiflora stabilized worn-out soil. It smelled sweet, looked good, and provided a natural protective fence for crops. Over two decades until 1960, farmers in North Carolina alone planted between fourteen and twenty million roses, and others in West Virginia set out more than fourteen million. Multiflora windbreaks even arose in the Great Plains.[18]

Multiflora's nineteenth-century virtues became—at least for the farmers whom the Soil Conservation Service was trying to help—twentieth-century vices. A single plant, according to James W. Amrine, Jr., of the West Virginia University Division of Plant and Soil Sciences, can produce a half million or even a million seeds in a good year. Multiflora's attractiveness to songbirds like robins and mockingbirds ensures that some of these seeds are spread when excreted. Songbirds do not grind up seeds as chickens, turkeys, and other gallinaceous birds do, but often pass them intact. The process actually makes the seeds twice as likely to germinate, according to one study. The government-sponsored overextension of this attractive and durable plant, coupled with its ability to spread and its tenacity once established, brought out its weedy potential.

Even in the 1930s, Kentucky and a few other states realized the hazard and limited multiflora planting. A leading Indiana botanist, alarmed by proposals to plant old cemeteries with multiflora, wrote to a correspondent that "[w]hen Gabriel sounds his horn, I am afraid some will be stranded and not be able to get thru the roses. Please do not recommend the multiflora rose except for the bonfire." In the 1960s his fears were realized. Multiflora had spread to hill slopes, roadways, and other marginal lands. First West Virginia, then Iowa, Illinois, Kansas, Maryland, Missouri, Ohio, and Pennsylvania branded it a noxious weed. In West Virginia, vocational agriculture teachers and county agents called it the state's most serious agricultural problem, according to one survey in the 1980s. In North Carolina, over two million acres of former meadow are now dense with multiflora. The living fences of multiflora, far from merely bordering and protecting fields, have overrun them; grazing and farming are impossible where it thrives. For the last thirty years, farmers have been trying to cut it, rip it, level it, burn it, poison it, and overgraze it with the last resort of herbicides, goats.

Unlike kudzu, multiflora has promising natural predators and diseases. There are tiny wasps, rose seed chalcids (*Megastigmus ac-*

uleatus var. *nigroflavus*), that lay their eggs in up to half the viable seeds, destroying them as they mature. Chalcids may eventually destroy up to nine out of ten multiflora seeds. Unfortunately, so many multiflora were planted as cuttings that control by the chalcids will have to wait until they disperse naturally—which means slowly, perhaps more than twenty years. Another insect, the rose stem girdler, does just what its name implies, damaging the stems of mature plants when its larvae emerge, but it does not kill enough plants for serious control. There are also mites that carry rose rosette disease (RRD), also known as witches'-broom of rose, often attributed to a virus but not really understood. The mites and RRD had spread to the East from the Western windbreaks by the early 1990s, by way of the Ohio Valley and West Virginia, and appear destined to spread on prevailing winds to New England.

RRD does not appear to threaten economically important fruit trees, but its use has potential revenge effects of its own. Because many plants with RRD show no signs of disease, the virus could easily spread inadvertently. Although some species appear to be unaffected, other rose varieties could be devastated by RRD. If the mites that carry it were transported to East Asia, they could infect the local multiflora rose populations. There multiflora is not a pest at all, but an important and appreciated plant. James Amrine fears the mites may—like gypsy moths, Dutch elm disease, and chestnut blight—become even more serious problems overseas than in their home countries.

Multiflora shows that given enough encouragement by experts, even an apparently well-behaved ornamental plant can turn into an officially condemned pest within thirty years. It also shows that biological control is neither fast nor free from revenge effects of its own.

Futureweeds: Cogon Grass

Because the garden seed and nursery businesses are as devoted to the impulses of fashion as they are to the demands of agriculture, and because the technology of transporting and propagating exotics continues to improve, new candidates for acclimatizing will continue to appear. Even as the environmental (left) arm of the federal bureaucracy points its accusing finger at nonindigenous species, the agricultural (right) arm will continue to extend at least a wary handshake. After all, one reason to encourage and maintain a diversity of life on earth is the opportunity this affords to study and introduce beneficial species. But acclimatizing studies can have revenge effects delayed by decades.

Cogon grass (*Imperata cylindrica*), which is starting to claim large areas in Florida and elsewhere in the Southeast, has tall, sharp, spiky leaves. When scientists at the beef cattle laboratories of the U.S. Department of Agriculture near Brooksville, Florida, first imported it, probably from China, its toothed edges discouraged cattle from grazing. The USDA dropped the project. The trouble was that seeds somehow were spread, and cogon grass turned out to flourish along roadsides, at the edges of pine plantations, and in other places where it was not as conspicuous as kudzu but displaced more valuable native vegetation. Other animals evidently found it as unpalatable as did the cows, and Florida's rich insect fauna would not touch it.[19]

The underground shoots or rhizomes of cogon grass are massive and durable. Beneath a single acre of the plant may be three tons of rhizomes, which retain water efficiently and may work their way up to four feet within the soil. Botanists believe the rhizomes produce chemicals that limit the growth of other plant species; whatever the reason, few other kinds of plants are found in dense patches of cogon grass.[20]

By the mid-1990s, cogon grass was joining the list of Florida's problem imports. "Once we became aware of it, we began to see it everywhere," said one Florida state land manager, "up and down the highways, everywhere. The Citrus Tract in the Withlacoochee State Forest is just loaded with it. If you keep your eyes open, you'll start finding it everywhere." A state task force has been studying the plant. The agronomist who heads it believes cogon grass is already widespread in the Southeast. In the absence of biological control, Florida officials can only apply massive doses of herbicides with names like Arsenal and Roundup. And the rhizomes of cogon grass demand unusually thorough treatment if the plants are not to grow back rapidly.[21]

Exploding Trees

While herbaceous plants and vines may spread to become nuisances on a regional and even national scale, new trees can be more troublesome locally. Trees can be weeds, too, especially when removed from their native growing conditions and insect predators. In areas of rapid development with disrupted landscapes, like the foothills of California and the waterways of southern Florida, trees have been among the most controversial of plant introductions.

Of all the developed world's landscapes, southern Florida's has been changed in as many ways and for as many reasons in as short a time as any other. It is actually a recent landscape; much of it was below sea level five thousand years ago. The biologist Daniel Simberloff has called the region "a habitat island, bounded on three sides by water, the fourth side by forest, and typified, as are oceanic islands, by an impoverished native flora and fauna." Lightning strikes and fires are frequent. Underlying soils are nutrient-poor and easily modified by human activity like agriculture, tree farming, and housing. Lakes and rivers are abundant. The relative absence of stream channels can expose plants to new water conditions when ditches and dikes are built miles away. These and other changes can alter the structure of biological communities and weaken their competitive advantage against newcomers. And like other disruptions, they prepare territory for weedy plants. Generations of canal building have changed a seasonally flooded wetland southeast of Lake Okeechobee into a million acres of agricultural and developed land. Florida's economic boom has been an ecological disaster for the wetlands of the Everglades, where vital water sources have been diminished and polluted by fertilizer runoff.[22]

Environmental scientists have been increasingly concerned about the kind of fragmentation Florida has undergone. When native vegetation is cleared away, even the areas not directly touched are subject to stress. Agriculture and building tend to increase the range of temperatures, with hotter days, cooler nights, and more frequent frosts. Boundaries between types of vegetation change wind patterns, exposing native plants to more weather damage and enabling seeds, insects, and diseases to migrate into undeveloped zones. Isolation reduces the number of species each remnant can support. Many animals specialize in living where two natural zones meet, areas that border on development and are biologically poorer and more stressed than those in the interior.[23]

Florida itself is a border state, a port of entry for tourists, business visitors, immigrants, and migrants from all of North and South America and Europe, all potential importers of new life-forms. The Port of Miami is also one of the busiest East Coast shipping centers, with double the exports of Baltimore and catching up on the value of its imports; plenty of the merchandise, too, is living. Florida is the major center of U.S. tropical fish importing and an exotic vegetation supermarket. Vendors of aquatic plants, including some from as far away as North Carolina, until relatively recently used Florida's public drainage canals, ditches, and other waterways to plant and grow their stock without maintaining expensive nurseries. And despite Canada's embargo in the face of the Africanized bee hazard, Florida is still a prime wintering ground for hives from the northern states of the United States. Not surprisingly, Florida now has over nine hundred introduced plant species, over a hundred of which have been classified as environmentally hazardous invaders by the state's Exotic Pest Plant Council, an organization of state and federal biologists studying the invaders.[24]

Of the troublesome newcomers, a number of the most disturbing were brought not by naïve householders or optimistic nurserymen but by professionals—public officials and academics—who were trying to improve and restore the environment. One plant explorer from the Department of Agriculture introduced Brazilian pepper (*Schinus terebinthifolius*) in 1898, and the USDA Plant Introduction Station in Miami made it available to amateur gardeners. The railroad builder Henry Nehrling, aided by friends in the nursery business, also distributed thousands of the plants. By the late 1980s it occupied thousands of acres of south and central Florida, the American Southwest, and Hawaii. This evergreen, growing up to forty feet high, has drupes that turn bright red late in the year and has been popularly known as "Florida holly" and "Christmas berry." It also grows aggressively, and can produce rashes on human contact. It turned out to be closely related to poison ivy.[25]

Probably because Brazilian pepper affects wetlands and their wildlife more than agriculture, the Florida legislature has been slow to appropriate funds for the plant's control, even though state law classifies it among the prohibited aquatic species. Brazilian pepper is now rampant in central Florida, especially along the drainage channels and ditches in Brevard County, and has become a serious concern of residents. In the Gulf Coast mangrove forest of the Everglades National Park it has been spreading from disturbed former agricultural lands to the previously untouched zone. Flood control measures, hurricanes, frost, and even large flocks of migratory robins eating berries and dispersing seeds have combined to make the plant a major problem. (The robins may have regretted their feast; the berries can be as unfit for avian as for human consumption.)[26]

It was into the disturbed wetlands of south Florida that *Melaleuca quinquenervia* was introduced. Melaleuca was another pest tree transplanted by professionals with the best of intentions: John C. Gifford was not only a real estate developer and nurseryman but a forester at the University of Miami. He brought the first trees to Florida in 1906, and soon melaleuca was spreading to coastal wetlands. But it was Gifford's colleague Hully Sterling who decided to restore the trees of the Everglades by scattering melaleuca seeds from the air. Just as aviation was to broadcast the "safe" pesticides of the postwar period over millions of acres, it was able to transform a landscape overnight in 1936. (There is no record of any federal, state, or local opposition to the airdrop.) Farmers also planted melaleuca as windbreaks and fence rows, and its rapid growth rate made it appealing for home landscaping. By the 1970s, melaleuca had spread into most of the water conservation areas of the Everglades, where Florida authorities have called it "the most serious threat to the Everglades ecosystem." In 1980 it covered 6 percent of south Florida land, including almost 13 percent of undeveloped wetlands. By 1993 it was estimated that melaleuca covered at least 380,000 acres, with some estimates exceeding 1.5 million, with a growth rate of fifty acres a day. Some believe melaleuca could eventually increase to ten times its present population.[27]

Melaleuca is an Australian and Malaysian tree that has no native insect enemies in Florida. It grows at an average rate of more than six feet per year, reaching heights of a hundred feet, and may live for seventy years or more. Stands of young trees reach exceptional density, choking out most native vegetation and supporting little wildlife even after they thin out. Melaleuca is sensitive to frost—only isolated trees have been seen in north Florida—but flourishes in soils that experience some flooding. With several times more leaf surface than native saw grass, it loses much more water to the atmosphere.

Melaleuca's Australian heritage helps account for its literally explosive growth. It belongs to a group of plants that propagated themselves with fire. Like Australia, Florida has a high rate of lightning strikes which often set fire to vegetation. In melaleuca, these fires not only spread readily to the flammable bark but can trigger an oil in its leaves that makes the tree's crown burn explosively, releasing seeds. In Florida, the extreme heat at which it burns (1,500 degrees F.) explodes native hardwoods nearby and can set off intense peat fires that change soil concentrations irreversibly, causing alterations in one season that otherwise would take thousands of years.[28]

Melaleuca does not need fire to reproduce; wind and other disturbances can also break open the seed capsules. Wintering bees from the North love melaleuca flowers and may have done their part to help the trees spread by pollinating them. (In fact, the only important benefit of melaleuca appears to be that honeybee hives attracted by it contribute millions of dollars to the value of other crops pollinated and cross-fertilized by the bees.) Once melaleuca covers more than 75 percent of an area, native fish, amphibians, birds, and reptiles decline in number, and invasions of frogs, rats, and house mice increase.[29]

In the 1990s, biologists are trying to save the Everglades from the airborne mission to restore them. Like many other chronic problems, melaleuca can be spread easily but managed and controlled only painstakingly. Contrary to the expectations of the foresters who seeded it, melaleuca is a poor source of wood and expensive to harvest. The equipment that could remove it leaves ruts in the soft soil, changing its topography. The best-known methods are pulling trees by hand, and girdling trunks and applying a herbicide. But after removing melaleuca, conservation officials must return seven to nine months later to pull up seedlings; from a few hundred to as many as eighteen thousand can germinate after one tree is disturbed. And no herbicide that can treat large areas of soil has been approved for standing water. State and federal entomologists are testing imported melaleuca-eating Australian insects, including a weevil and a sawfly, which don't appear to find any native American vegetation palatable. But it is not clear that funds will be available to continue the tests and introduce the insects—or that any insects will be able to multiply quickly enough to keep the rampant melaleuca in check. The experiment in restoring the Everglades may yet lead to its devastation.[30]

Melaleuca is a localized pest. It needs such a hot, wet climate that it can thrive only in a small area of North America. The genus *Eucalyptus,* another Australian emigrant tree, is as versatile as melaleuca is restricted—and in different ways, just as problematic. Most people who live with eucalypts don't consider them pests. In fact, eucalypts are among the planet's most versatile and valued trees. Cultivated carefully, the deep roots of some species can tap underground water sources unavailable to other plants, increasing the productivity of a managed environment. But some of the very qualities that people prize in them can have unexpected and unpleasant consequences.[31]

The environmental historian Stephen J. Pyne has called the eucalypt "the Universal Australian." The eucalypt is a scleromorph like melaleuca, one of a family of evergreen trees named for the small, hard leaves that retain nutrients and let the group adapt to many soils, including poor ones. The genus was once limited to a small group in Australasia, but the breakup of Gondwana that began thirty million years ago started it on its way to dominate the vegetation of the new continent: a scleroforest, as Pyne puts it. Fire was both a consequence and a cause of this process, as vital as precipitation to a rain forest. Fire selected plants for their adaption to it, and the spread of the scleromorphs in turn ensured that future fires would be well fueled by plenty of survivors. Eucalypts, thanks to their deep root systems, could find nutrients in a wide variety of soils, extracting phosphorus from sites where more specialized plants would have perished. Equipped with structures called lignotubers, the root systems can store nutrients for up to a decade. The whole tree is a marvel of frugality, optimized to withstand poor soil and drought.[32]

Eucalypts coevolved with the fires that regularly swept Australia. From the roots to the crown, every part of the tree came to withstand combustion, and to turn this to its advantage. Its roots were safely buried and could recycle the nutrients that fire deposited, as well as send out new shoots. Fire not only prepared the soil for new trees, it opened clearings necessary for their growth. And since 95 percent of Australia's forest trees belong to the genus *Eucalyptus*, its birds and other wildlife came to depend on eucalypts and on the fire regime that sustained them. All this was usually noticed too late by the tree's admirers.

The admirers were many. The Acclimatization Society of Paris helped establish experimental stations in Antibes and Algeria as a remedy for deforestation and, it was thought, as a swamp-draining malaria fighter. Famous physicians praised the external disinfecting and internal healing powers of eucalyptus oil—especially against malaria itself. Tolerance for drought and soil varieties helped the tree's popularity wherever frosts did not threaten it. Most of the great colonial powers encouraged planting one of the twenty or so species that appeared economically valuable. Less than a hundred years ago, under Emperor Menelik II, Ethiopia established a fixed capital for the first time at Addis Ababa ("New Flower") after the introduction of eucalypts assured a steady fuel supply. Today eucalypts thrive throughout Mediterranean Europe and the Middle East, much of Africa, India, and China. In the twentieth century, eucalyptus products are staples of the developing world. Hated by many environmental scientists there as pathologically thirsty nuisances, and sometimes as colonial vestiges to boot, eucalypts are nevertheless prized by rural dwellers for their lumber, bark, and leaves.[33]

Americans first planted eucalypts in California, where early settlers had burned vast stands of timber in their impatience to start farming and then confronted a treeless landscape. By the later nineteenth century, ruinous logging in the forests of the eastern United States, too, led some Californians to campaign for massive eucalyptus plantings in their state to promote local forestry. The president of Santa Barbara College, Ellwood Cooper, led the crusade. Cooper imported seeds from Australia and reprinted lectures of Australia's leading botanist, Baron Ferdinand von Mueller, on eucalyptus growing. The trees were beneficial to the soil, and profitable. Without their Australian enemies, they grew at record rates. Cooper reported that in three years, a seed would become a forty-five-foot tree with a nine-and-a-half-inch diameter. Some eucalypt species failed to withstand California conditions, but one lived up to Cooper's promotion: the fast-growing blue gum (*E. globulus*), which soon became and has remained a California landmark on roadsides and in windbreaks. The California Department of Forestry distributed thousands of trees of this and other eucalypt species. Arbor Day in California became a patriotic festival of blue gum planting.[34]

By the early twentieth century, federal and state governments were joining the boom. The Department of Agriculture warned in 1904 that Eastern forests had only a sixteen-year supply of hardwood left, and industrial users of oak, hickory, ash, and other trees appeared eager to accept eucalypt substitutes. The construction of the Panama Canal would make it possible for California wood to reach the East Coast cheaply. In 1908 the University of California College of Agriculture published a bulletin on eucalyptus. Read closely, it had a few warning signals. Most of the wholesale price of eucalyptus oil, praised as an antiseptic and medicinal ingredient, was used up in distilling it. Eucalyptus was better than nothing on mediocre land, but it needed good land to become a profitable tree. It could be hard to cut. Still, the College Experiment Station recommended eucalypt groves as a more stable source of income than fruit orchards. Authors and promoters sang the profitability of eucalyptus forestry with doubtful data, extolling its virtues as a source for everything from tannin and charcoal to bee nectar and paving blocks.[35]

Businesses and private speculators responded. The Joseph R.

Loftus Company of Los Angeles sold two-and-a-half- and five-acre tracts to absentee eucalyptus growers. ("The trees keep growing night and day, year in and year out, while you may be devoting all your time and attention to something else.") Farmers converted cropland to eucalyptus forest. Rancho Santa Fe, now an exclusive San Diego suburb and resort, was originally a blue gum plantation for the Santa Fe Railroad. The Pullman Company's own groves were to supply luxurious finishes to decorate sleeping cars. Jack London, hounded by creditors, at the same time was paying twenty Italian laborers to plant 100,000 trees on his Sonoma Range land. "Everything I can raise and scrape I am sinking into the planting of eucalyptus trees," he explained in a letter to a job seeker.[36]

Neither railroad capitalists nor social writers realized that the rapid growth of blue gum and other eucalypts in California would have a revenge effect. The wood, no matter when it was cut or how it was treated, defied cutting and working. Posts would split across a diameter; boards would curve out of true while being cut from logs at the sawmill. Cracks would propagate. Worst of all for the railroads, eucalypts were so riddled with fissures that it was impossible to bolt rails to them. Even as windbreaks, the trees were blamed for taking up the water that neighboring fruit trees needed. As the bubble burst, eucalyptus-based industries closed, and imports of Australian eucalyptus oil helped put distillers out of business.[37]

The unruly behavior of eucalyptus wood was a mystery at the time. Wood from the same species of trees was serving their native Australia. An Australian forest scientist, Max Jacobs, discovered the real revenge effect decades later. Eucalypts in North America were growing too fast for their own good. In Australia, leaf-eating insects and some fungi slow their rate of growth. Sent around the world as seeds, eucalypts left these predators at home. The trouble was that they had evolved with them. All trees, including eucalypts, develop stresses as they grow. The structure of eucalypts had apparently been optimized for the slow rate of growth with the normal complement of pests. In their absence, the fast-growing trees were booby-trapped by the laws of physics. Under great strains, parts of logs would pull away at the sawmill. The miracle tree simply could not be used for lumber in America. The eucalypts had become problem trees because they had left their own pests at home in Australia.[38]

Blue gums were not all uprooted and sold for firewood before the First World War. To the contrary, the tree may have failed as a crop, but it continued to thrive as a suburban ornamental. Towering over the Berkeley and Stanford University campuses, shading the lawns of many of California's costliest neighborhoods, it marks so many landscapes that the state would seem a different place without it. Even in suburbia, the failure of the eucalyptus boom left its stigma. Some communities banned eucalyptus trees as attackers of municipal water lines. Authorities condemned the trees not only for crushing pipes but for shedding branches on passersby. Some developers' deeds banned blue gums outright. But the blue gum grew too rapidly and too handsomely to stay repressed for long.

Eucalypts draped the costly houses of the Berkeley and Oakland hills overlooking the cities and San Francisco Bay. But planting these trees densely among shake-roofed buildings failed to take into account their most basic behavior. They had evolved with the Australian climate not just to survive fire but to help kindle and sustain it. Eucalypt branches regularly embrittle, crack, and drop off the tree. The leaves last only an average of eighteen months before falling off—not long for an evergreen. Not only branches, twigs, and leaves but bark, flower buds, and fruit capsules fall regularly to the surrounding earth. This "slash," as foresters call it, decomposes quickly in Australian forests. It can accumulate, though, in soil lacking the microorganisms and fungi that can process it efficiently.

If all this sounds like preparation for a giant bonfire—one that not only the roots but the trunks of the trees are adapted to survive—it has been just that, over and over, in California. The Berkeley fire of 1923, fed partly by eucalyptus litter, destroyed half the city and spilled over to the University of California campus. A smaller brush fire in 1970 swept through thirty-seven homes in the hills above the town. It could have been much worse. Just a few years earlier, a visiting Australian forester was so appalled by Berkeley's assemblage of wooden housing, eucalypts, and accumulated slash that he left his group tour. He retired to his motel room, dazed by visions of an urban firestorm in the making.[39]

The conflagration did come, on October 21, 1991. Residents had begun to forget the report that followed the 1970 fire, which recommended (among other things) removing eucalyptus trees and building firebreaks. And in the 1970s and 1980s, officials issued a growing number of permits for houses in densely wooded canyons with narrow roads. Most of the new construction had shake roofs, but residents and authorities believed that new fireproofing treatments made them safe. The revenge effect of the chemicals was that they protected individual houses from small-scale fires, but by prolonging the use of shakes, they made a large-scale fire, which overwhelmed them, even more likely. Already in 1976 another wildfire specialist, a professor of fire ecology at the university, was predicting a new catastrophe from the accumulation of combustible material. Yet when he and former students began to cut eucalyptus and Monterey pine with chain saws to turn the potential hazard into firewood, residents complained to the university and forced them to stop.[40]

Of course, neither shake roofs nor eucalyptus trees caused the 1991 fire. California was in the fifth year of a drought, and winds were strong. Construction workers probably started it by burning debris with gasoline, and the Oakland fire department may have let the fire re-ignite after having put it out. We have already seen the interface between suburbs and wildlands as a potent source of fire. The fire burned out power lines serving pumping plants, making it impossible to fill water tanks in the area. Meanwhile the heat became so strong that emergency generators and portable pumps were useless. By the time 370 engines had contained the fire, twenty-three people had died, 1,700 acres were burned, and 3,400 housing units were destroyed or seriously damaged—an estimated $5 billion loss. There were 148 injured and 5,000 homeless people. The alarmed observers had been all too right. One study by a Berkeley student calculated that eucalyptus trees had contributed fully 70 percent of the energy released by burning vegetation in the fire.[41]

State and local codes and zoning laws were soon revised to

reflect the fire's apparent lessons. Some cautious and affluent residents even covered their rebuilt houses with costly copper roofs with estimated life spans of hundreds of years. New laws restricted the species and locations of eucalypts and other rapidly burning vegetation, now stigmatized as "pyrophytes," literally "fireplants." It will still be hard to keep eucalypts out of the reconstructed Oakland and Berkeley hills for very long, if only because they grow so well and so fast on burned-over land. So far, violations of the new code have been rampant. Many residents treasure the shade the plants provide, and recommended replacements like redwoods take a long time to rise. Since many of the new houses are significantly larger than those they replace, there is scant room for the fifteen-foot distance between house and trees that the law requires. In the mid-nineties, collective memories of the disaster persist, reinforced by frequent fire patrols and inspections. Yet the Oakland Hills remain a wildland interface subject to frequent fires. It is not certain that vigilance can last another generation. If it does not, the fire will almost certainly repeat itself.[42]

Most of the unintended effects of eucalypts have nothing to do with fire. They arise from the trees' remarkable positive qualities: rapid growth and efficient use of scarce water. Circumstances can transform these into serious problems not so much for owners as for their neighbors. On the more affluent hillsides of suburban North America, and no doubt in comparable communities elsewhere, one person's shade tree can grow up to be another's nuisance. Residents of Beverly Hills have been known to hire crews to lop neighbors' trees illegally when they obstruct prized prospects; many of the targets are eucalypts that have crept up inconveniently since the neighbors moved in. Vandals have used chain saws to cut potentially lethal strips around the trunks of eucalypts—a variation of an ancient technique called girdling—and drilled holes to inject deadly chemicals. (There is even a Seattle agency called PlantAmnesty that fights these "crimes against nature.") The blue gum has survived its bizarre history in this country, from its introduction as a miracle worker and wildland colonist to its apogee as fuel for brush fires, and now is viewed as both a lynching victim and a candidate for protection.[43]

Outside the U.S. West Coast, others are following in the footsteps of Jack London. In much of the world it is poor agriculturalists and not rich homeowners who are trying to destroy eucalyptus trees as pests, and for more understandable reasons. The tree is beginning to take over world forestry. Acreage has expanded from 1.4 million in 1955 to 10 million in 1980, and the UN Food and Agriculture Organization (FAO) estimates that almost a half million acres are added each year. It already accounts for a third of world pulp production: a eucalyptus tree yields double the wood pulp of a pine tree. But people fearing the loss of olive trees and hardwood forests are unimpressed. Spanish villagers have uprooted thousands of seedlings in new eucalyptus plantations. Thousands of Portuguese farmers have fought police to attack trees. In Iberia and elsewhere, the thirsty eucalypts are blamed for drying up the water of olive and walnut trees, and other vegetation. An FAO report on the environmental consequences of eucalypts in the 1980s reached the sensible but not terribly specific conclusion that it all depended on which species, where they were planted, and what other trees, crops, or natural habitat they replaced. More recently the economics of eucalyptus planting have turned out to be disappointing in Iberia, as they earlier proved discouraging in California. Landowners have stopped planting the tree in Spain and in fact have bulldozed hundreds of thousands of hectares. Thus a hundred years later, Europeans are repeating something of the U.S. experience.[44]

The passionate advocates and enemies of eucalypts—they inspire stronger feelings worldwide than any other group of trees—disagree about a genus of organisms that developed long before modern agricultural science or technology could intervene. New varieties will undoubtedly be bred for specific uses: energy farming, pulp, hardwood. Selection or genetic engineering might reduce the fire risk of eucalypts as ornamentals and might optimize their growth rates and water consumption in particular settings. But since between 550 and 600 eucalyptus species are presently known, depending on the classification rules botanists use, we already have an exceptional choice.

The blue gum and other eucalypts may be the best harbingers of the conflicts and dilemmas we might expect from genetically engineered products. Even if eucalyptus-based medicine has lost its cachet, the trees don't appear to threaten human health. And nobody denies how much fiber a stand of eucalyptus can produce, at least for decades. The promoters of the early twentieth century were right after all, though not about where it would be profitable to grow the tree and for what purpose. While the forests of the Northeast and South grew back in the twentieth century as agriculture receded, worldwide deforestation (and soaring paper consumption) renewed the eucalypts' prospects. Even so, the divergent opinions of growers and opponents show how different the aesthetic response to eucalypts can be. For the American chairman of a Portuguese cellulose company that has been planting the trees, "Walking in a forest we build is like visiting a cathedral: vaulted, hushed, peaceful." Another American expatriate finds "sheer exaltation" in the eucalptus groves of southern India, with their "tall straight trunks, the loose bark spiraling off in long strips, the sickle-shaped leaves that glint in the sunshine." To others eucalyptus is a noxious weed, especially to people who "have seen their olive plantations and vineyards replaced with 20-meter high, blue-green eucalyptus plantations," as a Swedish professor of environmental forestry put it. A Portuguese environmentalist regards the hillside terraces of trees as a criminal transformation of mountainsides into "monumental staircases of eucalyptus." And elsewhere eucalypts don't just replace marginal agricultural landscapes; they endanger native plant life and wildlife habitats. The American in India acknowledges that local environmentalists detest the eucalyptus, first planted by the British and now ubiquitous, as the enemy of the *shola,* their ancient and endangered forests of tangled evergreens.[45]

Each of the technological advantages of eucalyptus also has an environmental or social cost. Their natural insect repellents leave little food for birds or other wildlife; beetles have no appetite for fallen eucalyptus leaves. The tree may reduce local employment with its rapid rate of growth and easily mechanized harvesting. A tree that originally was introduced in North America to help small farmers as well as railroad companies has instead become a worldwide symbol of the intervention of international corporate agriculture and the displacement of the peasantry. The Portuguese left has branded it the "fascist" tree.[46]

Eucalypts, kudzu, and melaleuca show how easily the vigor and adaptability of an organism—the reason for the enthusiasm of experts who sponsor it—can turn one person's (or generation's) miracle plant into another's pest. They also demonstrate that small-scale experiments don't necessarily reveal the potential of an organism. Kudzu was a popular plant long before massive planting brought out its weedy nature.

The revenge effects of the recent past have chastened today's professionals. But it's much too soon to say that we are beyond the days of expert-guided pest introductions. The garden columnist Allen Lacy has found in current catalogs some plants that may take their place with the weeds that nineteenth-century seed catalogs helped distribute. Oriental bittersweet, one of the most aggressive weeds, is still for sale. Lacy finds especially ominous the vines *Akebia quinata* and *A. trifoliata,* which can grow to forty feet and are praised in the catalogs as "easily grown," "vigorous," and "trouble-free"—all warning signs to the connoisseur of pests. Hybridized, they conquer all surrounding vegetation that much faster. A friend of Lacy's, having innocently planted *A. quinata* as an ornamental vine on a fence that happened to face an unpleasant neighbor, returned several years after a move. Akebia, without the friend's planning but to his satisfaction, was already overwhelming the neighbor's property. His friend has a new name for akebia: "revenge vine."[47]

8

The Computerized Office: The Revenge of the Body

It is not only in the environment that technology tends to replace life-threatening problems with slower-acting and more persistent problems. Nor is it only outdoors that nature fights back. Changes in our business surroundings have been equally surprising, and often as frustrating. On one hand, business is now conducted at a pace and with an accuracy that nineteenth-century scientists and visionaries from Charles Babbage to Jules Verne did not dare to predict. (In fact, microprocessor power has overtaken many of the grandest expectations of twentieth-century computer pioneers.) On the other hand, the benefits of automated data and word processing have not been quite as expected. The revenge effects are physical: what had promised to make work painless unexpectedly attacks muscles, tendons, and vertebrae. The revenge is also financial: what had promised to make services more efficient has returned astonishingly low net benefits. These matters are of course linked; disability, health, comfort, and stress are reflected in financial statements, too. But they are parallel stories, and each deserves a chapter of its own

I/O, I/O, It's Off to Work We Go

Over the last fifty years, Americans have moved in huge numbers from the hot, often dangerous factory floor to the relative comfort of the office. This shift marks not only a movement from production to administrative work, but also a deep change in the style of production. We have already seen the difference some historians of technology recognize between tool use and tool management, between acting *with* material and acting *on* it. By substituting microprocessor- and software-controlled machines for older mechanical linkages, we have added yet another level between us and the action. Originally all the controls of an aircraft were directly linked through cables to flaps, ailerons, and rudder. An airline pilot today, commanding the latest in aircraft design, except in a few emergencies has no mechanical contact with the plane's systems at all. The pilot is no longer even managing tools but managing the tools that manage other tools.

When I first came to work at Princeton University Press in the late 1970s, my most memorable experience was a tour of one of America's last major hot-metal typesetting plants with its small corps of linotype operators. The linotype is a managed tool, but it is still a hundred times more directly physical than the laser-driven electronic typesetters that have replaced it. The lintotype melts lead, arranges slugs with the reverse impressions of characters in rows, and then casts one full line after another. The force needed to work the keys, the sound of the mechanism, the smell of the lead—all make the experience a form of shop work similar to the hand-setting of type that it replaced over a hundred years ago, and in fact more industrial. Today's keyboarding can be done anywhere, including the Third World, and the space where codes are converted to a typeset image is likely to be quiet and cool. In fact, writing itself has become far less physical since the Victorian invention of the steel-nibbed pen relieved teachers, students, and clerks of the bother—and of the skill—of sharpening goose and turkey quills. With some reason, contemporaries called the steel pen "the knitting needle of civilization."[1]

Even where tons of goods must move physically, human agency is at greater and greater removes, based usually in an office rather than a shop. It is possible to manage all the freight in an entire region, in fact in all of North America, from a single giant bunker with no track in sight, as though it were the subterranean redoubt of the Strategic Air Command. (Not that centralized control always works with military precision; crashes of new systems have tied up passenger and freight trains for hours in the Western states and near Hamburg, Germany.) This manipulation of the world behind protective windows has found its way into popular culture, as Homer Simpson, a cartoon employee in a nuclear power plant, fumbles his way over and over into near-catastrophe. Homer is funny partly because tool-managing technology can insulate him from his own incompetence and inattention. Yet his hellish boss, Montgomery Burns, is removed by still another level of windows: the video cameras and the monitor bank in his office that together turn the plant into a vast panopticon.[2]

The twentieth-century office itself shows the same growing distance from production. My father's Depression-era Underwood typewriter, still graceful to use, reveals its bars, shafts, gears, and even its bell for all to see. Margins are set by releasing tiny levers and positioning metal blocks along a rod. With an assortment of replacement springs, screws, and other hardware—and perhaps a lathe and some ingenuity for really difficult repairs—it could last forever and produce works as good as the user's brain, just as untold nineteenth-century hand presses are still serving hobbyists, artists, and museums. A sequence of typewriters recently exhibited at New York's Cooper Hewitt Museum made it clear how industrial designers (and their customers) have been increasingly determined over the years to conceal these mechanisms. First a massive beveled-glass plate, worthy of a reliquary, was introduced; then the whole chassis vanished behind gracefully sculptured metal. Messy ribbons were packaged in cartridges and cassettes, and durable cloth yielded to disposable film. As the new models were finished in almost every color but black, typewriters were already on their way to being black boxes in the technological sense: mechanisms opaque to the user. Meanwhile, developing countries—with educated clerks and skilled mechanics but computer-unfriendly electric service—have been importing massive numbers of castoff American manual machines.[3]

Similarly, the photocopiers and laser printers that have replaced mimeographs not only eliminate the bother of cutting a stencil but also conceal the application of ink to paper. In fact, ink has nearly vanished as a visible fluid, once available in gallon jugs, and has been reincarnated as powdery toner in sealed housings stenciled with warnings against exposure to sunlight. Even most luxury fountain pens accept ink cartridges, and disposable low-end models are non-refillable. Likewise audio- and videotape, as well as computer diskettes, have retreated into hard plastic shells. The tortoise strategy has indeed won the technological race.

Shifting workers to offices and reducing direct physical contact with the mechanisms and materials of office work is part of the same process: an apparent insulation of men and women from the physical dangers and discomforts of earlier stages of industrial society. Margaret and Robert Hazen, science writers and cultural historians, have examined the retreat of combustion from the nineteenth- and twentieth-century household: from the open hearth to the cast-iron stove and thence to a basement furnace fueled by materials the householder never expects to see, touch, or smell. Electric power, whether generated by oil, gas, or nuclear fuel, represents the ultimate isolation of heat sources from the consumer. The barriers are imperfect; even natural gas lines, otherwise the cleanest and safest sources of residential energy, may leak dangerously. But who would deny that we are safer than the Victorians with their open fires and stoves?

Just as central heating has improved not only comfort but safety—residential space heaters are still far more dangerous than furnaces—the shift to automated monitoring, administration, and distribution in the modern office once seemed to promise a new age of more healthful and satisfying work. Within the workplace, the prophets of automation foresaw a golden age of human creativity—once electrical and electronic devices had successfully replaced drudgery. IBM's ubiquitous Think signs encapsulated Thomas J. Watson's mantra "Machines should work. People should think." Human liberation not only from gross physical danger but from stultifying routine still inspires the prophecies, prospectuses, and profits of hardware and software producers.

In the 1950s, American notables helped celebrate the twenty-fifth anniversary of *Fortune* magazine by painting the year 1980 in tones that rivaled the most radiant forecasts of the Soviet Politburo. David Sarnoff of RCA predicted: "Small atomic generators, installed in homes and industrial plants, will provide power for years and ultimately for a lifetime without recharging." John von Neumann of the Institute for Advanced Study and the Atomic Energy Commission speculated that energy might even "be free—just like the unmetered air." Henry R. Luce himself foresaw the global stewardship of consciousness by "High Organization" as represented by the multinational American corporation, bureaucracies, and labor unions. Making a living would no longer be an issue; bread had already become "a drug on the market."[4]

By the 1970s, prophets of the future workplace had learned that the safest bet was to predict unremitting and bewildering change. One could be right no matter what happened, unless, of course, a long period of stability ensued. As the world's romance with atoms-for-peace came to an end, futurists shifted their sights toward a new utopianism: visions of a knowledge-driven information economy in which "altering the position of matter at or near the earth's surface," in Bertrand Russell's phrase, would become almost a memory. Business magazines once adored the giants of smokestack industry; they have since shifted their admiration to entrepreneurs whose products either are largely intangible (Microsoft) or are produced in myriad overseas plants (Nike).

John Naisbitt's *Megatrends* (1982) delivered "a brief history of the United States" in the succession of occupational titles most frequently reported by the U.S. Census over the course of this century: "farmer, laborer, clerk." Even in work like farming and trucking, air-conditioned cabs and improved suspension have increased the distance between the operator and the immediate environment. Simultaneously, clericalization and computerization have gone hand in hand. According to one estimate, three-quarters of all U.S. jobs will involve full- or part-time computer operation by the end of the century. Already the paperwork-laden police ride with laptop computers mounted in the front seats of their squad cars as though they were partners.[5]

In the new workplace, everything seems to be under far more control. The air is warmer or cooler depending on the season. The lights are brighter. The incidence of traumatic accidents is far lower.

Tragedies still happen in manufacturing and food processing—for example, the fire at the poultry processing plant operated by Imperial Food Products in Hamlet, North Carolina, which claimed twenty-five lives in September 1991. But this was far from the record toll of 145 in the fire eighty years earlier at New York's Triangle Shirtwaist Company; and while Triangle's owners were never charged, the owner of the North Carolina plant accepted a twenty-year prison sentence for involuntary manslaughter. Between the 1930s and the 1990s, worker death rates fell 75 percent. In California, there were only eighteen deaths per 100 million hours worked in 1985, down from 127 in 1939.[6]

None of this implies that workers are now adequately protected from potential trauma. This still depends not only on worker practices but on the technologies, conditions, and policies set by their employers. Injuries rose sharply in some industries in the 1970s and 1980s, sometimes wiping out decades of gains. But steadily the problem becomes less technical and more social. We know more and more about making work safe. The obstacle is not the availability of technology; it is having the money and will to deploy it. The question is whether some combination of employer self-interest, collective bargaining, and legislation can require using what is already there. Rewards or pressure from supervisors sometimes lead workers to evade safeguards, increasing risk of injury. And of course there is an ethical question of how much risk a rational worker should be allowed to take voluntarily for a higher income—a question familiar to anyone who has seen the denouement of *The Magnificent Ambersons.*[7]

The reduction of fatal accidents does not mean an end to the health hazards of working. As the old catastrophes have receded, complications of chronic exposure to substances new and old have come to the fore. A report of the National Safe Workplace Institute in 1990 concluded that up to 10 percent of cancer deaths in 1987 and up to 5 percent of deaths from a variety of neurological diseases were work-related.[8]

But these are not what has driven up the cost of workers' compensation insurance. It is another kind of physical problem, one that results not from viruses, microbes, or toxins but from activities that in themselves appear harmless: sitting, typing at a keyboard, handling files and papers, looking at text or other data on a screen. Some of our concerns arise *because* we are in an office or home office—worries that don't trouble us when we are not using technology identified as business equipment. Consider electromagnetic fields (EMFs). Some physicists insist they have to be harmless because we already absorb so much electromagnetic radiation from the earth itself and even from our own bodies. Swedish and Finnish authorities in the early 1990s became convinced, on the other hand, that EMFs from video monitors could promote miscarriages and leukemia. Even the noted protechnology futurist Paul Saffo exclaimed in a 1993 column in *Byte* that the information revolution should carry a health warning, not because the case against EMFs had been proved but because manufacturers and the U.S. government were still refusing to deal with the evidence that already existed. Saffo's alarm is understandable, especially since there is no evidence of *benefits* that might balance possible risks of EMFs. I bought a low-radiation monitor and mounted a radiation-absorbing filter on it. I also wondered about my bedside electric alarm clock, which I learned probably emits more EMFs than the monitor. (I moved it a few feet away.)[9]

The most famous problems (but not the most common or necessarily the most serious) are the cumulative trauma disorders (CTDs). Whereas traumatic injury arises from a single, massive impact, the repetitive disorder is thought to arise from hundreds or thousands of small motions—or even from a position that seems to require little motion at all. The trauma is a spectacular event; repetitive injuries may appear suddenly and painfully, but they develop slowly and imperceptibly. A lay observer can identify the victim of trauma from blood, bruises, or the bandages and crutches that follow treatment; a physician can map traumatic damage through X-rays and electronic scans. But neither can always pick out a man or woman with a cumulative trauma disorder.

The white-collar workplace has its own potential catastrophes, from tipping file cabinets to robbers and deranged coworkers, and may harbor known (and suspected) microbes and toxins from Legionnaire's disease bacteria to the carpet chemicals we have already encountered. But much more likely are back and upper-limb problems. Once they appear, symptoms may persist for years. They are painful, sometimes disabling, but usually not visible on the employee's body. Consequently, the typical sufferer from them may need to fight for recognition of a chronic condition that is all too real.

In fact, physicians and scientists still have no way to measure pain directly and unambiguously either as nerve impulses or brain waves. Some have drawn up scales and proposed units of pain; others have tried to infer levels of pain from other evidence. But as yet there is still no reliable technique to establish the existence of suffering. Insurance company lawyers may call automobile whiplash injury a license to steal because no objective imaging procedure or test can diagnose it, but a strained muscle or sprained ligament in the neck is an extremely painful condition that can arise from even a minor collision. As often happens, such injuries may easily be faked. And perhaps there is a borderline region in which honest people, knowing that even minor symptoms are grounds for compensation, subconsciously focus on their pain, amplifying it. Yet the most recent research on whiplash injury suggests that there is no relationship between "neuroticism" and time needed for recovery.[10]

Office hazards share important characteristics with whiplash: potentially catastrophic consequences from relatively small impacts, psychological complications with charges of malingering and neurosis, skepticism of many physicians and insurers. Nothing illustrates the tendency of contemporary technology better than the prevalence of back pain, one of the most economically serious illnesses of the industrial world. It affects 31 million Americans and is estimated to cost $16 billion annually in medical care and disability payments. Physicians may explain sciatica, the often intense referred pain in the legs, by pointing to a herniated disk on an X-ray or a CAT scan, but 20 to 30 percent of all patients examined for other conditions also turn out to have disk bulges, yet feel no back pain at all. Some doctors claim that computerized motion analysis can distinguish organic from psychogenic pain, or pure malingering, but the tests are costly and have not been rigorously validated. Nor do they serve to distinguish psychological illness from conscious cheating. What tests can reveal is that, when asked to apply increasing force until pain compels a stop, a patient does not exert a consistent maximum force; over time the same tests can also document the effectiveness of physical therapy. But the body is a black box for these techniques. Degenerated disks, sprained ligaments, the sacroiliac joint, facet joints, lumbar instability—all plausible causes of pain—have vocal medical defenders, but there are too few controlled studies to make clear the origins of pain.[11]

Most of us would link back problems to heavy lifting and might expect that as men and women moved from farm and shop labor to the repose of the office chair, the nation's backs would have straightened up. Watching modern rail construction equipment put down long stretches of welded track as though it were slowly closing a giant zipper, today's train passengers may rarely see the the kind of backbreaking labor that built the original roadbed, made the cuttings, and dug the tunnels. But there are still far too many back injuries in manufacturing, transportation, and distribution, even though in general we understand industrial back injury, and have the technology needed to reduce it, but don't always use it. In addition to elevators, conveyors, and carousels, there is an impressive series of devices that can reduce stress on the lower back, even if some introduce new risks to the back from vibration, as do forklift trucks. Where lifting is necessary, industrial safety specialists have devised posters, charts, and well-established courses for employees. To many users, support belts appear to reduce back injuries significantly in warehouse work, nursing, and other occupations with high-risk lifting, though some studies have questioned their benefits.

Back care specialists have learned more about lifting in recent years, especially the dangers of twisting motions. Performed correctly, some heavy labor is actually safer than one might expect: a reverse revenge effect. Dr. Christopher Michelsen, director of the Orthopedic Spine Service at Columbia-Presbyterian Medical Center, has observed that longshoremen have a below-average rate of back injury, possibly because their work keeps them in better shape and aerobically fit.[12]

Office workers turn out to be at surprisingly high risk for back pain, even if they lift nothing weightier than the odd ream of paper. There are forty times more back pain cases among office workers than cumulative trauma disorders. The Western style of sitting may be less healthy than the floor-level living that prevails, or once prevailed, in Asia and the Middle East. Most chairs ought to carry warning labels. (The original Chinese word for chair means "barbarian bed.") Sitting in one, as opposed to standing or walking, puts dangerous stresses on the spine; in fact, office workers have as high a rate of back injury as truckers. Perhaps this is why even in the West, some tailors worked squatting until relatively recently. What makes the chair-bound life so hazardous—like the origin of other chronic conditions—is elusive. Part of the answer is negative. A person who sits all day and gets no other regular exercise is more likely to suffer back muscle disorders when at last doing something strenuous: the "weekend warrior syndrome." Yet some back pain patients are otherwise fit.[13]

Sedentary work can exhaust hardy men and women. Jack London's *John Barleycorn* recalls unforgettably how the office technology of the early twentieth century appeared to a laborer struggling to write in the evenings on his brother-in-law's typewriter: "How my back used to ache with it! Prior to that experience, my back had been good for every violent strain put upon it in a none too gentle career. But that typewriter proved to me that I had a pipestem for a back." London's experience, we should recall, had included seafaring, jute-mill labor, and time at a municipal railroad job shoveling two men's previous quotas of coal. London would have been unimpressed by the claim of an early advocate of the typewriter that the machine "banishes . . . cramp in the hand."[14]

High-backed, heavily upholstered "executive" chairs, long on status but short on support for the lower back, still pass in some circles as badges of corporate eminence. But even scientifically designed seating can confuse hierarchy and function. The designer of one of the earliest and most commercially successful ergonomic models once confided to me that the high back of top-of-the-line models is a secular variant of the nimbus of medieval religious art, a halo to frame the sitter's head—even though many chairs with head supports do not even allow enough backward leaning to make use of them.[15]

At the other social extreme are secretarial and "task" chairs offering little more than a seat and a contoured backrest. Until the 1980s, psychologists studying workstations pictured the ideal body position as a stack of rectangular blocks with their surfaces parallel to or perpendicular to the floor and work surface: a head/trunk cube, a thigh cube, and a lower-leg cube. Many office chairs still reflect this outmoded "cubist" model with its rigid ninety-degree posture.

A series of independent studies by groups led by E. Grandjean and by A. C. Mandal gave office-chair design new dimensions. Truly upright sitting, while it may look healthy and alert, has the revenge effect of rotating the pelvis backward, straightening the lordosis (curvature) that is part of the spine's natural shape in the standing position. Leaning backward 15 degrees (Grandjean) or tilting the seatpan and the backrest slightly forward to approximate the position of a horseback rider (Mandal) were both ways to cut stress on the spinal disks and muscles. The ergonomic psychologist Marvin J. Dainoff and his collaborators have argued that the choice of positions depends on the work at hand: reclining is better for reading text from a terminal with its large characters, while leaning forward is more appropriate for paper copy and other work involving fine detail.[16]

Grandjean's and Mandal's research appeared just as architects and manufacturers were beginning independently to produce furniture that not only looked stylish but responded to the user's motions and postures. The Bauhaus movement of the 1920s and 1930s, for all its homage to science and to functionality, appears never to have sponsored such expert studies of seating. Some of its aesthetic high points, like Marcel Breuer's tubular-steel-and-fabric-strap "Wassily" chair, are among those objects that are better for the mind to admire than for the body to use. (Ludwig Mies van der Rohe probably spoke for many another International School architect when he described a chair as more difficult to construct than a skyscraper.) Other modernist designs, especially those from northern Europe, are more stylish versions of old-fashioned desk chairs shaped more accurately to the human back. Others, like the Vertebra chair codesigned by the architect Emilio Ambasz in 1987, renounced the idea of a single correct position in favor of maximum freedom to change position, whether by inclining, reclining, flexing the back, or just shifting around. No doubt under the influence of Grandjean and Mandal, aggressive lumbar support yielded to a new, laid-back philosophy.

In the 1980s and early 1990s, the movement for safer office conditions collided with a drive to downscale the workforce—to reduce the number of employees and the money spent on them. Ergonomic seating may reduce expenses in the long run, but it is also costly: unit prices of $500, $600, and even $1,000 or more are common, compared with the $150-to-$250 range of mass-produced office furniture. But price is only the beginning. The real challenge in using ergonomic chairs to reduce back injury is that they require time to learn to use properly. Most employees never learn to make the adjustments needed to take advantage of the costly features. It takes little time or effort to do so, but the use of such chairs does have to be learned. Training videos are one way to impress their features on new users. So far, a handful of self-adjusting ergonomic chairs have lost in the marketplace; kneeling chairs like the Balans, the Earth Shoes of seating, delight some users and cut off the circulation of others. Even comfort apparently requires study and vigilance. (The most talked-about office chair of the mid-1990s, the membrane covered Acron by Bill Stumpf and Don Chadwick of Herman Miller, is said to mimic the body's natural motions and may become the long-awaited exception. It is one of the few chairs designed for body sizes, not corporate rank, but petite or king-size, and it is not cheap.)[17]

Even the finest of chairs suffers from another revenge effect of office technology on the back: the fact that more work is done without leaving the workstation. The networked office of the 1990s has reduced the need for trips to supply rooms, filing cabinets, printout racks, printers, fax machines, and other computers ("sneakernet"). Just as word processing eliminates the breaks that conventional typing requires, networked communications, especially document imaging, eliminates the walking from working. Each single instance of time or task saving may improve productivity in itself, yet taken together they can promote dangerous semi-immobility. And the division of labor among industrial psychologists is such that communication researchers study ways to reduce walking to face-to-face meetings while ergonomists work to offset the health problems of reduced motion.[18]

It is even hard to say how much back pain a chair can cause or prevent. There is reason to believe that frustration has more to do with back pain than the mechanical effects of seating. Even if research showed a lower rate of back pain in companies with top-of-the-line ergonomic seating it might not be easy to tell whether the credit should go to the chairs or to the enlightened management style that resulted in their purchase. Since consultants generally can't tell bosses to stop making people feel helpless and abused, the chair is the stand-in for the company.

When thinking of the physical side of the computerized office, we might as well start with the window the computer presents: the video display terminal (VDT) or simply the monitor. "Terminal" and "monitor" both hark back to an earlier day of interactive computing, when instructions and results appeared on a teletyper roll—traces of which survive in such computer commands as "echo" (repeat an instruction) and "type" (display the contents of a file, no longer necessarily in print). Now we relate to our monitors differently, not only visually but qualitatively. Interaction is no longer periodic; central processing units (CPUs) usually execute commands within mere seconds if not milliseconds, whether they are part of a mainframe computer, a network server, a desktop machine, or a portable laptop or notebook unit. It is the CPU that wastes millions of cycles waiting for us.

In the popular culture of computing, and especially where computers are being sold, the monitor is a window. It is not just an aperture that makes the machine operations visible; it is also a stage light that plays over the user's face. For computer marketers, and for computer enthusiasts in education, the glow of the screen is literally visible enlightenment, the user soaking up the rays of information.

Monitors are clearer and less expensive than ever, but they have dropped more slowly in price than most other computer components. Theoretically they don't have to be big or expensive, being little more than television sets outfitted with fancy electronics that let them receive more kinds of signals and produce tighter images. But only the costliest video cards can approximate the number of colors (and thus the degree of realism) we observe on an ordinary TV set. The large monitors that people really need in order to work simultaneously with several applications are still so costly that only 5 percent of all monitors sold in the early 1990s exceeded fifteen inches. The largest sizes, twenty inches and above, still cost more than $2,000, sometimes more than $3,000, because so many cathode ray tubes are imperfect and must be discarded in manufacturing. They may weigh sixty to eighty pounds, and may qualify as a space heater. High-quality flat active-matrix color screens are beautiful and legible, but they too have daunting reject rates in manufacturing, which result in corresponding shortages. While prices are declining slowly in the mid-1990s, and 10.4-inch color displays are becoming standard, the big, flat, moderately priced display once announced so confidently by industry futurists still seems years away.

The computer monitor is little more than an awkward compromise, usually too small to display a full page, let alone allow work on two or more documents at once as software vendors encourage users to do. The headers, menus, and multiple button bars of graphic-interface applications require some of the space, though it is usually possible to remove them. The optimal speed for repainting the screen is 100 times per second—faster refresh rates can actually make text harder to read—yet most monitors in use still don't support even the industry standard of 70 to 72 Hz (cycles per second). An ordinary book or magazine is typeset at 1,200 dots per inch (dpi); a computer display is only a tenth as sharp in a single dimension, which means only a hundredth in two dimensions. Earlier, office work required the physical *strain* of hard manual operations; now electronic offices (as well as schools and libraries) have added a new kind of visual *stress.* As the geographer Jean Gottmann put it, we have moved from hot sweat to cold sweat.[19]

In the mid-1990s we still know very little about the health effects of electrical and magnetic fields in general. The U.S. National Institute for Occupational Safety and Health (NIOSH) has consistently failed to identify significant risks. Even heavy computer users may get more exposure from electric shavers and clocks and microwaves than from their video display. One study found that ambient AM radio signals induce stronger currents in the human body than a terminal does. The greatest exposure is at the back and sides, not when seated in the normal operator's position. In any case, electromagnetic radiation from displays is relatively cheap to reduce. Most premium monitors already meet Swedish government antiradiation standards, among the highest in the industry. Effective antiradiation screens generally don't cost more than $100 to $150, a small amount relative to the cost of an entire computer workstation.[20]

Eyestrain is common among computer users. One survey of optometrists suggested that ten million Americans seek professional help for computer-related eyestrain and related problems each year. It also reveals that symptoms related to video terminals prompt 14 percent of all eye examinations. Excessive use can produce myopia requiring prescription lenses, although computers usually merely accelerate an existing problem or tendency. In fact, no study appears to have compared systematically the effects of equally intensive work with paper documents. Of course, it is easier to control the lighting of a conventional desktop than to adjust windows, overhead lighting, monitor positions, and filters to avoid glare. This part of the eyestrain problem is new, but it is growing not so much because of computer technology but because more people are working in offices.[21]

For most users, working with a monitor simply means following routines and practicing exercises, each of which is simple in itself, but which together demand much more attention than old-style paperwork ever did. There is not only the proper positioning of the monitor with respect to windows (which should be at right angles), but the adjustment of the monitor's height for the best angle of vision, the adjustment of brightness, contrast, and colors for different applications, and the need to take regular breaks from work. The best way to reduce eyestrain, in fact, is simply to focus from time to time on a distant object. All these are simple precautions, but they still require attentiveness that was unnecessary in the classic paperwork age of file cards and vertical files that began in the late nineteenth century and ended in the late twentieth.

If the risks of using monitors are relatively easy to minimize, keyboard disorders are still under a diagnostic cloud. A report published by Herman Miller Research and Design notes that the United Kingdom, Australia, and other Commonwealth countries prefer to call them "repetitive strain injuries," Scandinavia and Japan "occupational cervicobrachial disorders," and the World Health Organization "work-related disorders." There are a dozen or more muscle, tendon, or nerve problems included in the cumulative trauma disorder (CTD) category. Carpal tunnel syndrome (CTS) and tendinitis are only the best-known of the CTDs.[22]

Carpal tunnel syndrome, like other computer-related health problems, is hard to document. Similar to automobile whiplash injury, CTS appears both real and difficult to verify. Once more, the ability of medical imaging to isolate and map the site of a condition turns out to have unexpected limits. In 1993 an English judge rejected the diagnosis as a basis for compensation, infuriating CTS sufferers in the United States as well by calling the diagnosis of repetitive strain injury a "meaningless" one with "no pathology" and "no place in the medical books." In the United States, too, some analysts insist that fitness, personal habits, and stress at home and work have a much greater role than technology itself. The *Wall Street Journal* lists the following proposed explanations among others: "diabetes, weight, menopause, hysterectomy, wrist size . . . relationship with supervisors, job pace, posture, length of workday, exercise routine . . . decision-making authority, job security and contraceptive use as well as pregnancy." But while every popular diagnosis may magnify pain by focusing attention on it, the underlying disorder is too real to be dismissed as malingering or somatizing.[23]

The anatomy behind CTS is well known. As I enter this chapter on my computer, the nine flexor tendons that control the movement of the fingers of each hand are bunched in an opening surrounded by the eight carpal bones of my wrist, which are connected by a strong ligament. Through this tube, called the carpal tunnel, the median nerve also connects with the hand's sensory cells. This marvelously compact design is sensitive. A disease or injury that compresses the nerve can produce pain and limit the functions the hand can perform. Medical researchers have found that repeated flexing and extension of the wrist can expand fluid-filled protective sheaths that surround the tendons. But this protective reaction creates its own revenge effect by putting pressure on the median nerve in the carpal tunnel, causing pain and numbness. What makes a CTD such a painful and chronic condition is the thickening of the tendon sheath after repeated acute inflammation, which maintains constant pressure. Once more, the long-term effect of many small insults to the body can be more severe than the consequences of treatable major injuries.[24]

The human being did not evolve to perform small, rapid, repeated motions for hours on end. Upper-limb disorders appear to have been rare in traditional societies where workers made their own tools. In fact, John Napier, a physician specializing in hand mechanics and evolution, pointed out that the "modern" look in hand tools often made them less rather than more suited to human needs. "Early people in the throes of constructing hand-axes never made that mistake; their lives and livelihood depended on it." And as we have already seen, the peasant societies of Eastern and Central Europe used made-to-measure implements that showed an intuitive sense of the unity of tool and body.[25]

But we should not assume that cumulative trauma was entirely absent from preindustrial technologies. Luther Sowers, a self-taught North Carolina master armorer renowned for the accuracy of his historical reconstructions, wondered in a newspaper interview whether his medieval predecessors shared his tendon pain from working steel and his hand cramps from assembling chain mail. He expected that "[i]f they became apprentices at 12, they'd be cripples by middle age."[26]

The first medical reports of upper-limb pain appeared in the early eighteenth century. If by then it was already a well-established problem, it doubtless reflected rationalization of craft processes. Adam Smith's praise of the division of labor in *The Wealth of Nations* (1776) owed much to his belief that the changing activities of country workmen promoted "the habit of sauntering and of indolent careless application." Alternating tasks and tools may nevertheless have been healthy. On the other hand, the minute subdivision of activities Smith observed—as many as eighty persons were involved in making a button—may have been pushing workers to a dangerous pace. ("The rapidity with which some of the operations [in trades like button and pin manufacture] are performed, exceeds what the human hand could, by those who had never seen them, be supposed capable of acquiring.") The nineteenth-century division of labor and the intensification of small but fast-repeated tasks created such ailments as "weaver's hand," "sprout-picker's thumb," "stitcher's wrist," and "cotton-twister's hand." By 1840, the French physician D.M.P. Velpeau had devoted a whole monograph to upper-limb disorders.[27]

Were nineteenth-century information workers exempt from cumulative trauma disorders? The eleventh edition of the *Encyclopaedia Britannica* (1910) recognized "writer's cramp" and "scrivener's palsy" as marked by a spasm that appeared during, and only during, the act of writing, musical performances, or other learned activity. But, failing to recognize any source of these maladies in actual work conditions, the article described them as "an affection of the central nervous system," occasionally hereditary but more often resulting from "a history of alcoholism in the parents, or some neuropathic heredity." Because of the differences between writing with a pen and using a computer keyboard, it is difficult to say whether writer's cramp and computer-related carpal tunnel syndrome are closely related. Nor is it clear whether or not the "operator's cramp" or "glass arm" of telegraphers was any closer.[28]

What the twentieth century contributed was not the symptoms but the unifying diagnosis. Of course, imaging technology and clinical investigation played a part. But CTD is as much a social as a physiological condition. CTDs among blue-collar workers may have been underreported for decades. Evidence from the back injuries of forklift truck operators suggests that a high rate of injury among young workers may conceal the true scope of the problem by natural selection: workers with excessive pain may simply switch jobs without filing workers' compensation claims. Similarly, some mature blue-collar workers may be CTD-resistant survivors.[29]

Even with the survivor effect, cumulative trauma disorders are still undoubtedly more serious in blue-collar work. They have become alarmingly common. In 1993, over 300,000 new cases were reported, a more than tenfold increase since 1982. They now constitute more than 60 percent of all workplace injuries, but they are not evenly distributed across industries. The U.S. Bureau of Labor Statistics reports that in 1992 there were nearly 1,400 CTD cases per 10,000 workers in meat packing and 860 in auto manufacturing, but only forty-four in newspaper publishing. The difference is that white-collar computer users are not limited to data-entry clerks putting in up to eighty thousand keystrokes a day. Elite journalists and financial professionals now are at risk for proletarian aches and pains. Even the rising generation of computer-socialized senior managers has not been immune; some spreadsheets just can't be delegated. Add the growing number of men and women at all levels who spend most of their time at keyboards, and possibly using other entry devices from mice to checkout scanners, and the ingredients of an epidemic are all present.[30]

Carpal tunnel syndrome appears to be a repeating revenge effect. The revenge is that computers have reduced even the physical exertion of typing: of striking keys, raising blocks of metal for every capital letter, throwing back the carriage for each new line. If anything had the potential to injure hands and wrists, it would seem to be a manual typewriter. It had a nondetachable keyboard, and wrist rests were almost unknown. Yet there is remarkably little printed evidence of CTS among typists. The pain that Jack London felt on first using the typewriter was in his back, not in his wrists. Some older women typists recall pain that strongly suggests the presence of carpal tunnel syndrome, but contemporary reports of the problem are hard to find.

It may be that earlier office workers saw no point in reporting their problems to employers. It takes a certain social and legal framework to define a pain as a symptom. Only recently, perhaps, have safety legislation, medical and disability insurance, and workers' compensation made it possible to file such claims, which in turn appear more often. It is also likely, though, that old-fashioned typing was safer just because it was harder and slower. Some specialists believe that the need for extra pressure distributed more force to shoulders and arms. Manual keys literally had a spring to them that prevented the sharp impacts that firm strokes on a computer keyboard can bring. (Try this comparison with your computer and an old manual machine.) The mechanics of manual keys also encouraged straight wrists, making less likely the arched and suspended hand positions that seem to be risk factors for CTS. The need to keep keys from jamming may have helped maintain a slower work rhythm, though ergonomists now consider uninterrupted entry a far more serious work factor than rapid entry alone.

Electric typewriters may similarly avoid many of the problems of computer entry. Most of them limit the number of keystrokes per second, and they enforce the same brief healthful interruptions as did manual models. Electric machines have eliminated the hand carriage return, but inserting and removing paper, and even old-style erasures and white-outs, remain short but probably healthy breaks in routine typing. Even a study of office mechanization and automation published by the International Labor Organization in 1960, calling attention to complaints of "muscular fatigues, backache, and other such ills" in modern offices, did not identify hand-and-wrist pain as a specific problem. (The report emphasized instead the psychological stress of close attention to monotonous processes.)[31]

We have already seen that CTDs are far from restricted to the office, although they are thought to affect up to 10 percent of office workers. They happen wherever automation and the division of labor have led workers to perform manual tasks thousands of times a day, whether sorting letters, making air filters, or binding books. They can also arise from child care, hobbies, home repair, and sports. There appears to be a background rate, unaffected by activities, of about one reported case per thousand people per year. All in all, according to U.S. government sources, upper-limb disorders (including but not limited to CTS) now represent half or more of occupational illnesses. Carpal tunnel syndrome alone is said to have a lifetime incidence of 5 to 10 percent in the United States.[32]

Unanswered is the question of why rates of CTDs are so different in organizations with similar work. Worker age, weight, posture, lifestyle, hormone levels, arthritis, herniated disks—all have been linked to CTDs. It is also becoming clear that workers under greater stress, whether clerical or professional, are more likely to develop them. One NIOSH study of telephone operators suggests that even with the best-designed equipment, excessive demands on employees and job insecurity appear to promote carpal tunnel syndrome and other computer-related injuries. In another case, workers who had participated in exercises to reduce CTDs had the same rate of injury a year later. But nobody really knows why some newspapers have significantly higher rates of carpal tunnel syndrome than others with similar equipment. It is also not clear whether new ergonomic equipment deserves most of the credit for reduced cumulative trauma disorder rates, or whether most of the benefit has come from more enlightened work rules and policies. It seems to be one of the rules of revenge effects that just as their causes are multiple and uncertain, their cures can be equally hard to understand.[33]

Whatever the ultimate factors behind CTDs, some simple mechanical and behavioral changes can reduce them. Some keyboards permit lighter touch settings that are thought to be healthier; but a too-sensitive keyboard can be dangerous if it encourages users to suspend their fingers above the keys rather than maintain constant but light contact. Wrist pads can help people restrict a tendency to bend the wrists upward or downward in typing, but they can also lead users to slide their hands over to hit side keys rather than to lift the arms and hands. Larger ergonomic keyboards may permit more comfortable positioning of hands while typing, but they can also increase the distance from keyboard to pointing device. And this movement of the arm has health consequences of its own.[34]

Even the simplest device can turn out to have complex and unexpected results as part of a human routine. As computers have relied more and more on graphic displays, designers and users have shifted attention from the cursor and entry keys to pointing devices that employ sequential arm motions to navigate a cursor on the screen. Instead of remembering or looking up strings of commands to enter, users can pull down menus and change settings in dialogue boxes to highlight and perform operations on text and graphics. Most of these input devices are hand-moved mice. Manufacturers have also offered trackballs, pens, and various-sized levers from eraser-shaped knobs in the middle of the keyboard to full-sized joysticks, and generally mediocre adaptations for use with portable computers. Nearly all these pointing devices, designed to make navigation easier, have the revenge effect of requiring removing one hand from the keyboard for each operation of the mouse and then making the user return to the keyboard. (A miniature knob located in the keyboard can interfere with normal typing, which can be confusing enough; Apple chose to put tactile marks on its *d* and *k* keys, while IBM computers and compatibles have positioning ridges on their *f* and *j* keys.) Of course, there could be an unexpected benefit from all this shifting; like the carriage returns of old, it could break up and slow down routines that otherwise might be dangerously fast. But the mouse does add a problem: ergonomists warn against operating it from the wrist rather than from the forearm.[35]

Mice have escaped most of the early controversy about carpal tunnel syndrome probably because major cases have arisen among workers entering and processing data in text-oriented systems where mice were uncommon. But by 1991 a California medical journal was reporting a new syndrome, "mouse joint," observed in a married couple both of whom used mice at work and at home. The male patient recovered in a month of conservative treatment with splints, rest, and ibuprofen, but the condition returned when he resumed working with the mouse. He apparently preferred to work in pain than to shift to other computer equipment; possibly there was no keyboard-controlled alternative to the software he was using. (Until recently, many Macintosh programs, unlike Microsoft Windows programs, offered no keystroke equivalents to the commands in their pulldown menus.)[36]

Ten years after the first release of the Apple Macintosh and the rise of mouse-oriented software, there is still no public, agreed-upon standard for mouse safety. If there are important proprietary studies, they are well-kept secrets, and corporate safety research has the potential revenge effect of risking heavier damages from later lawsuits as plaintiffs search for evidence that a manufacturer was aware of risks. Thus when Microsoft introduced a new model of mouse in 1993, its comfort level was more certain than its health implications. Higher and longer, it was raised in the center to fit in the hollow of the user's palm and curved to fit the (right) thumb. The left and right switches clicked more precisely and sloped away from the center. Microsoft reportedly spent millions to develop the new mouse, and a long file on proper mouse use was included as part of the software. The product is the most comfortable device I have tried, but there is no published claim that its design is any safer to use than the original, Chiclet-shaped Microsoft mouse or competing models from Logitech, Mouse Systems, and other companies. In fact, while the mouse sits comfortably under the cupped hand, it does not follow that it is actually safer to use than other designs. While nearly all new personal computers are equipped with mice, and many applications do not fully support function keys and other keyboard alternatives, research on pointing devices and health is only beginning.

As of 1994, the major manufacturers of mice, as well as of keyboards and other input devices, were still avoiding specific health claims for their products. And wisely so. One of the most frequent (rearranging) revenge effects is that a sense of safety in one part of a system leads to problems in another. For example, a mouse more comfortable to grip could possibly affect the rate of carpal tunnel syndrome by encouraging more intensive use. A trackball, really an upside-down mouse that controls the cursor as the user manipulates a small sphere in a stationary socket, may solve some wrist problems while encouraging overuse of the thumb. A touchpad, a square of material that responds to the light pressure of the edge of one's thumb, may do the same. And for both keyboards and pointing devices, the patterns of use required or encouraged by software may affect the rate of injury as much as hardware design does. On closer examination, what seems to be science and technology turns out to be equally aesthetics, comfort, and a good dose of guesswork.[37]

Technological optimists, always ready to dismiss revenge effects as "transitional," have seen in voice-activated computing the remedy for the hazards of mouse and keyboard. Since many people already spend much of their day on the telephone with no ill effect, what can be the harm in an occasionally muttered "open," "save," or "exit," except perhaps the stares of one's coworkers at first? In fact, now that voice-response systems are commercially available, reports of voice strain are growing. Some of its sufferers are programmers who adopted voice input after losing full use of their hands to carpal tunnel syndrome.[38]

Like other overuse injuries, voice strain may arise from habits that can be avoided. Some users court injury by shouting their commands, but tense muscles can be hazardous even at low volumes. And like carpal tunnel syndrome, voice strain may be promoted by factors like posture and allergies. Users are now seeing speech pathologists who coach them on how to speak, relax, sit, breathe, and take frequent sips of water. It is paradoxical but not unusual that some of the most advanced and apparently effortless technology should require vigilance, practice, and sometimes professional advice in performing the simplest actions. One programmer with voice strain observes how under work pressure, "all of a sudden you find your voice has tensed up." In the late twentieth century, staying relaxed may demand ceaseless effort.

From easily reversible eyestrain to crippling back, hand, and wrist pain, the physical problems of computing have important things in common. They are incremental. They develop slowly, often without a noticeable onset. There can be a sudden crisis of disabling pain, the result of conditions that have persisted for weeks or months. These are only indirectly measurable. X-rays and other imaging can show anomalies consistent with pain, but men and women with similar physical images may not have similar feelings.

Above all, these conditions are shaped socially. Political conservatives usually insist on the existence of an objective reality and deny that scientific and even technological knowledge is socially constructed. But partly because there is no "dolorometer," no recognized test for pain—only at best devices that may reveal suspiciously inconsistent responses—conservatives and especially neoconservatives deplore the economic cost of computer-related claims and see neurosis at work, if not fraud. Liberals, who otherwise perceive the self-interest of medical providers in "new" syndromes and diagnoses, consider computer-related illness to be objectively real. Both sides would agree that injury rates tend to be higher where work is more stressful. Organizations using similar hardware and software have had such different experience with injuries that social processes must be responsible for part of the difference. But nobody understands yet what the processes may be and how they operate.

In the outbreak of reported cumulative trauma disorders in Australia in the 1980s—"the largest, most costly and most prolonged industrial epidemic in world history," according to one medical critic there—there was agreement that medical attitudes were part of the problem and actually helped make it worse. But who was creating the unintended consequences? Was it the office workers who were reporting it, or their labor and feminist allies who helped to promote oversensitivity to minor symptoms and even encourage outright malingering? (Australian trade unions are among the world's most socially and politically active, and workers' compensation laws reflect labor's political influence.) Or were sympathetic physicians helping "the powerless and dependent, and those who cannot otherwise express their righteous rage at their supervisors, employers and spouses," to use their "exquisitely symbolic pain and incapacity" to communicate distress, as one Australian doctor has suggested? Or were skeptical physicians helping to create chronic symptoms by refusing to take early reports seriously and putting the burden of proof on patients, as other analysts have argued? Either way, the epidemic was in part an unintentional consequence of medicalizing what the Australians called (following Commonwealth practice) repetitive strain injury.[39]

Not paying attention leads to injuries. But focusing on the physical problems of computing or anything else might have amplified the symptoms. Are sufferers hard workers who have driven themselves too far, coming forward reluctantly only when the pain is unbearable? Or are they consciously or unconsciously trying to escape from responsibility by medicalizing their problems? Questions that begin with seatpans and backrests, forward and backward tilts, microswitch clicks and wrist supports turn out to be have answers that are psychological, organizational, and even political. The question is whether the ethical burden is on employers to control stress even at the expense of profits and "competitiveness," or on workers (whether data tabulators or editorial writers) to stiffen their upper lips as well as their lower backs.[40]

In the last thirty years, the office may have grown more quiet but it has also become more tense and lonely. Shoshana Zuboff, a social psychologist, has explored the unintended consequences of the automation of office work. Electronic mail has replaced not only face-to-face conferences but even telephone conversations. One clerk standing and asking a question at another's station more often than not signals a problem to supervisors. Software-driven application of rules has supplanted decision-making based on personal experience; sequences of screen prompts leave little room for judgment. The implications of all of this for the body are apparent. Zuboff's fieldwork among clerks suggests that reports of physical problems owe less to equipment design than to social constraint:

> Automation meant that jobs which had once allowed them to use their bodily presence in the service of interpersonal exchange and collaboration now required their bodily presence in the service of routine interaction with a machine. Jobs that had once required their voices now insisted they be mute. . . . They had been disinherited from the management process and driven into the confines of their individual body space. As a result, the employees in each office became increasingly engulfed in the immediate sensations of physical discomfort.[41]

One of Zuboff's subjects looked back nostalgically to the arrays of clerks she remembered from an earlier era as she described her new situation with graphic metaphor:

> No talking, no looking, no walking. I have a cork in my mouth, blinders for my eyes, chains on my arms. With the radiation I have lost my hair. The only way you can make your production goals is give up your freedom.[42]

Published just as cumulative trauma disorders were becoming a major issue in North America, Zuboff's study challenges the vulgar Platonism of computer studies that assumes a frictionless and disembodied world of information processing. Bodies were never noticed much in computing until they started to take revenge in the form of lost time, lawsuits, and workers' compensation claims.

There is an important positive consequence of the new agonies of computerized offices. Without the complaints and the potential litigation that eye, upper-extremity, and back problems have brought, the health and comfort of all computer users would receive much less attention. Whether reclining is better than tilting forward; whether clicky keyboards are really healthier than "mushy" ones, or vice versa; whether trackballs are safer to use than mice, everyone benefits when such questions are taken seriously. We have seen that some complaints were old ones that had been ignored, in part because they affected mainly women, and non-elite women at that. Even the male bosses didn't seem to understand much about comfort in the office, to judge from their own furniture. It took the specter of legal damages to help make them take their own well-being seriously.

It is true that this real improvement may not immediately promote satisfaction; it may have the revenge effect of increasing the number of reported problems. As medicine tends to make people feel worse about their health even when objective measures are improving, better workplace design may focus people's attention on what is still less than optimal. The office *can* be much safer, though. Specialists believe that by the late 1990s, changes in behavior will complement improved machines and furniture to reduce eyestrain, back problems, and cumulative trauma disorders of the modern office.

The price of better conditions, however, may involve new complications and higher levels of maintenance. Ergonomists call this the "degrees of freedom problem." To respond to changing users and tasks, for example, a chair may have numerous adjustable features, including seatpan and backrest angle, seatpan height, backrest height, tension, and armrest height. The height and angle of keyboard and monitor may also be adjustable. As Marvin J. Dainoff and James Balliett have observed, too many variables may discourage users from trying to take advantage of all the ergonomic features of their equipment, while there are no effective ways to prevent unsatisfactory combinations of adjustments. In combating the repeating effect of office routines, we have found ourselves with a recomplicating effect.[43]

Once more, technology usually fails to solve chronic problems in the direct, automatic way we have been led to expect—or hope. This does not mean that technology fails; quite the contrary. It means that it requires skill and vigilance. A study in Finland, for example, showed that even with state-of-the-art workstations, employees at video display terminals still had significantly higher musculoskeletal complaints than other office workers until a physiotherapist explained why and how office equipment should be adjusted. The rate of shoulder pain dropped by 70 percent. Nobody predicted that automating intellectual work would make it necessary to pay attention so constantly to the body. Nobody thought that just sitting in a chair would become a skill one had to learn. Even taking short breaks from work turns out to require more planning than anyone expected. On their own, workers may return to the office prematurely or plow on until they are too fatigued to continue at all. There seems to be an unexpected benefit of frequent brief interruptions: productivity improves though time spent at a given task declines. Even abolishing electronic mail in favor of telephone use and face-to-face meetings seems to have positive effects.[44]

The irony of computerization is apparent. The pastel-walled, air-conditioned, computerized office turns out to be a far more complex, possibly dangerous place than even the most hardbitten Luddites of the 1960s had dared to predict. Just as new devices let men and women complete mental work with less physical exertion than ever before, new complaints and new scientific thinking showed that some of the old inefficiency was actually beneficial and should be reintroduced into the workplace, by corporate decree if necessary.

That often-deplored metaphor the Information Highway turns out to be surprisingly apt in considering health problems of the computerized office. The massive increase of the power given to the operator (truck driver or white-collar employee) leads to health hazards from rapidly repeated small motions (seat vibration or finger and wrist movement). Focusing occasionally in the distance, maintaining a light touch, maintaining proper posture, and taking frequent breaks can reduce the stress and the likelihood of injury. The similarities can be taken only so far; a software crash or blowup is not the same as one on the road. Still, the price of health may require the reduction—if not the avoidance—of the intensity that newer technology makes possible. Of course, it helps to have monitors that maintain a tight focus, chairs and work surfaces that adjust, high-quality keyboards and pointing devices, and proper ambient lighting, but many ergonomists emphasize how much can be done without the latest equipment.

As usual, revenge effects have an unexpected positive side: computer-related complaints, as interpreted even by imperfect legal and insurance systems, set natural limits to authority. Physical symptoms have financial consequences that force managers to take stress seriously. Unfortunately, they are not the only obstacles to the profitability of computerizing. It is time to turn to the others.

9

The Computerized Office: Productivity Puzzles

The subtle risks of computers have not been their only unpleasant surprise. Their greatest paradox may be that it is still hard to evaluate their benefits where these promised to be greatest: in the most rapidly growing parts of the economy. There is little debate about the value of electronics in manufacturing and distribution. Robots still have limited applications, but nobody doubts that they sometimes can do things faster, better, and more cheaply than human workers. It is unlikely that any old-fashioned purchasing agent or warehouse manager could match the performance of good software. For the widget-masters and metal-bashers of the world, computers usually improve speed and quality and reduce waste.

But in manufacturing too, consequences can be unexpected, especially the results of improved communication. Consider the U.S. garment industry. It has always been cheaper to make textiles and clothing in Asia than in rural America, let alone in New York City. What protected American manufacturing as much as tariffs and quotas was the flow of information. Even with airmail, it might take too long to send designs and patterns overseas. Fax and electronic mail have changed that, taking weeks off turnaround times. With air freight an option for higher-priced merchandise, a major advantage of North American and European manufacturers—closeness to the end customer—has been seriously eroded. By buying time cheaply, new technology works against those who used to have time on their side.

The introduction of the IBM PC as the standard for personal computers in 1981 (older rivals soon faded, though the Apple Macintosh made its debut as an alternative standard in 1984) launched a new era for office work and the service sector. Mainframes had long controlled the operation of large corporations and government bodies, from manufacturing control to the processing of tax forms. Microcomputers promised to transform medium-sized and small businesses that could never have considered buying a minicomputer costing tens of thousands of dollars. Book-length documents could be edited with ease. Complex financial calculations, previously demanding hours of work, could be completed, and when necessary revised, in seconds. Records that only a few years earlier would have needed a variety of awkward colored tabs or punch cards plus long needles for sorting could now be stored on inexpensive disks or tapes. Programming no longer required months or years of apprentices hip in the old FORTRAN and COBOL languages; schoolchildren were mastering BASIC. And for those who worried that BASIC as an introduction led to bad programming habits, there were PASCAL, Turbo-PASCAL, and a host of more exotic entries from ADA to XENIX.

Throughout most of the 1980s, it was hard to imagine services and white-collar work in general *not* becoming more productive. The cost of everything from random-access memory (RAM) to disk storage space seemed to be shrinking by half every eighteen months. One of the first things that computer users have discovered is that not trading up to the latest processor means forgoing many new products and software versions, and even losing support for older software. WordPerfect 5.1, introduced in 1989, runs fast on 286-class computers from the mid-1980s; WordPerfect 6.0 almost demands a 486-class machine, generally four or more times as fast, as many Windows 3.1 applications did in the early 1990s. Microsoft's Windows 95 needs at least a *fast* 486 machine to run efficiently.

If the automobile metaphor is apt at all, it would probably be best expressed as a salesperson having to take on a larger and larger territory, paying the same amount every three to five years for a vehicle (ultimately an airplane) with greater and greater range, but getting an ever-smaller trade-in price for the last model. In 1992 I spent about the same for a Hewlett-Packard LaserJet 4 printer capable of eight pages per minute that I had paid seven years earlier for a "fast" C. Itoh daisy-wheel printer with less than one-eighth the speed. But unlike other consumer hardware (automobiles, cameras), my printer had lost nearly all of its resale value in seven years, even though it was still in excellent working order and could produce Courier 10 text that was actually sharper and clearer than copy in the same font from my LaserJet. Even the retail trade-in value of the most unpopular automobile models is a bastion of stability by comparison—all the more surprising because aging cars need costly overhauls that most mechanical printers don't. (Heavy-duty daisy-wheel printers are the massive, almost immortal elephants of office electronics. A school is now using mine.)

My experience was hardly isolated. The revenge effect of the explosion in computer performance is, as every excited new system buyer soon discovers, an *im*plosion in value. In February 1994, a Compaq 486/66M machine with a list price of $4,654 was projected to keep only 17 percent of its retail value over two years and only 6 percent after three years, 3 percent at wholesale. The wholesale and retail value in three years was listed as "salvage." Meanwhile, as of 1995 the Internal Revenue Service still required computers to be depreciated over five years.[1]

Not only does resale value decline precipitously; the price of replacement systems actually does not go down. There was a saying in the 1980s that "the system you want always costs $5,000," and for all the breakthroughs of the 1990s this still seems to be the case for state-of-the-art products, especially portables. A new entry-level system still costs around $1,500, as it did in the days of the beloved Apple II, except that the product has many times the processor speed, memory, and storage capacity. Of course, cheap computers can still do a lot. But they can be the most expensive of all because they can become socially obsolete—unable to run new releases of important software efficiently—within a year if not months. Some popular on-line services are already unavailable for DOS-only computers. Upgrades may be disappointing. The price of additional memory chips, for example, has not declined as much as the prices of other components. It fluctuates with demand and is most likely to increase when users need it most, to support more powerful operating systems.[2]

By the mid-1990s a 340- or 528-megabyte hard drive is a near necessity for handling bulky program and graphics files. These numbers in turn demand a high-capacity tape backup, and many new software packages are now written to be delivered on CD-ROM, requiring yet another hardware addition. The wish to run several programs concurrently using graphic interfaces like Microsoft Windows or OS/2 prods users to upgrade from 14- and 15-inch to 17-inch and larger monitors. Meanwhile, business travelers who at first demanded smaller and lighter notebook computers can now add portable printers (with batteries good for only about thirty pages), wireless transmitters, CD-ROM drives, and speakers, bringing total weight and price back up to the days of the original Compaq luggables and beyond.[3]

The huge investment in computing in the 1980s and early 1990s, then reflected one of the great cultural inversions of our time: North American and European corporations, and millions of professionals, small businesspeople, academics, and students, learned to stop worrying and love planned obsolescence. If the watchword of the 1970s was "survival," that of the 1980s was "empowerment." People felt autonomous, in control, more powerful, and absolutely more productive. But toward the end of the 1980s, the sentiment grew that something was not right. Throughout the decade, scholars like the sociologist Rob Kling and the political scientist Langdon Winner were pointing out that computerization was a movement that reflected social conflict and organizational infighting as much as technical change. By the 1990s, a new set of critics had joined them. Unlike the social scientists, philosophers, and technophobic humanists who had long challenged the cult of productivity at all costs, the new skeptics approved of capitalism, higher output, technology, and all the rest. They had emerged from within technocratic culture: from economics departments, business schools, and consulting firms. And their message was that things weren't happening according to plan. The service sector, where the potential for gains was greatest, was strangely lagging.[4]

Intuitively it is hard to believe that huge investments in improving the quality of anything will not pay off in the long run. There is a brilliant, well-supported argument that we are actually in the early stages of a more fundamental revolution. In a famous article, the economist Paul A. David has compared the early results of microcomputer use with the introduction of small electric motors to an industrial world formerly dependent on mills requiring massive power sources driven by water and later steam. Decades after the introduction of electricity, most plants still relied on networks of shafts, pulleys, and leather belts that were not so far from the world of the late-eighteenth-century millwright. With large investments in functional older systems, manufacturers often added electric motors controlling a group of machines while keeping the centralized power plants and drive trains in place. This may have stretched the useful life of equipment, but it prevented the most efficient use of either old or new technology. Even managers optimistic about the ultimate benefits of powering each machine with its own electric motor may have rationally preferred to introduce electric power in a series of smaller-scale, exploratory applications.[5]

By the 1880s, the advantages of powering each machine with its own electric motor were apparent. As a result, factories could run efficiently on a single, skylit level. The (literal) overhead of noisy, oil-splattering drive-train equipment could be eliminated. Materials and semifinished work could circulate with less delay, as heavier machinery no longer had to be placed near the main shafts and drive belts. Workers could control machine speeds more precisely. Factories could make new goods that would have been difficult to produce with shaft-driven power. All these advantages were known. Yet it took forty years or more for all the elements to fall into place and for electrified factories to transform industrial processes and raise workers' standards of living and conditions in the 1920s. Replacing a big steam engine in the basement with a large electric motor was only a small first step toward more productivity. Not only the investment in older plants but utility rates and the price of both electric motors and machinery helped determine the curve of investment in, and benefits from, electricity. And it took time to bring new, single-story plants on line. All this leads David to conclude that as different as information and electricity may seem, we may be looking too closely at the early problems of shifting from one set of structures to another.

In manufacturing, electronically controlled processes have supplanted earlier systems much more swiftly than electric power replaced steam. Microprocessors and other solid-state components have improved reliability of many goods and lowered unit production costs by simplifying manufacturing and in turn raising sales. While electricity meant little more than lightbulbs even for the fortunate minority of consumers enjoying service in its early decades, the consumer electronics of the 1980s have brought radical changes at a faster pace. As recently as the early 1970s, a working four-function calculator was still an industrial-grade business machine that looked and cost the part—usually hundreds of dollars. Some electric adding machines did clunky semiautomatic multiplication, and there were exotic European handheld gizmos that looked like pepper mills, but nothing more. By the 1980s, solar-powered calculators no bigger than a credit card were selling for less than $10. The lag that David noted in reconfiguring industrial power to exploit the advantages of electricity does not seem to have occurred in the almost instant, universal dissemination of microelectronic hardware.

Statistics have nevertheless been discouraging. According to one study by the economist Stephen Roach, investment in advanced technology in the service sector grew by over 116 percent per worker between 1980 and 1989, while output increased by only 0.3 percent to 1985 and 2.2 percent to 1989. Two other economists, Daniel E. Sichel of the Brookings Institution and Stephen D. Oliner of the Federal Reserve, have calculated the contribution of computers and peripherals as no more than 0.2 percent of real growth in business output between 1987 and 1993. Meanwhile, manufacturing productivity was growing rapidly. One possible explanation, so far not systematically studied, is that the mediocre performance of the service sector is due to the average of results from companies that had done well along with others that had lost ground. Precisely because service companies now depend so much on computer systems, there is more to be lost as well as gained if technical bugs get out of hand. This is consistent with the shaky job security of chief information officers (CIOs), natural scapegoats for computer-related debacles even when other managers may be more at fault. Some banks and brokerage houses seem to fire their CIOs almost every year, and more than a third of CIOs surveyed in the early 1990s reported that their predecessors had been forced out (or down). But even if the problem has been high gains offset by high risks, rather than overall mediocrity of performance, something has not gone according to plan.[6]

Could the problem be system failures? Could hardware defects and software bugs be responsible for much of the loss in production? Computer catastrophes are definitely worrisome, especially for computer professionals. The details of known electronic disasters have filled whole books.

Software code, like laws and sausages, should never be examined in production. Indeed, it seldom is—if only because today's developers write code in teams, each member producing a small part of an increasingly bulky and complex whole. There is no software equivalent of Tracy Kidder's hardware odyssey, *The Soul of a New Machine.* There is, however, an excellent contrast in Lauren Ruth Wiener's *Digital Woes,* the best general-interest explanation to date of why and how computer software is inherently unreliable. Some of the stories she and others tell are chilling. In the mid-1980s a computer at the Bank of New York began overwriting data on government securities transactions. Before the bug was discovered, the bank owed the Federal Reserve $32 billion without knowing who had bought what securities, and it ultimately lost $5 million in interest on a $23.6 billion loan it had to take from the Fed, pledging all of the bank's assets as collateral. At about the same time, cancer patients in Texas, Georgia, Washington State, and Ontario, Canada, received lethal X-ray doses from a software-controlled radiation therapy machine that failed after a specific sequence of commands to change its mode, as instructed, from a high-intensity X-ray beam to a low-intensity electron beam—while it nonetheless removed the tungsten shield that protects patients during X-ray sessions. Death and serious injury were the consequences. The *Risks Digest,* an Internet bulletin board expertly moderated by Peter G. Neumann of SRI International, leaves little doubt about the hazards of software as readers throughout the world contribute cautionary tales.[7]

The ghastly consequences of some of these failures, and the existence of computer codes that might automate nuclear strikes, make complacency about electronic risks nearly criminal. But talking to computer support and maintenance professionals changes the picture. They believe that at least microcomputer hardware has if anything become steadily more reliable over time. Floppy disk drives and even hard drives fail much less often, even if they always seem to crash at the worst possible time. Some manufacturers in the early 1990s have announced MBTF (mean time between failure) rates equivalent to thirty years of continuous use—not bad for equipment that may be socially obsolescent in three years because of new standards and capacities. As computer parts get cheaper, it is easier to protect systems by building in two or more of all crucial hardware items, especially hard drives. A computer with a RAID (Random Array of Inexpensive Disks) stores two or more copies of data. At a price, companies can regularly back up data to secure offsite storage services. Even if an office has to be evacuated for days or weeks, as happened after the World Trade Center bombing in 1993, employees can resume work in temporary quarters—a resilience impossible in paper-based offices.[8]

For both corporate and individual users, software failure is still not only a likelihood but a certainty. Yet here, too, technology has taken the edge off the sudden disaster. The cost and ease of backing up large data sets continue to improve. Few computer buyers pay the $180 to $300 that a tape drive costs, but this is a trifling price for the security of being able to store hundreds of megabytes of data on a $25 tape. It's easy to program a computer to back itself up unattended, during a lunch break or in the evening. There is still, of course, a burden of vigilance: shuffling floppy disks for the majority of users who still don't have tape backups, maintaining the discipline of backing up every single day, comparing backup data regularly with original files, and performing all the little rituals that are part of the technological strategy against catastrophe. Most network administrators now routinely back up files that reside on individual computers.

Even viruses are less menacing now than they at first appeared to be. Not every virus is a serious threat to every computer user; most users exchange programs and data intensively with a relatively small group of other users. According to researchers at IBM, few viruses reach the level—very similar to the infection threshold in medical epidemiology—at which they become stubbornly established in a population. To this social barrier is added the protection of antivirus software. This may be far from perfect, it's true. Once more, the price of protection is chronic vigilance. Here it means regularly downloading definitions for new viruses—dozens are identified every month, and there are even underground toolkits, like legitimate software development packages, for creating new ones. A few viruses have been designed to attach themselves to popular antivirus software. It is also true that locating and eradicating a known virus on every single node in a corporate computer network might cost tens of thousands of dollars, because each workstation must be disinfected individually. Still, deleting and overwriting files accidentally, spilling beverages, forgetting one of those proliferating passwords (some managers must memorize dozens), and other mundane mistakes do far more damage (as yet) than all the world's diabolical code. One of the biggest computer outages so far, the crash of the AT&T telephone network in January 1990, resulted from a single ambiguous statement in the C programming language that controlled the system's switching center—not the criminal conspirators sought by police investigators at the time. The world's angry hackers will continue to be a chronic nuisance and from time to time the source of local disasters, but not the great enemies of productivity some have portrayed them to be.[9]

The same can be said of power spikes. The two-strand telephone wires used in our homes and businesses were originally designed for voice communication, and only with great ingenuity can they be used for data transmission. Similarly, electrical networks were built to heat tungsten filaments and to power motors, not to regulate sensitive electronic equipment. Most people aren't conscious of voltage spikes and fluctuations unless their lights flicker visibly, but salespersons for surge suppressors do a brisk business at trade shows by plugging instruments that record each potentially dangerous spike into the exhibit hall power. These usually appear as though on cue.

While operating a computer during an electrical storm is never advisable, the main danger is not a sudden power surge that fries the computer—though these certainly happen—but repeated smaller excess voltages that gradually reduce the reliability and life of components. Once more, what appears to be an acute threat is in reality a surprisingly incremental and chronic one. And this in turn creates a rearranging revenge effect for hardware protection. Most surge suppressors use varistors, zinc-oxide wafers that react each time an excessive voltage passes through them, acting as electronic shock absorbers during transient high-voltage peaks. Unfortunately, the equipment itself generally reveals no visible signs of these repeated blows the components have taken, or of the life left in the system. It would be as though there were no way, either from mileage or from its behavior (or even from an inspection), to tell the condition of a car's brake pads.[10]

To make things worse, when a powerful surge does come along, most protectors don't really suppress it; they send it into the ground with (logically) the grounding wire. The computer and other equipment, especially modems, are nearly always on the same grounding circuit, and it is not uncommon for the surge to be shot right back up into the equipment that is supposedly protected. I bought an expensive and well-reviewed unit that used a different technology, only to learn in a later review that it was susceptible to in-house surges that were no problem for good conventional units. The same review also asserted that the degradation of metal-oxide varistors was not really a problem after all. The rating of my expensive surge suppressor had dropped to unsatisfactory. Surges are important enough to be a significant source of trouble and expense, and another demand on vigilance, but not important enough to affect computer productivity significantly.[11]

As always, no technology (including manual and mechanical systems) is entirely safe. Nor do people invariably use the safest systems, or use them with sufficient vigilance. But considering our dependence on electronic hardware, the important thing is not that it fails but that off-the-shelf hardware has become more and more stable. We can rule out system failures as a big component of the productivity paradox. The problem is really in the software—and its use and misuse.

Computers tend to replace one category of worker with another. There are two ways to get something done. You can find one group trained to accomplish things the old-fashioned way. Or you can pay another group to set up and maintain machines and systems that will do the same work with fewer employees—of the older category of worker. You are not really replacing people with machines; you are replacing one kind of person-plus-machine with another kind of machine-plus-person. When IBM persuaded corporations to modernize their bookkeeping in the 1950s, businesses were able to get along with far fewer accountants, as they expected, but they had to hire more programmers than they had anticipated. Automatic teller systems also require programmers and technicians paid four times as much as bank tellers. If things go well, banks need less than a quarter of the staff, and they come out ahead. But it is notoriously difficult to predict all problems, or their levels of difficulty, in advance. And one mark of newer technology is that while it is cheap in routine operation, it is expensive to correct and modify.[12]

We have all seen the sign "The Difficult We Do Immediately; the Impossible Takes Time." Computerization turns this manifesto on its ancient head. Software can devour highly complex tasks with ease if they fit well into its existing categories. But even a simple change illustrates the revenge effect of recomplicating. The scientific typesetting program TEX, developed by the computer scientist Donald E. Knuth and now the standard in many branches of physics and mathematics, makes short work of the most fearsomely complex equations that once cost publishers up to $60 per page to typeset. An author proficient in TEX—and I have had the good fortune to work with several of them—can prepare camera-ready copy that stands up to most commercially available systems. But making small changes, alterations that might require dropping in a metal slug or pasting in a new line in traditional systems, can sometimes take costly programmers' time. A hairline rule can take more time and money than pages of author-formatted proofs brimming with integration signs, sigmas, deltas, and epsilons.

Minor incompatibilities between authors' TEX programs and publishers' typesetting equipment can delay book-length manuscripts for weeks and run up costs well beyond those of conventional typesetting. Worse still from the publisher's point of view, some inexperienced and unskilled TEX-using authors—including distinguished scientists—blame the publisher and typesetter when their work is held up.

As editors of conventional manuscripts, my colleagues and I could identify problems early and request changes before texts went into production. Even experienced electronic manuscript specialists cannot evaluate a TEX manuscript reliably just by looking at the author's laser-printed version. Messy or nonstandard coding may fail to reproduce the same beautiful output when fed into professional typesetting equipment. Consequently, there are real hidden productivity costs associated with an "inexpensive" TEX manuscript; it may require open-heart surgery rather than a haircut. Publishers and typesetters discovering such insurmountable glitches have been known quietly to set the author's electronic manuscript aside and dispatch the hard copy to Asian compositors for conventional keyboarding. This may be speedier than waiting for the author to learn the fine points of TEX, but it inevitably delays production, embarrasses author and publisher alike, and introduces new errors. What computerization offers—or simplifies—with its right hand it can withdraw—or recomplicate—with the left.

TEX demonstrates the additional burdens of vigilance that advanced technology imposes. It may slash production times and costs for a scientific or engineering publisher, but only if either (1) the whole burden of typesetting is shifted to the author, who then has to be knowledgeable and vigilant about levels of detail that copy editors and typesetters otherwise would supervise, or (2) the author's editor is prepared to spend hours learning the fine points of TEX, adding technical support to his or her job description.

What about less esoteric programs—the word processors, spreadsheets, and databases that ordinary office workers use? Surely these have become easier? Text appears onscreen more or less as it will on paper (What You See Is What You Get). Graphic user interfaces (GUIs) with colorful icons have replaced most of the mysteries of DOS command lines, which usually required constant reference to a manual—a good thing, because new documentation is notoriously skimpy. If everything is now so easy, why have there not been more gains in productivity?[13]

In fact, another set of revenge effects is at work. Graphic interfaces like the Macintosh System 7 and above, Microsoft Windows, OS/2, NextStep, and others actually do not simplify underlying programs. What they do is add all kinds of programs and files behind an organized facade. If things go right, as on a luxurious cruise ship, nobody worries about the engine room. When a problem does arise, though, the sheer number of elements and the unforeseeable ways in which they interact means that a high price may have to be paid for the seeming ease of everyday use—just as the levees, locks, and channels of the Army Corps of Engineers may tame the fair-weather river while shifting and even magnifying the destruction of floods when they come.

In the world of computers, there are two solutions to this problem. One, adopted by Apple Computer for the Macintosh, is to police the system closely and enforce high compatibility and integration. Consistency has made Macintosh computers the easiest to learn and use, even when they have been somewhat slower in performing tasks than PCs running DOS with comparable processors. Apple's refusal to license the Macintosh operating system until the mid-1990s kept prices high but also helped it maintain a more consistent user interface. On the other hand, it has always been more expensive to develop applications that meet these standards than to program for the original DOS, so there has been less choice as well as less grief for Macintosh users. (Graphic artists and scientists are the exceptions; their Macintosh applications are more abundant and powerful than DOS or Windows counterparts.)

In the mid-1990s, processor speed and storage capacity continue to grow exponentially, but operating systems and programs are multiplying their demands, too. IBM's OS/2 2.1 took no less than 40 megabytes for full installation—more than many users' whole hard disks. Its successor OS/2 Warp 3 needs 65 megabytes. Microsoft Windows 3.1 plus MS-DOS needed 20 megabytes; Windows 95 requires 60. Rarer, more powerful systems like elegant NextStep 3.2 demand 120. The inflation of code, also reflected in 25-megabyte application programs, illustrates a chronic problem of computing that contributes to the productivity paradox: as swiftly as resources grow, the demands of software tend to expand even faster. Programmers and developers are understandably working for the future, preparing for the next generation of machines. Most users are living in the past because either their budgets or their bosses won't let them upgrade hardware to the current standard. Meanwhile the planned obsolescence of software means that earlier versions of major programs may no longer be supported. Computer managers are moving users to Windows without always giving them the processor power, memory, or disk space to work effectively. For many users, computer software seems related to hardware as Thomas Malthus believed population responded to food resources: software does not just match but eventually outstrips hardware in their respective rates of growth. Users who do not have enough computer power spend more and more time waiting for their systems to finish work or struggling with messages that protest, accusingly, "out of memory."[14]

Even users with memory and disk space to spare are starting to think twice about upgrades. It is sometimes the applications that don't have the power to run under the new version of the operating system. Even the ardently pro-Windows *PC-Computing* had to acknowledge in late 1994 that many favorite applications simply would not work under the next release of Windows. By the mid-1990s it appeared distinctly possible that operating systems were reaching a point of rapidly diminishing returns.[15]

Even when one's computer has power to spare, the techniques that make computers more "user-friendly" have revenge effects of their own. Take the icons that are replacing typed commands in controlling computers. Supplanting words with symbols is one of the great goals on the agenda of scientific idealism. Otto Neurath, a philosopher of the Vienna School, even founded an Isotype Institute, continued by his wife, to supply humanity and especially education with a universal iconic language. One of his disciples, a lawyer named Rudolf Modley, emigrated to the United States and founded the Pictograph Corporation, designing charts that conveyed information vividly and almost without words. Modley later collaborated with the anthropologist Margaret Mead and influenced generations of graphic designers. With growing regional and global movements of population, roadsides and airports sprout signs using standardized, international pictographs in addition to (or entirely in place of) texts in one or more languages.[16]

Computer icons have the advantage of being more readily recognizable and distinguishable than strings of text. They also require less space. Pointing to an icon is also defended as more "natural" and "intuitive" than spelling out a request. In the normal course of education one is taught to spell and not to point, but much of the appeal of icons (and even more elementary graphic interfaces like Microsoft's Bob) is their playful innocence. The trouble is that simplification can have its own recomplicating effects. At first, with only a small number of icons, and a single organization (Apple) largely controlling their use, the idea proved a smashing success.

The real problem surfaced when open standards were adopted to encourage independent software producers, while manufacturers (especially Apple) lost part of their control over the interface. Meanwhile, applications of software swelled in size with new features as developers sought to please users in many occupations with each program. Today there are hundreds and perhaps thousands of icons in use, not to mention popular packages for creating and modifying icons. There are even animated and sound-enhanced icons, and no doubt some will soon be able to sing in 24-bit sound and four-part harmony. The recomplicating effect is that while some commands and programs are much clearer as symbols than as words, others are resolutely and sometimes inexplicably nongraphic. (The world has at least two serviceable Stop designs, but still no decent Push and Pull symbols for doors.)

"I love standards; there are so many to choose from" is a computer industry joke that applies to many icon designs. It probably does not matter too much that Apple Computer has copyrighted the Macintosh icon of the garbage can as a symbol to which unwanted files can be dragged for deletion. Windows software uses a wastebasket instead. But what does a shredder mean, then? Does it discard files, and/or does it do what paper shredders are supposed to: make the original text impossible to recognize or reassemble? And some symbols mean different things in programs written for the same operating system. A magnifying glass can call for enlarging type as it does in some Apple software, but it can also begin searching for something or looking up a file in Macintosh as well as in Windows applications. A turning arrow can mean Rotate Image, but a similar arrow can denote Undo; Microsoft tried a hundred icons for Undo and finally gave up. People might think of Stop or Greetings when they see an extended hand, and in some countries this might be an obscene gesture, but it is supposed to mean Move Through Document. No wonder software producers are starting to add text to icons, sometimes in the form of pop-up balloons that appear when the cursor is moved over the icon—a recomplicating effect if ever there was one. And while the computer industry is rightly proud of the special features it has provided for people with disabilities, current reliance on graphic interfaces is at least a temporary setback for vision-impaired users, who will probably have to rely on voice synthesis. A survey by London's Royal National Institute for the Blind revealed that nearly three-quarters of software firms producing text-based software were planning to shift to graphic interfaces.[17]

Color, like symbols, also seems to make computers friendlier, although we have already seen that its introduction has been of doubtful benefit for user comfort. Moreover, spending more on color displays can actually reduce understanding. Most medical people prefer a gray scale for reading digital images. The eye can distinguish more shades of gray than of any other color—as many as 128 with special training and experience. While a color image appears richer in information, it actually may be misleading though more interesting. Even professionals may focus on the wrong data because, like the rest of us, they attach special importance to certain colors that are coded for urgency in our culture: red traffic signals and yellow highlighting, for example. The prominence of some colors can be equally deceptive. Yellow areas look larger than equal areas in blue because of yellow's higher luminance. It is hard to assign colors to data without subtly and unintentionally distorting information.[18]

Even following cultural conventions can be dangerous. If hot is coded as red and cold as blue, water-faucet style, the brain has difficulty judging intermediate shades. Our perceptual systems let us see gradations along red-green and blue-yellow axes. A scale that mixes the two may look richer but is not so clear. More information (in the shape of a richer color palette) may reach the brain as less understanding. Psychologists like Bernice E. Rogowitz at IBM Corporation and computer scientists are beginning to collaborate on tools for clearer presentation of data, but it probably will be years before these are adapted in common application software packages like spreadsheet and graphics programs.[19]

Until software incorporates good psychology, not only professional tools but essential everyday information sources like maps may be compromised. The geographer Mark Monmonier has observed that "there is no simple, readily remembered and easily used sequence of hues that would obviate a map reader's needing to refer back and forth repeatedly between map and key." Since many successful programmers of mapping software have no cartographic or design background, users of their products are "as accident-prone as inexperienced hunters with hair-trigger firearms. If you see one coming, watch out!"[20]

Whatever the perils of color, it is gaining—and driving up the costs of computing. Color inkjet printers may be relatively inexpensive, but the color ink they use is costly. As the prices of standard laser printers fall, corporate buyers are eyeing color laser printers—a recomplicating effect that should also multiply the sales of color photocopiers. But just as video distributors have stopped colorizing classic films and art directors are exploiting the surprising impact of black-and-white in a polychrome media world, a stalwart minority of computer users are staying with monochrome monitors. The renowned economist Fischer Black, a managing partner at the investment banking firm Goldman Sachs, was still using his Compaq 386 as of early 1995, and was proud of having the only monochrome screen on his firm's trading floor.

Adding the bugs and glitches of software to the limits and quirks of hardware is a recipe for problems that can easily wipe out the productivity gains in any system. This need not happen, of course, but the triumphs of computing are understandably reported more frequently than disasters that most managers would rather keep from customers as well as from the competition. The problem is usually not catastrophic but—once again—gradual. What occurs is a slow leakage of staff time that does not show up in most statistics, but is nevertheless a drag on output.

Sociologists of science and technology have long realized that getting both experiments and machinery to work takes skills that don't appear in textbooks or manuals. So-called high-technology professionals learn these as artisans always have, by working with masters of the craft, and of course by trial and error. Some companies are lucky enough to have staff specialists who know how to coax programs, machines, and people into working well together, just as manufacturers have millwrights who, with nothing more than an ordinary toolbox, can get moving a stalled production line that would baffle managers and even most engineers. Despite twenty years or more of electronic diagnostics, automobile repairs still depend primarily on basic mechanical know-how and real-world expertise. Yet many organizations cannot find—or afford—the technical support they require. As the cognitive psychologist Thomas K. Landauer has observed, software is written by programmers whose logical, spatial, and mathematical skills exceed those of most users, and whose experience often blinds them to user needs. Some features, like the Windows games Minesweeper and Solitaire, have become major corporate timekillers, and conscientious employees sometimes struggle for hours to get columns to align on their printers—behavior that a tongue-in-cheek survey by one software producer, SBT Accounting Systems, dubbed the "PC futz factor." And according to the 3M Company, 30 percent of business users lose data every year, costing an estimated 24 million business days to replace.[21]

As both software features and the numbers of users have multiplied, support by manufacturers has not kept up. Long waits and high charges are becoming more common, even when a bug in the product is at fault. Nearly 80 percent of companies increased their support costs between 1990 and 1995, but even in-house support cannot be ubiquitous. Filling the gap is peer support, the person down the hall who becomes a resource without any amendments to the job description. Many of these people can beat outside specialists. After all, they are likely to work with similar hardware configurations, using the same software for the same purposes. But there are also hidden costs involved. To help other users, relatively high-paid executives and professionals are spending time as peer consultants, time they have to take from their real jobs. Although the time thus lost never turns up as a line item on any financial statement, it means that these Good Samaritans may never get around to refining a strategy further, to making an extra client presentation, or to doing any of a number of other income-producing things.[22]

The Boston consulting firm Nolan, Norton & Co. has documented what happens when organizations—among them AT&T, Bell Laboratories, Ford Motor Company, Harvard University, and Xerox Corporation—add workstations and networks without increasing support staff. Peer support expands to fill the gap. While journalists continue to celebrate the ever-lower prices of computers, the total costs of *computing* suggest that there is no such thing as a free menu. Nolan, Norton found that a single PC workstation can cost up to $20,000 a year. Of this, $2,000 to $6,500 appears on the organization's information services budget as the cost of equipment, supplies, and in-house technical assistance. Peer support turns out to cost two to three times as much, or $6,000 to $15,000. The hours needed to help end users learn things and resolve problems seem to reach a natural equilibrium. Networking multiplies the need for support. And even among professional end-user computing staff, these needs change priorities. They spend most of their time reacting to problems and have less time for planning and implementing more effective enterprise-wide computing. Most peer-support time is reactive, too.[23]

Peer support often turns out to be a rearranging effect. We save time by having computers accomplish things for us, but before they can do them we take time from our colleagues. A local area network (LAN) may help a department work more effectively together, but it can also take hours of time as employees become administrators and assume the care and feeding of cables, interface cards, and networking software. The problem is more serious for those who do not have a formal role. Most amateur gurus do not mind interrupting what they are supposed to be doing. They may like computers more than their real jobs. Even if they love their work, they probably find peer help psychologically rewarding and politically valuable in accumulating allies and favors owed. Professional computer managers appreciate the volunteers for relieving their budgets and of course for knowing their colleagues' needs better than a full-time technical person could. Since some peer helpers are world-class scientists, engineers, and other professionals, the information services department is getting a bargain—and the organization is losing a part of the high-priced time it is paying for. Yet without them, as William M. Bulkeley wrote in the *Wall Street Journal,* "millions of U.S. workers might turn their PCs into expensive hatracks."[24]

One obvious solution is to bring support costs into the open by expanding budgets for information services to reflect the hidden demand. In the mid-1990s, however, corporations appear to be moving in the opposite direction: buying more and more "bargain" hardware and expecting the same staff to support it. Managers think that because local area networks are cheaper than mainframes they should cost less to maintain. Actually, one consultant for network strategy, Janet Hyland, points out, "For client/server, ongoing support is much more costly than a mainframe." And she adds that "no one will ever know how expensive it is" to support networks throughout a corporation.[25]

Despite industry call for usability, then, the computing world of the 1990s turns out to be a patchwork of stand-alone machines and networks, professionals and amateurs, always in a state of tension between the productivity benefits greater power brings and the learning and support costs that it requires. Whether by planning or by chance, some organizations have the people needed to make computers work brilliantly. These become the legends of the movement. Other organizations groan under the weight of their new machinery, because they don't have people with the right skills in the right places. It comes as no surprise that computing, like most other forms of contemporary technology, is neither a miracle weapon nor a dud, but a set of tools that need constant attention and maintenance. The software crisis is yet another example of how the mastery of great tasks increases rather than reduces the chronic burdens of vigilance. Some say that this is a transitional situation, that computing is approaching a new stability and reliability. Users' experience with Windows 95 and applications written for it will be a good test of that claim. "Technology is on a collision course with marketing," *Forbes* magazine warned a few months before the system's debut, as new home users fumed about their unruly machines; but early users were more positive. Meanwhile, something else is happening to software that calls the productivity of computing into further doubt.[26]

Well into the 1970s, computing was the rugged, spartan territory of electronic data processing. Some people punched cards cut to the size of 1890s-era U.S. currency. Others fed the cards into readers and removed bulky fanfolded oversized paper from giant, clacking printers. A few lucky people, generally scientists and engineers, had terminals that could communicate with mainframes and minicomputers and even over networks. Some organizations had fancy daisy-wheel printers that could approximate the quality of an IBM Selectric typewriter—itself a spinoff of an early computer printer program. But there was nothing visually exciting about the vast majority of computers. Only a handful of pioneers saw that with enough processing power and memory, they could produce images that would speed use, aid analysis, and clinch sales.

By the early 1980s, programs like Visicalc and later Lotus 1-2-3 showed that computers could display information much more powerfully than anyone had thought. Large corporations have always invested heavily in graphic design and the production of charts, even for internal processes. Now executives and professionals need no longer depend on artists and draftspeople. They can enter data and see it in tables, line charts, or pie and bar charts with the click of a mouse. They can print data as letters, numbers, or as real graphics. By the 1990s, organizations could buy equipment for $5,000 or less to make high-quality color slides and prints, and four-color laser printers in the same range have now come on the market in the mid-1990s. For $10,000 and up, dye-sublimation printers can produce even more vivid images.

All this represents a stunning achievement. Even in the late 1980s, a laser printer that could come near typeset quality (600 dots per inch) sold for as much as $20,000, and early desktop publishers considered them bargains. A page that might have cost $15 or $20 to set conventionally could be generated in-house for little more than $1 in paper and toner. Even when maintenance, power, and labor were counted, the savings were impressive. In 1992 my Laserjet was already capable of 600 dpi at under $1,300. Equally to the point, it had dozens of standard typefaces built into it, making costly type cartridges unnecessary. Meanwhile, software had not stood still. Programs like Harvard Graphics, Lotus Freelance Graphics, Microsoft PowerPoint, and WordPerfect Presentations let users mix texts with images to create near-professional-quality transparencies and slides. What, I wondered, is the connection to productivity?

Among otherwise matched competitors, whoever has a powerful technology first has an obvious advantage. In war, it was horse-drawn chariots, then cannon, rifles, Maxim guns, submarines, decoding devices, atomic weapons. Technology does not always decide the contest, and sometimes the efficiency of shoe factories can count for more than the design of tanks, but nobody doubts the power of being first. In peaceful contests, the prize may go to the more powerful analysis—which is why mathematicians and economists can earn seven-figure incomes on Wall Street—but may also go to the more effective presentation. Not only marketers but technical and medical researchers must learn to play the presentation game. In fact, scientists are so entranced with visual display that one of them, John Rigden, has satirized their fixation with a hilarious account of Lincoln presenting a revised Gettysburg address to accompany a set of slick transparencies.[27]

For years, economists have studied how competitors learn to

fight against the successful early adopters of a new idea or technique. The value of powerful new instruments shrinks over time as more people have them and learn how to use them. When no other civilians had carrier pigeons, these birds were literally the highest financial technology; in 1815, they gave Nathan Mayer Rothschild several hours' advance notice of the outcome of the Battle of Waterloo, letting him make a fortune in stocks depressed by Blücher's recent defeat. No matter how much faster communications have become since then, the person with an on-line service can trade securities (short-term at least) better than one who scans the next day's financial paper. In turn the subscriber to a quotation service with a fifteen-minute delay will beat the occasional user of the electronic news, the subscriber to a real-time service will have an advantage over the reader of delayed trading, and of course where open-outcry trading exists, somebody who has bought membership on the exchange can feel and hear the pulse of trading in the pits. Similarly, even powerful spreadsheet programs will not perform as well as applications developed for a single purpose or industry; already venture capitalists are tired of seeing business plans drawn up by a few standard software programs. No matter how efficient or cheap the technology becomes, a hierarchy remains. Retaining an edge remains costly.[28]

The search for graphic superiority is not new. After the relative restraint of Enlightenment design, the nineteenth century spirit of competitive enterprise promoted documents ranging from the elegant to the bizarre. Victorian circus posters and corporate letterheads alike testify to a zest for visual hyperbole, typographic tinkering, and virtuoso flourishes. Steel pens and cheap paper encouraged grand stylistic profusion, and ambitious white-collar workers took lessons from masters' penmanship in their quest for advancement. Copying "all the letters in a big round hand" was an important skill for the aspiring clerk. The Coca-Cola logotype, created in 1887 in the exacting Spencerian style by the Atlanta bookkeeper who also suggested the name, and the Teutonic-baroque Budweiser label of the same period, have outlived war, depression, and sans serif functionalism.[29]

Modern presentations, like the Victorians' efforts, have a scale of values. The problem is that as elegant laser output gets more affordable and "mail-merge" programs make customized form letters almost effortless, handwriting has moved up to the top of the executive communication scale. Top-of-the-line Montblanc, Waterman, Parker, and Pelikan fountain pens cost more than many computer printers. The head of Montblanc's U.S. operations believes that when word processing removed the prestige once enjoyed by a flawless letter that only a top executive secretary could have typed, the handwritten note took its place. But many things still can't be handwritten, so graphic escalation continues, to the delight of hardware and software suppliers. The technology of the high end may change, but the cost of being on the high end apparently never goes down by much.[30]

Even the most enthusiastic fountain pen owners can write only a fraction of their communications by hand. Otherwise they must compete for recognition with everyone else. After investing in the newest generation of equipment, one is at best only months ahead of others. Meanwhile there is the cost of learning how to master the new capabilities. And here is where another chronic problem emerges. Even as the cost of equipment goes down, the increased power and flexibility of hardware and software make them take more time to learn and to use. Of course, they would probably be much quicker if one did the same old things. But there are always new things to be done. Some are automatic, like right-hand justification—which came to the office just when both psychologists and typographers were discovering that uneven right-hand edges are more legible and even more attractive because they permit the most consistent spacing. Most right-justified computer copy requires uneven padding between the words. It also requires minute variations of the distance between individual characters (letterspacing), a practice discouraged by traditional typographers. One of America's greatest, Frederick Goudy, observed years ago, "Anyone who would letterspace lower case would steal sheep."

As high-quality laser output has become increasingly available, and screen fonts like Adobe Type Manager and TrueType have made it easier to produce, the impact of computerized documents has declined. When Times Roman replaced ordinary Courier 10 as a standard, it became ordinary Times Roman. By March 1994 one type guru could even write a First Law of Fonts in a leading computer magazine: *"Never use Times or Arial* [the standard Windows sans serif font] *for anything, ever."* How long will it take for Baskerville, Garamond, and Palatino to share their fate? Laser printing has a way of turning even the best-designed types—and Arial, like Times Roman, is especially legible—into clichés.[31]

User problems hardly end once the proper type is at hand. The recomplicating effects of graphic interfaces continue to work their mischief when users share documents electronically. Any DOS text file can be read by any other DOS computer. A Windows file needs a copy of the original font on a recipient's computer to display the same carefully produced effects. If the font is not there (and if it is not one of the hackneyed, standard fonts it probably will not be), Windows looks for the next-best thing, which usually looks much worse than the default DOS font. On the other hand, including multiple fonts in a file bloats the file and may violate copyright laws. Some complex programs promise to convert fonts attractively, but these take time to install and learn, and may not be easy to use.

Despite all the fuss that graphic escalation brings, what makes it most impressive is that it sometimes has no direct economic or competitive logic. A colleague who helps produce a newsletter for a local private school has told me how the school once managed very well with simple Macintosh output until new desktop publishing and page layout software became available. The new features were too much for volunteer editors, and the school had to hire a designer to lay out an attractive publication. Even if a new specialist does not show up on the books, the hidden costs of graphic escalation can be immense. Executives and support people spend hours—taken from other work, not to mention the time of their peer gurus—learning to make slides, handouts, and newsletters. The habit seems to be entrenched. A colleague teaching at the Sloan School of Management at MIT confirms that even some M.B.A. students now feel that they are not sufficiently competitive if they fail to submit their course work in color.[32]

By the mid-1990s, just as everyone seemed to have learned the ropes at last, the graphic escalator lurched upward again. This time the new dimension is multimedia, which means adding sound and fury to the smoke and mirrors. The cost of playing back multimedia files has plummeted as much as anything in computing; before long most personal computers will have a CD-ROM drive, a sound card, speakers, and a microphone as part of standard equipment. But as usual, when anything seems to be getting cheaper, a recomplicating effect is around the corner. Voice annotation of spreadsheets and other documents is a perfect example. Just when software was finally making it possible to share documents across computer systems and business programs, the multimedia document raises its head. It hogs disk space, remains partly unreproducible on most of the world's computers, erects new (if no doubt temporary) obstacles to file sharing, and above all makes it impossible for anybody to get full information from a printout, no matter how many colors it may have. All this for what footnotes or an explanatory paragraph could accomplish at a fraction of the cost.

Soon we will no doubt be able to add video clips as well as sound to documents. Watching a video clip presents no problem, but producing video material is another question. The technological history of the motion picture industry suggests that every enhancement of the capabilities of a medium actually makes it more expensive. Refinements of science, far from ending the need for craft and for vigilance, create new categories of specialization and multiply their importance. Business people are not, in general, interested in animating three-dimensional dinosaurs or even two-dimensional ducks, but they share a problem with the George Lucases and Steven Spielbergs. More powerful hardware, even when it drops in price, does not run itself. The more awesome it is, the more skilled labor is needed to realize its potential, and in turn the more production costs. In the early 1990s, the most stunning films using computer animation, despite budgets well into the tens of millions, could not afford much more than ten minutes of real three-dimensional effects. Copyright owners will no doubt license some footage for corporate presentations at a price, as clip art is now sold as part of desktop publishing packages. But this will have its problems, as the bundled solutions will soon be overexposed. (We have already seen how extensive use weakens the impact of printer fonts and business-plan templates.) And as Thomas Landauer has emphasized, many computer uses are unproductive: "more concentration on operations and customer service applications and less on toys for professionals and managers is in order."[33]

What about analysis? Of course computers take time to learn, and some people may miss the point. But who can deny that people think better, work better, analyze better, and write better with computers? Actually there is growing evidence that computers may not be consistently better for real work after all. This suggestion infuriates many computer professionals. Industry columnists arguing vehemently with one another about almost everything else concur in denouncing anyone so perversely goofy as to doubt that computers really do make everyone more productive. (For computer magazine editors and writers, there is no doubt that a rapidly changing industry promotes productivity; they would have very little to produce without a monthly shot of new hardware, programs, environments, and even operating systems.)

From 1981 to the present, American businesses have spent hundreds of millions of dollars on software to help improve decision-making. Software producers are now disbursing millions of dollars on research to make their programs even better and easier to use. Spreadsheets and other decision-making aids are now ubiquitous in academia and the professions as well as in business. Thus it is all the more remarkable how little research has been devoted to the effects of computers on the quality of decision-making. Given more data and more powerful analytic tools, is it reasonable to assume that the quality of decisions will be correspondingly higher? This may seem plausible, and possible. It may even be true. Yet only a handful of studies exist. After all, why try to prove what everybody knows intuitively and what is built into the curriculum of business schools? The problem is that there is growing evidence that software doesn't necessarily improve decision-making.

Jeffrey E. Kottemann, Fred D. Davis, and William E. Remus, specialists in business decisions, tested the ability of a group of students to control a simulated production line under conditions of demand with a strong random element. They could add more workers with the risk of idle time on one hand, or maintain a smaller workforce with the risk of higher overtime payments on the other. They could restrict output relative to demand, possibly losing sales if orders jumped, or they could maintain higher output at the cost of maintaining possibly excessive inventory.[34]

The subjects, M.B.A. student volunteers and experienced spreadsheet users, were divided into two groups. One group was shown a screen that prompted only for the number of units to produce and workforce level. The other group could enter anticipated sales for each proposed set of choices and immediately run a simulation program to show the varying results for projected inventory. They could immediately see the costs of changing workforce levels, along with overtime, idle time, and nonoptimal inventory costs, all neatly displayed within seconds after entering their data. They had no more real information than the first group, but a much more concrete idea of the consequences of every choice they made.

If all we have read about the power of spreadsheets is right, the "what-if" group should have outperformed the one with more limited information, unable to experiment with a wide variety of data. Actually the what-if subjects incurred somewhat higher costs than those without the same analytical tools, although the difference between the two groups was, as expected, not statistically significant. The most interesting results had less to do with actual performance than with the subjects' confidence. The "non-what-ifs" rated their own predictive ability fairly accurately. The ratings that subjects in the what-if group assigned to their own performance had no significant relationship to actual results. The correlation was little better than if they had tossed coins.

Even more striking was how the subjects in the what-if group thought of the effects of their decision-making tools. What-if analysis improved cost performance for only 58 percent of the subjects who used it, yet 87 percent of them thought it had helped them, five percent believed there was no difference, and only one percent thought it had hurt. This last-mentioned subject was actually one of those it had helped. Another experiment had an even more disconcerting result: "decision-makers were indifferent between what-if analysis and a quantitative decision rule which, if used, would have led to tremendous cost savings." In other words, the subjects preferred what-if exploration to a proven technique.[35]

Why the attraction of what-if analysis? Perhaps the subjects, being M.B.A. student-volunteers, had a strong belief in their ability to manipulate numbers. But that does not explain why they also rejected a carefully formulated equation in the second experiment in favor of free-form interaction. The real reason, Davis and Kottemann surmise, must be what psychologists call the illusion of control: the way we can easily convince ourselves, given the proper setting, that we are making things happen when in reality they are chance events. Both the beauty and the risk of computerized analysis is the concreteness it can give our plans—even when our underlying data are doubtful and our models untested or even wrong. We have seen the theatrical power of computers; in the illusion of control, we turn it on ourselves to reassure ourselves that the powers we possess are indeed real.

Computer advertising appeals to this sense of control, and in fact it often *is* real control, at least in the right hands. But consider all the hardware and software producers who have failed or lost tens or hundreds of millions of dollars. Surely their employees and at least some of their executives were among the world's most proficient users of computers of all kinds—with IBM at their head. And surely they used the best models and techniques available. But the losers faced situations in which what-if questions are of limited value, in which politics, distribution, the evolution of standards, and sheer bluff matter as much as technical excellence. Of course, we can't go back to graph paper, slide rules, pencil marks, and eraser shavings, but we do not seem to be moving forward as rapidly beyond the present categories of software as we might, despite the extravagant early promises of artificial intelligence research. In fact, some old-style financial executives still astound younger colleagues by using slide rules and mental techniques at meetings to get faster results than the others obtain from their calculators and notebook computers.[36]

Spreadsheets are still valuable tools, but they are more like lathes than can openers. They need careful setup, adjustment, and oversight. The decision scientists Raymond R. Panko and Richard P. Halverson, Jr., found that while spreadsheet users have a low error rate at the cell level (0.9 to 2.4 percent), these errors cascade into disturbingly high error rates at the bottom line (53 to 80 percent). As the logical intricacy of spreadsheets increases, the error rate also grows—a recomplicating effect. When a bug in mathematical operations of the Pentium chip in 1994 alarmed users, few understood that their own mistakes in entering data and failure to debug their programs were a far greater risk to their results than the problems of Intel's processor could ever be. Spreadsheet users, like other beneficiaries of technology, have not realized that the way errors propagate from step to step in a program is yet another chronic problem of technology. The answer is that once more, a labor-saving technology needs a surprising amount of time-consuming vigilance to work properly.[37]

To return to Paul David's comparison of electric motors and electronic workstations: could it be that only now are we starting to see the beginnings of new structures that will make possible the kind of productivity increases that manufacturing enjoyed in the 1920s? Networks seem to be rebuilding organizations in new ways. If only there were at least basic agreement on what is happening. On the one hand, networks are said to inject grassroots democracy and equality, ushering in a kind of corporate New Age where rank and title matter less than the quality of a contribution, bare as electronic mail is of the usual trappings of executive office. Documents can be transmitted and processed with new speed, sent between employees with different computer types and operating systems. Electronic meetings can open the floor to ideas of those who might otherwise have been pushed to the sidelines by superiors or more aggressive colleagues in a conventional face-to-face session. Regrettably, an in-house computerized meeting room costs $50,000 to $200,000, and outside providers charge thousands of dollars a day for their use.[38]

The flip side of the innately democratic network is the malignant authoritarian network. Where some find inventiveness percolating up and correspondingly rewarded, others find discipline and punishment raining down and privacy trampled underfoot. If networks appear to open channels previously barred—and it's not clear how having to put ink on paper ever prevented sending a message to top management—they also make it possible to read files surreptitiously, monitor activities, and even trace message traffic to discover clusters of malcontents. Aside from such ethical lapses, a collegial style doesn't necessarily mean a flattening of power. The *Info World* magazine columnist "Robert X. Cringely" suggested not so long ago that the casual culture of Microsoft masked a management style that was at heart not so different from the hard-driving ways of the robber barons. ("You can work any eighty hours a week you want.") Of course, an authoritarian organization can be extremely productive, whether or not the iron fist is in a faded denim glove, but there is no evidence that networks as such make managers different. Rather, people seem to build networks in their own image.[39]

Apart from issues of democracy and authority, computers have encouraged a trend that may have unintended results for corporate efficiency: reducing support staff. The American Manufacturing Association, in a much-noticed study, found that staff reductions in general did not reliably increase profits. At downsized companies, profits increased for only 43 percent of firms, and actually dropped in 24 percent of those studied. Almost as many reported a drop in worker productivity as reported an increase.[40]

If computers really made it possible for a smaller number of people to accomplish the same amount of work, there would be little outcry about longer hours for middle managers and professionals. In fact, after examining twenty case studies in five major U.S. corporations, Peter G. Sassone, an economist and management consultant, concluded that computerization has helped reduce rather than promote the amount of time that these employees spend performing their highest and best work. Sassone found that many highly paid people were spending a significant amount of their time performing what amounted to secretarial and clerical functions, usually working with computers but often not doing what they spent years at college and graduate school learning to do. Why not? One reason is that companies tend to expand by adding professionals and to cut back by laying off support staff, giving them a top-heavy structure. Another is that computer systems make it possible for managers to do more things for themselves. Their jobs become more diverse in a negative way, including things like printing out letters that their secretaries once did. Sassone's analysis suggests that the increased productivity of office technology has an "indirect and unintended effect on staffing" that "may cause overall organizational productivity to decline." This he calls the "law of diminishing specialization."[41]

This is just fine for many people. Either you like the division of labor or you don't. Some social thinkers, and some people in business, believe that doing your own word processing, filing, and so forth builds character. But economic theory suggests that it doesn't do much for profits. The most rational way to deploy staff is to have the most skilled and specialized employees working as much of the time as possible at the highest level of specialization, while less skilled and lower-paid staff should take over the rest of the work. A famous precomputer textbook example is the story of the best lawyer in town who was also the best typist in town, but who would look foolish not employing a typist. Computers can create an illusion that the machine is doing all the work, but a surprising amount actually remains: formatting letters, replenishing and unjamming paper, replacing toner, and of course addressing and stuffing envelopes. Professionals who are doing these things aren't doing other, more productive things. Of course, top executives understand this pitfall—when it comes to themselves. They rarely cut back on their immediate support staff. Sassone claims that by reversing the formula and employing fewer managers and professionals and more support staff, corporations could achieve savings of $7,400 per employee. What keeps them from doing this is another form of the illusion of control, in this case the misconception that now support functions can take care of themselves.[42]

The relentless speed and efficiency promised by microcomputers and networks, their computation capacity doubling every eighteen months, has a catch. The more powerful systems have become, the more human time it takes to maintain them, to develop the software, to resolve bugs and conflicts, to learn new versions, to fiddle with options. Once again, the intensity of a given technology may bring repeating and rearranging effects in its wake. Early computer-minded historians were overjoyed at their new powers of quantitative analysis—until they realized that they would probably have to spend months or years entering data by hand before a machine could spend its few minutes processing and analyzing the data.

One approach to computers and productivity is to insist that many benefits of computing defy conventional measurement. If anything should change, it is the methods of economists, not the claims of the computer industry. A more radical form of this argument is that the point of the computer is the pleasure of mastering and using it, its responsiveness when things go well. In this view, that game of Minesweeper on an employee's monitor should not make the boss scowl; there probably is something to learn from it. Computerization is as much an end as a means. Nearly every computer user must feel such joy when things go right—seeing the first gorgeous page of text from a new program or printer, for example— that it seems crass to talk profit and loss. But this approach ignores the times of correspondingly great frustration. Some programmers are repeat keyboard smashers. And if personal experience is paramount, what of the countless people who still enjoy manual ways of doing things?[43]

For both technophiles and technophobes, the best, and perhaps the only, way to avoid the revenge effects of computing is to maintain skills and resources that are independent of the computer. We can learn back-of-the-envelope calculation to beware of misplaced decimal points. We can work on face-to-face and telephone relationships with colleagues and outsiders to avoid the misunderstandings of excessive reliance on electronic mail. In a way, the problems of computers, like so many other revenge effects, have reminded people of the value of other things. In the first decade of computing, handwriting and handwritten letters began to rebound from years of neglect. Bugs, glitches, and crashes have a positive side: they are the machine's way of telling us to diversify our attention, not to put all of our virtual eggs in one electronic basket.

10

Sport: The Risks of Intensification

Sport, unlike the office work we have just surveyed, is an ancient part of Western and indeed of human life. Today only a few scholars care about the operations of Roman and medieval administration, whereas images of the Greek Olympics are still vivid, thanks to the revival of the games by Pierre de Coubertin more than a century ago. The attraction appears almost universal. Even countries sharply critical of European influence have generally competed in the Games rather than abstained. For better or worse, Western concepts and customs rule international sport. (In fact, even judo is a European-influenced version of Japanese martial arts, and the system of colored belts was introduced only in 1927, in London.) Coubertin's revival of the discus and javelin reestablished these field events after millennia, and contemporary statues and vase paintings of Greek athletes have a striking immediacy.[1]

For all of the world's identification with the athletes of antiquity, we are also conscious of a distance from them. Part of it is religious; the church rejected athletics as a holy calling, making occasional and partial exceptions for medieval warrior-monks and nineteenth-century muscular-Christian schoolboys. More of the gap is ethical and technological. As the sports historian Allen Guttmann first observed in his book *From Ritual to Record,* ancient Greek athletics, including the Olympic Games themselves, lacked a concept basic to our enjoyment: the idea of a measurable level of performance that, once achieved, sets a benchmark for future athletes to surpass. While exceptional athletes could earn sums not far removed in real purchasing power from the winnings of top twentieth-century players, their prestige was based on stories of their prowess, not on recorded numbers. Greek runners did not attempt, for example, to achieve a new time for a given distance—not just because they lacked chronometers, but because the concept of a record was foreign to them. Roman athletes were conscious and proud of their numbers of victories, but apparently not of performance statistics as we know them.[2]

High-performance equipment was absent from Greek sport, partly because most of the events required almost none at all (even clothing was forbidden) and partly because there was no incentive to achieve a new record score. Yet sport as we know it today can hardly be separated from technology. The modern Olympics themselves depended first on railroad and steamship schedules and now rely upon airline services. The staggering expense of producing them today can be offset only by sales of broadcasting, sponsorship, and other rights at correspondingly high prices—justified to advertisers by potential worldwide television audiences in the hundreds of millions. Interestingly, sport as such a global force was unimagined in the predictions of bookish nineteenth-century political economists. Marx certainly made room for leisure in his post-revolutionary utopia, but the Eastern European sports machine that once awed the world in his name would surely have astonished him. Marx's own lifetime (1818–1883) coincided with what the American sports historian John Rickards Betts called a "technological revolution," embracing not only transportation of teams and spectators but electrical transmission of sports news to nascent mass media.[3]

There are two sides to the unforeseen consequences of sports technology. One reflects the multiplication of chronic health problems, the other the paradox that sports is the one field in which a technology can be *too* good.

In this century, coaches and scientists have rationalized professional and amateur sports with a vengeance, raising them to levels of system and performance that many other professions might envy. (In fact, the ideology of the amateur now stands largely discredited as a vestige of nineteenth-century middle-class prejudice against career athletes. Here the still-useful word *amateur* is merely a briefer synonym for recreational athlete, not an endorsement of bygone snobbery.) The quests for industrial efficiency and for sports supremacy have gone hand in hand. Promoted to neutralize the mental fatigue of school and factory work, athletics were soon subjected to studies of time and motion. In 1900 a French commission used advanced motion-recording instruments to judge "the precise methods and the effects of different sports on the organism, and to compare their value from a hygienic standpoint." Ever since, the scientific study of industrial efficiency and the pursuit of maximum athletic performance have been raising the bar, literally and figuratively. The physiques of today's actors show the results: compare Johnny Weissmuller with Arnold Schwarzenegger, or Kirk Douglas with Sylvester Stallone. Equipment is generally lighter and stiffer. Video recorders that would have made that early French commission pant with envy are honing the performance of athletes at all levels. While computer-assisted teaching has yet to prove its value in the classroom, computer analysis of sports performance has a solid record of success on the playing field.[4]

The advances of both technology and technique nevertheless have limits. Improvements of protective equipment have not stopped lawsuits claiming that the same equipment has been the cause of injury. New designs have led to multimillion-dollar claims against sports associations. As we will see, projectiles that can fly higher and farther have sometimes threatened the lives of spectators and judges. A flourishing athletic drug culture defies medical advice, official sanctions, and police investigation. Operations intended to restore performance can lead to agonizing disability. And while new technology can reduce the costs of a sport (aluminum-alloy baseball bats, for example), it can also multiply it (racing boats).

Danger, direct or vicarious, is part of the interest in most athletic competition. The fans of each sport have, at any point, a baseline of expected risk. Technology can do a lot to change the apparent danger of a sport. If the participants and spectators wish, it can affect the reality, too.

Consider the bicycle. With its high center of gravity, jolting ride on paving stones or dirt, and tendency to pitch the rider forward, the picturesque giant-wheel "pennyfarthing" cycle was a menace to its rider. But even falls from the far more stable "safety" cycle in use for the last hundred years can cause severe brain injury. Until the last twenty years or so, the only widely available head protection was the racer's leather helmet. More recently, helmets of lightweight polymer foam for shock absorption, covered with a thin, high-impact shell, have offered both greater protection and more comfort than their predecessors. According to some studies, wearing helmets can reduce the death rate from cycling injuries by 90 percent. Death and brain injury rates from bicycle accidents decline sharply where legislation and publicity have increased children's use of the helmets. While many children and adults still resist wearing helmets, there is little doubt that they work—that technological innovation, as we have seen so often, does mitigate dangers great and small.[5]

Recreational boating is another success story. In the water, as on land, technology and regulation (and perhaps some effects of congestion) have combined to make boating safer even as the number of boats has multiplied. Some worry that electronic navigation aids, especially satellite positioning equipment, may encourage boaters to ignore traditional navigation skills. A colleague in technology studies, himself a boater, recently raised this possibility in an electronic discussion group. But this plausible revenge effect has yet to appear. U.S. Coast Guard statistics show that while the number of U.S. pleasure boats more than tripled from 1962 to 1992 (from nearly six million to over twenty million), annual fatalities declined from 1,114 to 816, reducing the fatality rate per 100,000 boats from 18.7 to only 4.0 per year. While the total rate of accidents seems to have dropped less sharply, legally mandated technological improvements—and a campaign against drunken boating—have been a powerful force. Between 1972 and 1991 alone, the Coast Guard issued more than fifty new regulations and amendments, covering everything from flotation standards and standardized visual distress signals to devices (similar to the interlocks in automobiles) for preventing outboard motors from starting in gear.[6]

Other sports have also become truly safer. Luge, with eighty-mile-per-hour runs, was barred from the Olympics until the 1960s as too hazardous, but has improved crash safety by introducing lighter sleds, Kevlar helmets, face protectors, and redesigned tracks. Of course, there is always the chance that perceived safety will attract more novice athletes and thus multiply at least minor injuries. And in fencing, for example, better materials for weapons and clothing have led to more dangerous styles of play. But in general, new materials for protective equipment have consistently reduced sports casualties—that is, the catastrophic ones. If anything, professional and amateur athletes are probably too slow in adopting new protective equipment, especially polycarbonate eye and face shields in sports where flying balls or opponents' fingers are likely to cause eye damage. In the most dangerous sport of all, horseback riding—not coincidentally the least technologically changed—resistance to helmets is still strong. In the United States, more than 121,000 people sought emergency room treatment for riding injuries in 1989–90, but programs to promote the use of helmets have reduced injuries significantly among amateurs. Of course, competitive riding and professional racing are likely to remain potentially lethal; of America's two thousand jockeys, an average of two die and two suffer catastrophic injury each year. Their polycarbonate helmets and Kevlar vests, spinoffs of military gear, mitigate injuries but cannot prevent them, nor does such gear protect a rider's legs when a horse rolls over them.[7]

It is not clear that jockeys take any more risks with their special gear than without. As risk-taking professionals who find health insurance almost unavailable, they are already exerting themselves and their horses at the highest levels of which they are capable. Yet in other sports, especially contact sports in which a lack of protection does give athletes second thoughts about striking with maximum force, "safety" equipment can make a game more dangerous because it encourages rougher play. Or it may in turn lead to changes in rules, changes that will at least temporarily offset the gains of the new technology.

It is wrong to think that all safety measures are futile, that athletes will inevitably misuse them to maintain a steady level of risk. They sometimes do and sometimes don't. But rule-making counts as much as technology. By prescribing and proscribing equipment, association executives and professional athletes shape the future of their sports. The political scientist Langdon Winner has underscored the power that can reside in things—for example, in the bridge abutments of Robert Moses' Long Island parkways, which may have been designed in part to exclude buses and thus limit mass transit and low-income development on Long Island. He calls this power "the politics of artifacts." A livelier ball, a faster surface, or a new refueling system may either stimulate or reduce revenues of professional players, entrepreneurs, and manufacturers. Decisions about technology are political ones determining the balance of power among competitors.[8]

Boxing

Where players and spectators enjoy the direct display of force to begin with, there is less technology can do, and more room for revenge effects. The last hundred years or so of professional boxing shows this all too well. As the historian Elliott J. Gorn has noted, John L. Sullivan's national tours in the 1880s left no doubt that real men could fight with gloves and play by the Queensberry rules. "The ring continued to call forth images of primitive brutality, of lower-class and ethnic peoples venting their violent passions. But gloves and new rules *appeared* to curb the animality sufficiently to allow a titillating sense of danger inside safe and civilized boundaries." Electric lighting, as Gorn also points out, also helped transform the boxing ring from disreputable diversion into commercially raffish evening spectacle.[9]

The new conditions, of course, were intended for the benefit of promoters, not fighters. The gloves and lighting were parts of another scheme—to intensify the match, preventing boxers from wrestling or colluding to slow down, the illumination exposing them to better scrutiny. (Industrial employers of the same period, not coincidentally, were experimenting with production methods they hoped would discourage the workers' informal limitation of their pace, or "soldiering.") These continued dangers of the ring were intended, not inadvertent. Together, gloved hands, standard three-minute rounds with one-minute rest periods, and a reduction of knockout time from thirty to ten seconds quickened the pace of the sport. It was now definitely worth the far smaller risk of a broken hand to land a hard blow to the head; in thirty seconds an opponent would nearly always have been able to get up and resume fighting. Gloves also promoted roundhouse punches by weighting and protecting boxers' hands; a British Medical Association report has even compared them to the infamous hand weight, the *caestus,* of Roman gladiators. But there was another effect of gloves that neither promoters nor athletes understood, and that was indeed a revenge effect: they multiplied cumulative, chronic damage.[10]

Gloved play and the managed interest in the knockout blow encouraged boxers to land one hit after another on the side of the opponent's face. Bouts were sometimes as brutally bloody as the old bare-knuckle fights, but more often the greatest damage was not apparent. Today we know that the rotational blows of respectable boxing are far more harmful over the years than the straight punches of the bare-knuckle prizefight on the open countryside of the earlier nineteenth century. The friction of a leather glove against an opponent's skin rotates the head, gradually destroying layer after layer of nerve cells, killing the axons of others, and forming clumps of useless material in the brain. (The bill of a California acorn woodpecker decelerates at a thousand times the force of gravity, or 250 times the force on an astronaut's body at liftoff, yet because the bird's head moves straight forward, its braincase provides protection. You will never see a punch-drunk woodpecker.) Some physicians believe that long-term eye damage from the repeated blows is an even more serious problem than chronic brain injury. In fact, gloves or no, professional boxers appear to age twice as fast as nonboxing males, and physicians' groups regularly call for abolition of the sport.[11]

Boxing does not seem to be changing under the weight of criticism from doctors or the press; some say it is even more intense than ever. A former light-heavyweight champion, Jose Torres, described the experience of a really hard blow thus: "It is like one million ants get into your brain and into your whole body." And he added that while he took only three such blows in his whole career, "[t]oday, fighters get hit like that three times in one round." Still, it appears to be the repetition of lighter punches rather than knockout blows that causes Parkinsonism and other chronic neurological conditions in boxers. Gloves also are no protection against the serious eye damage that many boxers suffer; some of them blame the thumb of the glove for inadvertent injury.[12]

Boxing with gloves need not have these severe long-term consequences. College boxing is a relatively safe sport, and studies of amateur athletes do not show the same brain damage as that suffered by professional fighters. The professional sport's medical notoriety seems to have encouraged conservative calls at amateur events. But the pain of the sport, and its myth of social mobility bought with blood, are exactly what attract bourgeois authors no less than Middle American fans. Joyce Carol Oates has put the dilemma perfectly: without its excesses, the sport "becomes less satisfying on a deep, unconscious level, more nearly resembling amateur boxing; yet as boxing remains primitive, brutal, bloody and dangerous, it seems ever more anachronistic, if not obscene, in a society with pretensions of humanitarianism."[13]

Football and the Perils of Padding

Football shows as clearly as boxing how twentieth-century technology tends to convert catastrophic injury to chronic ailment. And as with boxing or any other sport, what matters is less the equipment used than the tacit understanding among players, coaches, and spectators concerning the application of force. A former college player, Brendan Kinney, wrote: "The 30-some pounds of equipment a football player wears are not for protection. The pads and helmet are made of hard plastics and steel. The helmet alone can weigh eight pounds, and when propelled by a 220-pound man at about 13 miles an hour, becomes a weapon of fantastic destruction."[14]

(Rugby, played without helmet or shoulder padding, is known to be less violent than football, though it is still as rough as it looks. It has a high rate of "major" injuries resulting in absence from multiple games. Some officials acknowledge that like football, it is a collision sport as well as a contact sport; its accident rate lies about midway between soccer and ice hockey. One English study of the professional game in the late 1980s found well over twice the amateur soccer accident rate per thousand participant-days. Waves of major injuries have beset British professional rugby as they have U.S. football. Still, there is no comparison with body-armored American football for major casualties. An Australian study underscored the virtual absence of major concussions and the small number of minor ones. Not adopting helmets clearly had a reverse revenge effect as far as the head, and probably the spine, were concerned.)[15]

Rough as American football is now, it once was still more violent—when it began as an elite diversion. President Theodore Roosevelt declared his "hearty contempt" for any young man who "counts a broken arm or collarbone as of serious consequence when balanced against the chance of showing that he possesses hardihood, physical prowess, and courage." And it was with more admiration than horror that the president of the University of California, Benjamin Ide Wheeler, described the game as "[t]wo rigid, rampart-like lines of human flesh," with the offense forming "a catapult to fire through a porthole opened in the offensive rampart a missile composed of four or five human bodies globulated about a carried football." The game had evolved from rugby to become a rigorously coordinated and meticulously choreographed team event. But its casualty rate became a national scandal, and while there were only a few deaths in intercollegiate play in 1905, twenty-three men died that year in intramural games and practice. Roosevelt threatened to outlaw the sport, and it took two sets of rule changes, in 1905 and 1910, to end the most serious carnage. The college presidents and coaches who worked out the new rules were intensely conscious of the importance of managing spectator interest. The forward pass, once a controversial "dream-like" play, helped attract a mass public following in the 1920s. Equally important was the acceptably smaller number of catastrophic injuries.[16]

It may have been the experience with the helmet in the First World War—the first major conflict fought with metal head protection—that led to rubber-lined headgear. Since steel was obviously too heavy, players continued to wear padded leather until after the Second World War. Plastic helmets were introduced in 1939 but were not used widely until after the war. Well into the 1950s, there were college holdouts. In fact, Haverford College players wore leather MacGregor helmets as late as 1967. Richard F. Malacrae, a Princeton trainer who was on the staff at the time, believes the school's conservatism reflected an enduring suspicion that hard helmets promoted more aggressive play. Where plastic helmets were adopted, players intent on using maximum force to stop an opponent began to use their headgear, with the mouth guard that soon accompanied it, as a battering ram. This intensifying tactic all too often had its own unintended consequence: spinal fracture and paralysis. Dr. Joseph Torg, an orthopedic surgeon at Temple University and later the University of Pennsylvania, began to analyze spinal injuries in the mid-1970s and discovered that broken necks and damaged spinal cords were not random accidents. Studying films of actual games, Torg found that stronger helmets introduced in the 1960s and 1970s for better protection enabled players to use their heads as battering rams. When they bent their heads forward, the spine straightened. Like a freight train hitting an obstacle on the track, the spinal column, propelled by the body's momentum, kept moving after the helmet struck its target. What seemed to be a technological solution had become an extension of the medical problem. College and professional officials had been trying to solve the same dilemma that Roosevelt faced: keeping the sport rough and bruising but free from disabling, catastrophic injury.

Once more, conspicuous disaster led to remarkable if incomplete improvement. The NCAA banned aggressive use of the helmet in 1976, and injuries dropped. Medical catastrophes in professional and school games combined dropped from thirty-six in 1968 to fifteen in

1989, and two in 1991. Even though standards officials and sports organizations have so far been unable to agree on a single prescribed design—there are now two competing standards—coaching, training, and emergency care have improved. And perhaps because there is still no universally accepted optimum design, litigation against helmet makers has continued. In 1988, Rawlings Sporting Goods stopped making helmets after courts had levied $39 million in judgments against it and other manufacturers in the eight years since 1980 alone. Spearing, the use of a helmet in place of the shoulders to knock down an opponent, is now banned but is still widespread, and not just in professional play. The middle linebacker Mike Singletary broke an average of four of the extra-tough helmets during each of his four years at Baylor; as the football critic Rick Thelander pointed out, these were helmets strong enough to withstand the impact of a baseball bat. Whatever courts may rule about the responsibility of equipment makers, the burden of constant vigilance is on officials, coaches, and trainers to prevent potentially lethal but tempting behavior.[17]

Artificial grass, first famous in the 1960s when the Houston Astrodome installed AstroTurf, was supposed to make football more widely playable rather than safer to play. Still, the choice of this surface had a revenge effect. Coaches relished the superior traction it provided. What they did not realize was that star players would be sidelined more often with patellar tendon injuries, torn ligaments, and damaged toes. One carefully controlled study of knee injuries in NFL games in the 1980s showed significantly higher rates of knee sprains on AstroTurf, even allowing for all other factors. Like other new technologies, it also takes much more care, maintenance, and vigilance than people expected. Stadiums with well-maintained artificial turf, like Cincinnati's Riverfront, have safety records much better than those without. But even the good safety statistics of some artificial-turf arenas may conceal yet another instance of damage from repeated stresses. Phil Simms, the New York Giants quarterback, told the filmmaker and former college player Robert Carmichael, "Playing on grass is a much easier game than playing on artificial turf. Grass is rarely frozen or lumpy. . . . The jarring of this artificial surface takes its toll. After a game on the turf my lower back is tight, and I can feel it in my legs."[18]

The real problem of artificial turf, though, is likely to be the same as the ultimate revenge effect of the armored style of American football. Intense recruiting, more strength and endurance training, greater conditioning, and illegal but ubiquitous steroids have made the players bigger and faster than ever, and impacts correspondingly more serious. Linemen of the 1990s would overwhelm their already massive counterparts of the 1960s. Medical and surgical technology, from helicopter evacuation to arthroscopy to hip replacement, and with superb rehabilitation equipment, have returned athletes who otherwise might have been permanently disabled to active playing careers and restored function to those who might have become paraplegics or quadriplegics. This improvement, however, has led to the biggest revenge effect of all: the prevalence of chronic conditions among former professionals and even many college players.

While there are fewer catastrophes, most of which result from spearing and other dangerous practices, serious injuries have actually increased with the spread of better protective equipment. From the First World War through the 1950s, only four in ten professional players per season reported injuries that needed surgery or resulted in prolonged absence from the game. By the 1980s, seven in ten were seriously hurt each season, according to a study by the NFL Players Association. Through the 1950s, only one player in three needed surgery; in the 1970s and 1980s, this jumped to two out of three. Technology, from helmets and padding to training aids, has helped intensify the game to the point that Dr. Robert Goldman, a leading sports physician, has compared a hit by a pro player to the impact of a small car. The game's "ballistic" style calls for brief but powerful bursts expressed as joint- and vertebra-jarring collisions far more severe than those of Theodore Roosevelt's day. The helmets, face shields, mouthpieces, and padding are better than ever, and deaths may be rare, but neither protective nor conditioning technology can prevent damage to the joints.

Since massive injections of anti-inflammatory drugs and painkillers make it possible for battered athletes to return to play, the new intensity means trading immediate relief for long-term disability. In fact, the more macho branch of sports medicine with its multiple injections fits in all too well with the play-with-pain ethic football players have come to accept, actually increasing long-term injury. Knee and hip surgery can extend players' careers, but usually only at the price of later pain, inflammation, and repeated rounds of surgery. During twenty years after leaving the game, the New York Jets quarterback Joe Namath needed four operations to repair tendons, cartilage, and ligaments in his knees; by 1992 his left knee was buckling, and surgeons replaced both knees with artificial joints. (Strangely, such experiences have not stopped even recreational athletes from demanding hip replacements to continue playing.) Shoulder and elbow injuries have left some players unable to raise their hands to brush their teeth. The better technologies for recovery from acute injuries become, the more severe the chronic consequences. One study of NFL players' long-term health showed that two-thirds of those who had played in the 1970s and 1980s retired with a chronic injury.[19]

Recreational Athletics

Of course, football and boxing are anomalous. There are few really recreational football players; even most college games are either professional or semiprofessional. The fans demand unusual intensity. And the athletes themselves are not only exceptionally large *and* agile, but also among the greatest stoics in sports, meeting pain with resolute denial. Long-term disabilities may be honorable battle wounds. It is not clear that many boxers or football players regret their careers. But what of sports that remain predominantly amateur in character?

The good news is that the health benefits of most forms of vigorous exercise more often than not offset the risks. A sedentary person is far more likely to suffer a chronic ailment in middle or later life than an active one. But it may not be easy to see in advance which technologies and practices are healthy. And our techniques for measuring the effects of exercise leave many questions open about where revenge effects might set in.

Running

Think first about the costs and benefits of running. Synthetic materials of the 1970s made running regularly on ordinary surfaces more efficient, by absorbing shocks and improving stability. In fact, running shoes became the comfort benchmark for all footwear. But these very advantages of better shoes also made it easier to run longer distances more often, multiplying stresses to joints, tendons, muscles, and bones. With each step, runners experience loads of three times their own weight on their bones and joints. Minor injuries are common, especially knee problems, even if with proper care these are not serious. Among long-distance runners, some women lose so much calcium that they are at risk of developing osteoporosis, although more moderate running can help maintain bone density.

Runners with proper technique and no physical abnormality do not face higher risk of developing arthritis and other joint diseases than do others. But many have skeletal or joint conditions that predispose them to injury—up to half of all middle-aged men. Nor are most physicians trained to recognize everything that may put a runner at risk. Systematic autopsies have shown that many people's knees have small congenital irregularities that might not be noticed in normal activity or detected by physical checkups but that could nevertheless lead to problems. Some combinations of anatomy and style, such as high arches and excessive pronation (shifting of weight to the inside edge of the foot), may increase knee damage. Sports physicians who deny that heavy running schedules lead to arthritis nevertheless acknowledge that chronic damage is possible if runners continue to ignore pain. Yet painkilling drugs make it easier to do just that. Comfort can also be unreliable. Running in expensive, heavily cushioned shoes may increase injuries by lessening the foot's ability to adjust to uneven ground and by encouraging pronation.

Finally, not even running surfaces are completely understood. Nearly everyone agrees that concrete is bad, but because of its unevenness, grass can be worst of all. What this shows is that even in a sport that has undoubtedly improved the health of millions of people, repeating effects can appear. It also suggests that especially for serious athletes, exercise demands more vigilance and maintenance than expected.[20]

Even the value of running for longevity is subject to revenge effects. Despite well-publicized tragedies like the fatal heart attack of the runner-author James Fixx, the sport probably does extend the lives of most participants. (It may have added years to Fixx's, too; he had a family history of premature death from heart disease.) Intensive running improves the heart's efficiency, often raising output to double that of a sedentary person. Many distance runners have pulse rates in the forties or even lower, though some outstanding runners have had faster rates. A lowered pulse may promote longevity simply by stretching the heart's useful life, though this has not been proved. But while running strengthens the heart's long-term fitness it also results in immediate wear and tear. Mathematical biologists have calculated that in theory, a vigorous running program should produce steady gains in longevity up to a point, after which the benefits of additional running time decline and eventually become negative. Even apart from analysis of the mechanisms involved, running clearly takes time that could be used for other activities. Running half an hour every day for twenty years probably adds time to one's life, but it has also probably taken a full year of one's time if transportation, warmups, showering, and dressing are included. Of course, few people run only for longevity; the point is that an excess of almost anything beneficial may have revenge effects.[21]

Data from the late 1980s support the idea that running and other regular exercise can, after a point, actually become hazardous to your health. Ralph Paffenbarger, a research physician, studied the incidence of heart attacks among men who used a stair climber and recorded their activity. The rate of all heart attacks and especially of fatal ones fell until it reached a certain level per week. Then it began to rise. Paffenbarger is on record as dismissing the increase as the result of some erroneous reporting, but there is other evidence that at a particular point exertion can become harmful. In the early 1980s, the English immunologist Lynn Fitzgerald found her health declining rather than improving when she became a star long-distance runner. She discovered signs of immunodeficiency in herself and later began to find immunosuppression among athletes with heavy training schedules. Since then, research of Fitzgerald and others has suggested that while moderate exercise strengthens the immune system, overtraining may compromise it by reducing the supply of the amino acid glutamine. (Of course, since sitting in a chair puts a heavy load on the spinal column and extended bed rest is a risk factor for osteoporosis, inactivity has revenge effects too.)[22]

Skiing

If more comfortable shoes can indirectly produce health problems as well as benefits, the consequences of more complex sports technologies are even harder to calculate. In skiing, new materials and techniques have transformed not only footwear but every item of equipment from skis to outerwear—not once but several times over since the Second World War. And while the results have made the sport faster and more enjoyable in many ways, they have equally and unintentionally changed both performance and the patterns of injury.

There is first the fact that better technological concepts are not always easy to translate into superior performance. While Americans believe in the power of technology, Timothy K. Smith of the *Wall Street Journal* has astutely pointed out that in most Olympic sports where advanced equipment provides a decisive edge, it is Europeans who have won more medals; Americans, ironically, excel in track and field, boxing, basketball, and other sports in which performance reflects more closely the intrinsic ability of players. Thomas P. Hughes has also noted another hidden advantage of Europeans: a corps of artisans skilled in the unwritten rules of high-quality custom production, arts neglected by Americans bent on the perfection of high-volume goods for the mass market. In the long run, it is difficult to keep any technology out of the hands of competing athletes, and some sports require that equipment used by any team be available for a reasonable time to all competitors.[23]

Profound as the changes brought by new equipment and materials to boxing, football, and running have been, the transformation of skiing and of skiing safety has been greater still. The replacement of wood by plastics and composites in the 1950s changed and extended the sport just as dramatically as lifts had done earlier in the century. Gone were the rituals of waxing. And on the way out, it seemed at the time, were the broken bones that once formed part of the folklore of skiing. At first, the new equipment shifted some of the injury from ankle fractures (common with lower prewar boots) to twisting fractures of the tibia. A fall often led to this spiral break of the bone. Then came further improvements. New, rigid plastic boots and bindings employing strong, lightweight alloys were designed to release the legs of skiers at a predetermined level of force. The boots encourage a crouching position that novices find unfamiliar at first, but they promised a superior level of protection.[24]

Once more there have been unforeseen consequences of new designs. For careful skiers, both downhill and cross-country skiing are safer than ever. In fact, skiing now has a significantly lower injury rate than tennis, a rate that includes beginners' accidents and probably understates its safety for experienced skiers. Still, the safety technology of skiing has revenge effects that can make some accidents more serious than they would be otherwise. These may be divided into two groups: voluntary risk seekers and the rank and file.

To the extent that skiers are risk seekers, they will respond to safer equipment and more carefully maintained slopes by seeking more dangerous runs and increasing their speed. For championship athletes, there is no doubt that speeds are higher; in World Cup skiing they have gone from seventy-five to ninety miles per hour for men and from sixty to seventy-five miles per hour for women, thanks to new ski and binding designs. (Even cross-country skiers can reach fifty miles per hour with rigid fiberglass skis on a downhill stretch of a groomed track.) When the Austrian skier Ulrike Maier lost control fatally in the 1994 World Cup at Garmisch-Partenkirchen, one official acknowledged that "we're pressing the envelope of what these bodies can do." Protection also leads to greater risk-taking in the slalom event, where skiers voluntarily use protective gear, including helmets, to take a straighter course down the slope, touching the flexible gates lightly with their bodies as they go instead of having to swing entirely clear of them. Among amateur skiers too, especially younger enthusiasts, there are signs that better equipment and greater perceived safety have led to riskier skiing. American resorts have invested millions in advanced and "extreme" runs for these customers without having a consistent scale for marking them, using slogans like "Make the mountain bleed." New resorts like Crested Butte, Colorado, offer slopes of up to 55 percent. Manufacturers promote "extreme skis" and even "extreme sunglasses." The result is that despite lower fatalities and accidents under typical conditions, and despite some of the most expert and best-equipped ski patrols on conventional slopes at Crested Butte, the number of misadventures on extreme terrain has increased. Catastrophic ski injuries jumped from a plateau of thirty per season in the 1980s to seventy-five per season in 1992. Thus, in downhill skiing there is at least preliminary evidence that some sports interests and some athletes are using safety technology to reintroduce danger.[25]

Since only a small minority of skiers, and mainly younger ones, are either professionals or temperamentally risk takers, the major revenge effects of better equipment lie elsewhere: in the kinds rather than in the numbers of injuries. In the days of wooden skis, the cast-encased leg was a cartoonist's cliché, but with some reason. The ankle fractures of that time were relatively common; though painful, they healed within a month or two. Between 1960 and 1980, the proportion of ski injuries to the foot and ankle plummeted from 45 to 10 percent of all injuries, and all lower-extremity injuries dropped from 80 to 55 percent. Knee injuries have remained constant at 20 percent, while arm, hand, head, and torso injuries increased relative to others. It is the change in knee damage that caused special concern.[26]

Robert Johnson, an orthopedic surgeon, has worked with Carl Ettlinger, an equipment consultant, for more than twenty years trying to solve the riddle of knee sprains. Skiers have been spraining anterior cruciate ligaments (ACLs) more often than medial collateral ligaments (MCLs). The small ACL, connecting patella and tibia, is essential for stable leg movement. ACL injuries are not only more immediately painful but have alarming long-term consequences if not properly treated. Knee cartilage can break down and arthritis can develop.

In the 1980s, videotape analysis made it clear to Johnson and Ettlinger that equipment was playing a major part in producing such injuries. Most recreational skiers with ACL sprains, it turns out, were twisting their knees during maneuvers to stop or to keep from falling backward. Hard, molded boots and more responsive bindings evolved together. Both were designed not only for easier skiing but to prevent that earlier scourge of skiers, the tibia fracture. But the boots, bindings, and more easily maneuverable skis with sidecuts can have a revenge effect when a skier falls backward. The boot and ski go off on their own (the "phantom foot"), with the ski applying a devastating load to the skier's knee—substituting a potential chronic condition for the old-style ski accident. ACL sprains now account for up to six injuries a day at large resorts and up to 100,000 annually in the United States. Surgeons can usually repair a torn MCL by stitching ends together; a sprained ACL demands much more difficult techniques, including tendon grafts. Specialized surgery can be effective but costly; its leading practitioners are among America's most affluent doctors.[27]

Equipment designers are already studying new designs to reduce the frequency of ACL injuries without bringing back tibia fractures. Meanwhile, the answer to revenge effects is once again not technology but vigilance—and, as Ettlinger and Johnson point out, instructors do not like to teach people how to fall, even though falling is part of skiing. Improved boots and bindings represent major advances—but only with more, not less, attention to technique.

Climbing

In mountaineering and rock climbing, technological revenge effects are of a different kind. In general the technology of climbing has few direct revenge effects. It is an intensely physical sport, but one without human opponents as in boxing and football. It is also a sport that is not intended to be safe and in which there has always been a high rate of injury. If drivers of vehicles equipped with antilock brakes have more accidents than those without, we may frown on their defeating what responsible people consider the purpose of these devices: to promote safety. If climbers use new safety technology to undertake more difficult and thus riskier routes, we can't say this effect is unintended. The designers are usually top climbers themselves, and they know they are not only making the sport safer for beginners but letting experienced climbers do what previously was impossible.

In his profile of the British climber Mo Anthoine for *The New Yorker,* A. Alvarez mentions that the highest of seven grades in British climbing, "Extreme," was once reserved for only a few routes but now has more numerical subgrades than all the other levels put together. While some top climbers like Anthoine are prodigiously powerful, climbing did not reach the Extreme level in the same manner that performance in many other sports attained new levels: by a search for previously untapped populations for athletes extraordinarily well endowed for the specific demands of an activity. (Much of the improved performance of late-twentieth-century athletes, in the West as well as in Eastern Europe, can be traced to early recognition of outstanding physical potential in a broad population; at least in the West, climbers are not generally recruited by colleges or professional teams.) Instead, it was a technological revolution that began after the Second World War and that let climbers take greater risks. At first new, harder alloys were introduced for the pitons that climbers hammered into the rock. Then it was braided-filament nylon and Perlon ropes that gave climbers an extra margin of safety by stretching just the right amount to break a fall. These improvements were followed by spring-loaded "friends"—lightweight devices with expanding, comma-shaped cams that could fit rapidly and securely in crevices—and other devices that could grip tenaciously. New synthetic uppers and high-friction rubber soles subsequently improved the traction of lightweight climbing shoes, a far cry from the hobnail boots of the 1930s. Plastic helmets and nylon body harnesses reduced the danger of falls. Meanwhile, synthetic materials were improving the weather resistance of outdoor clothing. Each improvement made it possible to contemplate more difficult climbs. On the highest summits, added lightness, strength, and weatherproofing were decisive: dryer plastic boots prevented frostbite, more aerodynamic tents kept out cold, stoves and freeze-dried rations were lighter, insulation dryer yet breathable.[28]

David G. Addiss and Susan P. Baker, public health specialists, interpreted accidents in U.S. national parks: "In contrast to the success of safe product design in preventing injuries in the home and workplace, the introduction of modern climbing equipment, designed for efficiency and safety, has been accompanied by a sharp escalation of climbing standards. Rather than decreasing, demands have increased dramatically to parallel the high performance made possible, in part, by better equipment." To public health professionals, the trend is indeed a revenge effect, but to climbers and perhaps to the designers of their gear, this is doubtless intentional. In fact, Addiss and Baker also note that in climbing as in other risk-oriented sports, the inexperienced novice is up to ten times more likely to get hurt than seasoned participants, but experts may be more likely to die. Many move through a series of increasingly challenging maneuvers. Himalayan climbers, exceptionally able as a group, seem to have the highest climbing death rates of all. (For similar reasons, racing drivers actually have significantly higher off-track crash rates than other motorists, despite rigorous training and superior skills. Their confidence prompts them to take greater risks.) The survivors among top climbers tend to be those who, like Mo Anthoine, resist pushing the outer limits.[29]

In football and skiing the weak link exposed by new technology is the knee and its anterior cruciate ligament; in climbing it is the hand and fingers. As the orthopedic surgeon and climber Mark Robinson reminds us, our hands have evolved for precision work. Monkeys have retained the powerful forearms and tough flexor tendons ideal for rock climbing; when one climber teased them with bananas left on a cliff, they made an ascent at the very high difficulty level of 5.13. Considering the inherent risks of scaling a sheer cliff face, today's ropes, shoes, cam devices, and other equipment use new materials to superb effect. But what they cannot do is compensate for weak joint cartilage that makes certain power grips dangerous. Not only joints but especially tendons are at risk. Just as the characteristic motions of keyboarding can inflame the carpal tunnel, repeated stress on the tendons can produce "climber's finger," a sprain or strain that partially tears the tendon or stretches it painfully. In extreme cases, the annular pulleys through which the tendons pass may be broken as well.[30]

Experienced climbers are at risk not only from the more difficult ascents they attempt—a conscious and rational balance of danger against accomplishment—but even more from the insidious risk of repeated flexion. Studies of climbers cited by Robinson show that half to three-quarters have "long-standing" finger injuries, with many also suffering from swollen joints or deformities. Many of these have conditions that do not appear as arthritic in X-rays but that, Robinson believes, may develop into full-blown arthritis. Carpal tunnel syndrome can appear among climbers, and inflammation of the tendons can scar the median nerve. Carefully supervised practice facilities with artificial walls may help climbers develop proper technique and confidence, but they, too, have potential revenge effects. Some veteran instructors are concerned that novices who learn mainly in indoor gyms may be vulnerable to accidents on less predictable real-world ascents. It is still too soon to say whether their concerns are justified.[31]

As in skiing, both conservative treatment and surgery can usually prevent potentially long-term injury from becoming chronic. It is even likely that clinical experience in dealing with knee and hand injuries in sports will lead to better procedures for treating similar conditions arising from activities at work and in the home; we have seen how military medicine has benefited civilian emergency treatment, plastic surgery, and other specialties. The study of sports injuries has also improved training and conditioning. As knowledge has increased, the attitudes of participants have also changed. Robinson observed in 1993 that within only five years, "climbers have become much more sophisticated and careful about training, and now routinely behave more like athletes than the anarchic maniacs they used to be in my glory days." Even so, climbers more than nearly all other recreational athletes have to accept chronic injury routinely. Robinson acknowledges that training and proper treatment can reduce but not eliminate these problems. Sooner or later, all serious climbers encounter injuries that probably will cause permanent damage. What care can accomplish is limiting these "to a level that can be regarded as a fair price for the pleasures of climbing."[32]

The ultimate revenge effect of technology in sports is not a health problem at all but a chronic one nonetheless. Not only for climbers but for cross-country skiers and backpackers, gains in both climbing and rescue techniques mean that more people feel confident in taking risks. Improved transportation and more leisure have made this growth possible, yet without a greater sense of security there might be as many sightseers but far fewer participants. In 1951, only 300 people climbed Mount Rainier near Seattle; by 1985 the number had increased to over 4,000 with nearly 3,500 others turning back before reaching the top. Mount Hood in Oregon has 10,000 climbers a year. Mount Fujiyama in Japan has 100,000. Mount Everest in 1953 tested the ability of two of the world's strongest climbers, their conquest bringing international acclaim, but 485 people had reached the summit by late October 1992—even though from 1953 to 1973 there had been only three dozen. Technology does seem to be taking much of the fear not only out of climbing, but also out of backcountry hiking and cross-country skiing. And this is just what alarms many park officials. The recreation officer of Mount Hood National Forest told a reporter in 1986 after eleven people, including nine members of a school expedition, died in two separate incidents: "Mountain climbing is a dangerous sport. If you made it so safe for everybody to get up there, you'd have a lot more fatalities because people wouldn't recognize the risk."[33]

One reason for continued high fatalities in some locations, despite unprecedented safety measures, is the illusion of control that we encountered in the paradox of computer productivity. To an experienced climber with respect for the mountains and weather, improvements seem to be mainly a chance to achieve greater success—and pleasure—with the same degree of risk. But others take risks without realizing their seriousness; vacationers overestimate the safety and predictability of nature. In fact, some of the technically easiest climbs can be the most treacherous. On Mount McKinley (Denali) in Alaska, the highest mountain in the United States and Canada, of one thousand attempts a year, about six hundred are successful. It is unlikely that there would be three or so attempts a day if the risks did not appear reasonable and controlled, at least to the extent that climbers are backed up by helicopter rescue. Yet eleven people died in the first half of 1992 alone, ten from falls and one from altitude sickness. The problem is that Mount McKinley's weather is unusually severe and unpredictable, with wind gusts of up to two hundred miles an hour near the summit and fiercer cold than Everest. Weather and the altitude of climbers often prevent use of the helicopter, which critics compute costs about $200 a year to operate for every person making the attempt. Some traditionalists have called for abolishing the rescue service. "If rescues weren't so easy," wrote the editor of *Climbing* magazine, "maybe some of these people would have the sense not to go on the mountain."[34]

AVALANCHES

Not weather itself but avalanches are the most serious threat to climbers and others on the mountain slopes, especially in winter. In the nineteenth century, avalanches were primarily a hazard of railroad workers and miners; now they are a hazard largely for outdoor vacationers and athletes worldwide. In 1990, forty-three climbers died in an avalanche on Mount Lenin in the Soviet Union; in 1991, nine skiers perished at a helicopter resort in British Columbia; and over Christmas 1991 there were a death and injuries at Albertville, France, during preparations for the 1992 Winter Olympics. In the United States, avalanches kill from fifteen to twenty people a year; in Europe, they kill about 120, more on average than earthquakes do.[35]

An avalanche is one of the most lethal natural disasters because victims engulfed by snow may have less than an hour to live. Prompt rescue is essential. Unlike earthquakes, avalanches tend to strike in the same areas with the requisite slopes and snowfall; they happen again and again in the same place. As a report of the U.S. National Research Council put it, "unlike other ground-failure hazards such as rock-slides, which once released are spent, snow avalanches automatically 'reload' with each snowfall and can 'fire' several times in a given year." We are only starting to understand avalanche prediction; the mathematics turn out to be exceptionally complex, an aspect of what is called "self-organized criticality." The only effective ways to deal with avalanche hazards are still to keep people out of zones at risk and to create and manage small avalanches. The U.S. Forest Service and other agencies provoke slides with military gear, including artillery shells from "avalaunchers," shots from recoilless rifles, and helicopter-delivered bombs, in hopes of preventing more serious disasters.[36]

May we consider avalanche a revenge effect of winter sports? Unlike an active fault or a forest fire, an avalanche zone, like a flood zone, is a hazard—an object of public concern—only if people live or travel in its path. We have already seen that the very features that make the forest edge and seashore attractive for living also make it dangerous. Similarly, in hopes of consistently good skiing conditions, several Austrian pensions were built in a known danger zone, with results fatal to a number of guests in the late 1980s. Typically, it is more common for skiers to die in avalanches they trigger than in structures in the path of the cascading snow. Locals, including local skiers, respect the snow as they respect other natural features. Of course, multiple deaths happened before modern sports tourism developed, often to railroad laborers and gold seekers in the U.S. West; but urban skiers, climbers, and hikers now bring a new attitude to the slopes. They are used to urban tempos and timetables. They have jobs and families in the city. If they cancel a climb or give in to prudent fear of weather or snow conditions, they may forfeit precious vacation time, and often money. While these vacationers may have had years of university science, they understand the mountains less than most local dropouts do. For them the revenge effect, the danger of technology, is precisely their feeling of immunity from natural events. The more reliable the control of avalanches is thought to be, the bolder some skiers and other winter athletes will feel.[37]

Even advanced avalanche monitoring, radio identifying signals, and public education programs have not reduced the number of accident victims. In Colorado an all-time record of twelve deaths was recorded in the winter of 1993, mostly of climbers, backpackers, and skiers who had wandered off from approved slopes. Far from satisfying the demand for risky experience, the well-maintained if steep slopes of areas like Crested Butte seem to draw a minority of skiers for whom even these are too tame, people who seem determined to use every new safety measure to get more excitement at the same risk. When seven skiers trapped by an avalanche near Aspen gained national attention in February 1993 (which turned to notoriety when they offered the story of their rescue to Hollywood), it turned out they were not ordinary fools but experienced backcountry skiers equipped with avalanche beacons that may have given them a false sense of security. Similar accidents have been reported from the mountains of Scotland and the Alps of France.[38]

Courting Risk with Safety Devices

The growing number of backcountry deaths, like the popularity of extreme skiing, challenges the view that Western society is timid and risk-averse. The opposite seems more on the mark. Knowledgeable and experienced skiers and climbers yearn more than ever for the most challenging experiences available. The crowds on the groomed slopes at the most popular resorts and the traffic on easier climbing routes drive some people farther afield to get the same mountain experience—a recongesting effect. What an international ski tour operator told the London *Times* after a tragedy in the French Alps also applies in North America: "The job [of a guide] has changed, it's now to take people into a risky situation. It used to be to avoid it. Clients often urge instructors to take risks now. An instructor will push the limits."[39]

Safety devices add risk not just because guides and clients feel safer but because it is easy to leave them in the wrong mode or not fully activated. An example from the high seas illuminates what is happening on land. The world-class American yachtsman Michael Plant died after his sixty-foot sloop capsized during a solo transatlantic voyage in October 1992; concerned to arrive in France in time for the beginning of an around-the-world race, he had not registered his Emergency Position Indicating Radio Beacon. Had its signals been recognized promptly, Plant might well have survived. His resourcefulness was legendary, and his sloop was built with multiple safety precautions, including a lifeline, a doubly protected living area, and watertight compartments. Plant's death shows that the greatest risk of safety devices may be that human beings retain the power either to operate them in the wrong mode or to disable them, as Plant did when he chose not to register the beacon. We have already seen how technology increases the need for vigilance in medicine, plant and animal introductions, computing, and driving. It is no less true in recreational sailing, even for the most skilled.[40]

Ordinary recreational sailors and backcountry skiers face smaller risks in relying on electronic devices. The Global Positioning System (GPS) gives the recreational market nearly the same ability to find coordinates that until recently only the U.S. military possessed. By 1994, 40 percent of one leading maker's GPS sales were recreational. At the same time, rescue organizations were starting to report a growing number of calls from people who were lost or perhaps hurt but not in grave danger. One of the problems of most warning technologies is that safety concerns will lead to many false alarms, which in turn may have the revenge effect of diverting attention from real emergencies. (Up to 99 percent of electronic residential burglar alarm calls to police departments are, as we noted in Chapter 1, false, and may actually weaken police efforts against crime by diverting patrol cars. My own municipality, Princeton Township, requires an initial response by a private guard service—yet another example of how automatic technologies demand more rather than less vigilance.)

Already in early 1994, rescue organizations were reporting multiple calls from climbers and hikers with cellular telephones requesting help in situations that response teams did not consider true emergencies. Sometimes these calls tied up lines that might have been needed for real emergencies. GPS devices might seem to help by making it harder to get lost, but they can also promote the idea that a quick rescue is always at hand. They also may remove the incentive to develop the basic wilderness skills that might still be needed if the new technologies do not work. It is too early to say whether this revenge effect is real or only hypothetical.[41]

New technology can enhance not only performance and enjoyment but safety. Yet often it does not. As we have noticed in medicine, the environment, and the office, what begins as an improvement all too often shifts a problem or even magnifies it. Injury, acute and chronic, may remain constant or actually increase. In part, these consequences follow from spectators' expectations of violence. Helmets, gloves, and padding can all be abused to intensify combat; in fact, the suspension system of the first plastic football headgear became part of the U.S. Army helmet of the Second World War. In part, the revenge effects of sports technology also arise because participants in sports like skiing and climbing appear to want a certain level of risk. In bicycling and baseball, where risk is largely incidental to participation, safety equipment like helmets and breakaway bases has been effective. Not technology but the values of participants and spectators determine the danger of sport.

Even where attitudes help reintroduce risk, revenge effects can be overcome once we realize that the benefits of technology take far more vigilance and attention than we thought. In contact sports this means that officials must be able to make and remake rules and enforce them for safety. And the measures they take, like football's prohibition of spearing, are only as effective as officials' determination to enforce them. It means that everyone from physicians to park rangers and ski patrols has to be more, rather than less, alert to abuses and dangers. Above all, it means that athletes themselves have to recognize that they are participating in more complex technological systems that need more understanding and watchfulness than they may have realized. Today, even leisure can take real work.

11

Sport: The Paradoxes of Improvement

We have seen that technology can both improve and undermine the healthiness of sport—preventing many formerly common casualties while creating new risks of acute and chronic injury. It has influenced even more powerfully the performance and economics of the games we play, and watch.

Before the Second World War, technological limits were as important as the rules themselves. The materials for the best sporting equipment still derived largely from plant and animal sources. Those substitutes that were available generally offered inferior performance and showed little promise of changing the games in which they were used. Known materials and processes were a kind of performance envelope that even skilled metallurgy could not exceed. With the major exception of steel shafts introduced to golf in the 1920s, few had changed much over decades. (Even in golf, steel woods were offered as early as 1919—"Metal Golf heads . . . they never rust!"—but failed to catch on.) There was limited room for new rules because few new designs changed sports significantly.[1]

Again there were exceptions. Recumbent bicycles, allowing cyclists to lean back and pedal far more efficiently, were invented in the 1930s, only to be excluded from official racing by international cycling authorities. Even here, it was not only official conservatism that kept new designs out of the mainstream. A stable recumbent cycle is heavier and more complex than the familiar safety bicycle that has changed so little in a hundred years. Fairings, aerodynamic shells that cut wind resistance, can make it even more competitive with cars and motorcycles—but also closer to them in bulk and cost. Rules could have been modified to admit recumbent cycles to competition, perhaps in a separate class, and public roads could have been opened to their use by providing dedicated bicycle lanes. But a commercially successful recumbent cycle still depends on lighter, stronger, and cheaper materials than are now available commercially.[2]

The fifty years since the end of the Second World War have witnessed a revolution in the materials used in sport, but these have also created new categories of potential revenge effects. Inexpensive plastic fabrics revived the short-lived late-nineteenth-century sport of hang gliding; some of the earliest gliders of the 1970s were made of nothing more than bamboo rods and garbage bag material. With hundreds of fatalities, gliding was for a time one of the most hazardous of known sports, and still is risky despite improved Dacron sails, aluminum construction, and relatively rigorous design and training.[3]

It might seem odd to talk about the improvement of a sport through better technology. Improvements available to one athlete or team sooner or later will be used by all. Does it really matter if everyone has a better score? On the other side, since everyone knows the contribution of technology to improved performance, what would be lost by staying with the old technology?

One answer might be that, as we have seen in the last chapter, many sports technologies have improved safety as they have helped performance. The fiberglass pole would be nearly useless in vaulting if athletes had to land in the old-style sawdust pit; modern foam cushioning is as important for today's twenty-foot vaults as the pole itself. Rubber-coated weights relieve weight lifters of the psychological and physical stress of lowering barbells to the ground; they can now be dropped without worry.[4]

Safety, though, is not the main point. There is a more fundamental side to the role of technology in sports. Bernard Suits, a philosopher, has defined it elegantly. Sports, and all other games, are "goal directed activities in which inefficient means are intentionally chosen." If the only point of an automobile race were to complete so many circuits of a track as rapidly as possible, the Indianapolis 500, Formula One racing, and the Soapbox Derby as we know them would fail. Each limits what participants can do, yet permits unlimited ingenuity within these limits. Some events prescribe spartan simplicity, and others encourage or require prodigious equipment costs and payrolls. But in no race is a competitor entirely free to choose the most efficient way to cross the goal line ahead of rivals. Marathon runners are not free to deploy motorized roller skates, no matter how much more suitable these might be to the task of completing the course quickly. They obey this rule not because they consider externally powered locomotion wrong in itself, but because the absence of a rule against it would turn a marathon into a different event. A recumbent bicycle is more efficient in every sense than a conventional one, and it was banned from international competition *because* of its efficiency.[5]

Inefficiency is not enough. Devising rules is a craft for the "gamewright," who must avoid both looseness and excessive restriction. Technological change may encourage a style that players and spectators feel either enhances or trivializes their effort. Many social rules, and even economically important technologies, appear to be founded on what are essentially games: social rules that create interest and competition independent of ethics or functionality. Think of the fortunes spent on producing no-iron and wrinkle-resistant garments, on all the chemical and mechanical inventions that depend on the convention of looking pressed. Yet this look is socially conditioned, a game with changing rules. In the nineteenth century, when mass-produced trousers were shipped tightly folded in bales, a sharp crease was proletarian and gentlemen preferred the tubular look. Automatic transmissions and stick shifts are now about the same in fuel economy and performance, but some drivers prefer the control of determining a precise shift point while others don't want to bother. Some want the instantaneous numerical readout of a digital watch; others value the graphic representation of elapsed and anticipated time that is built into an analogue watch. Neither alternative is clearly more functional than the other. One makes experience more interesting than the other by increasing one kind of information and reducing another.

Technological change heightens the role of style in sport by increasing choice. Cable television multiplies the number and variety of events, including overseas events, that are available to spectators. George Will notes that rising living standards and the suburban environment have reduced baseball's share of potentially outstanding athletes, with more youth preferring to play basketball or soccer, or participate in other activities. (This is probably the reason for the inability of today's top pitchers to match the velocity and skills of the stars of earlier generations, who as teenagers had no such competition for their time. Elite pitchers are the great exception to the general improvement of athletic skills over time.) Sports throughout the industrial world rise and fall in public esteem and social connotations. Equipment is not the only reason for these shifts; national rivalries, ethnic pride and prejudice, and contrasting personalities are still mainstays of the interest in sports.[6]

Technological change still is crucial. It can do three things. It can increase the potential audience for sport through extending media; 500 million people around the world watched the 1992 Winter Olympics at Albertville, and the number of those who viewed the 1994 games at Lillehammer was probably significantly higher. It can multiply opportunities for watching live sports by controlling the environment; floodlighting transformed baseball, and artificial turf made enclosed multiuse stadiums possible. And it can draw participants by changing the skills needed to play. Sometimes this means a more forgiving design, but at other times a more complex one. In the 1960s and 1970s, narrow-tired derailleur-equipped bicycles, though more difficult to maintain and use than the three-speed variety, helped promote a new bicycle boom in the United States. Of course, the purpose of the derailleur was to achieve the greatest possible variation in mechanical advantage according to terrain, along with the lowest possible weight. But the point is that simplicity and complexity can be equally effective in building interest.[7]

Interest also depends on the profile of fans and participants.

Violence may be a great draw for a sport, but also an obstacle to attracting a genteel clientele either as spectators or as participants. The present rules of boxing began in the studios of "professors of pugilism," who taught the noble art of sparring—as opposed to the brutal spectacle of prizefighting—to the gentry. Gloves, a very simple technology, marked the boundary between upper-middle-class sportsmanship or self-defense and the ruffian ways of the bare-knuckle fighter and his plebeian audience. But the prizefighting fans of the time had little interest in sparring. Even today, college and professional boxing are almost two distinct sports. In England, American football became popular as a middle-class family sport unvisited by football hooligans, while generations of elite American males have celebrated rugby football as "elegant violence" and "a ruffians' sport played by gentlemen."[8]

The elites who set sporting rules understand how important the implications of changes can be. That does not mean that they always act in the interest of either amateurs or professionals, simply that they are aware of the complexity of issues of interest. Better than most people, they see that issues presented as technological usually turn out to be social. A change in equipment or rules usually benefits one group of players over another, whether its value is speed or stamina, brute force or subtlety, distance or control. Any "improvement" in equipment or surface may not only threaten existing records; it may temporarily or permanently replace a class of record-holders. It may be an example of bad efficiency in Suits's sense.

The philosopher Paul Weiss saw sport as the pursuit of excellence, but the dilemma of sports management is using rules and records to maintain interest. Sports is more than people trying to run as far or as fast, to jump as high, or to swim a certain distance as fast as their bodies will allow. It is setting standards that define and redefine excellence and that will encourage men and women to continue training. Allen Guttmann has suggested that an extraordinary performance can actually lead to a decline of interest in a sport; a Japanese archery game was abandoned in the seventeenth century after one master set such a spectacular record that others gave up hope of ever equaling it.[9]

The inventor of basketball, James Naismith, lived to deplore the rigor that expert coaching had brought to a game he had devised as a casual amusement to pass the time between outdoor seasons. While the technology of basketball has changed little since Naismith's day—except, of course, for the shoes—the technological apparatus surrounding the game has been immensely transformed in the last decade. In fact, constant statistical feedback is no longer limited to the coaches and players. To many spectators it has become an important part of the game. The tennis writer David Higdon, lamenting the "stagnant and dull" state of his sport in 1994, summoned officials to follow the National Basketball Association (NBA) and its "marketing strategy geared toward generating, educating and entertaining fans." This meant, among other things, intensifying the flow of information. Higdon and countless other fans of the Portland Trail Blazers feel their enjoyment of the actual play enhanced by "a board [that] flashes up-to-date shooting percentages, free-throw percentages, number of turnovers, etc." Portland has a " 'Hustle Board' which compares the number of blocked shots, rebounds, and steals between the competing teams." On the other hand, Higdon favors keeping human line judges over electronic line systems lest tennis turn into "a Nintendo game." Interest is thus a strange thing. People who spend their time analyzing numbers for a living find their recreation incomplete without a steady flow of numbers to analyze.[10]

But if Higdon is right, they are not so fanatical about the accuracy of those numbers; they would rather have from time to time what the late Herman Kahn called a "warm human error." In fact, few professional sports still use instant video replays to help make or verify official decisions. The NFL dropped it in 1992. Officials resented its challenge to their authority and decisiveness; fans and broadcasters resented the time it added to games. The video record may well turn out to be as controversial as whatever memories the officials and spectators may have of the original call—as the Rodney King and Reginald Denny cases showed in very different contexts in the courtroom.[11]

In spectator sports, a combination of extraordinary performance and superb media technique may paradoxically decrease interest in the sport. According to Russell Davies, an English sports journalist, the sweeping view of today's cameras eliminates the blur and lost motion that was part of the excitement in older television newscasts of skiing. Camera angles also make challenging runs look far flatter and easier than they are. To make things worse, the fastest, most efficient skiers usually can't afford to have a distinctive style, any more than the fastest sedans can deviate much from an optimal aerodynamic profile.[12]

On the positive side, machines can enhance a contest where drama has been flagging. Consider chess. In this game, as in others, changing the guard is not a bad thing. When competitors dominate their sport for too long, the interest of fans, and ultimately money, is lost. Bobby Fischer challenging the Russian grandmasters was Cold War theater at its best. Had Fischer not dropped out of tournament chess to spend years as a recluse, but instead maintained his championship without a serious contest, he might have harmed the game in the West as much as he had once helped it. By the mid-1990s, Gary Kasparov, widely considered the best player who has ever lived, seems to have shut out most other grandmasters from the world title, raising once again the specter of stagnation through excellence. The answer to this dilemma is another paradox: the machine as contender. Chess computers with names like Deep Thought and Big Blue, rather than human players, are adding new and essential uncertainty to the game. With computer power doubling every eighteen months and the human game improving only marginally, the ultimate contest may well turn out to be machine versus machine and not machine versus human.

The strongest models have now moved up in rank to the master level and will soon be able to defeat international masters. The threat of a computer as world chess champion, far from reducing interest in the game, gives it a new poignancy. It is true that the strongest chess machines and software programs still win mainly by brute-force tactical computation, by calculating more moves ahead. They still cannot think creatively and strategically. But this very flaw, if it is one, enhances interest in the personalities of masters and grandmasters, who thereby appear all the more human.

The most exciting scenario for chess, and a kind of revenge effect of the chess computer, would be a series of increasingly powerful and sophisticated machines matched against a world champion of Kasparov's stature, with its postindustrial echoes of John Henry's fatal contest with the steel-driving machine. Even weaker chess computers and programs for the consumer market may also make the game more interesting by allowing human players without access to a club to match their skills against an expert opponent. The rise of the power of computer chess in the early 1990s did not seem to hurt serious participation; membership in the United States Chess Federation grew from 58,000 to 72,000 between 1991 and 1995 alone, and the new importance of the electronic game encouraged the semiconductor manufacturer Intel to become a major sponsor of professional matches.[13]

Technology can increase the interest of other sports provided it retains or reintroduces the vital inefficiency. Old record-holders, like silent screen actors after sound came in, must learn to play a different game. When equipment changes, a new group of athletes may seize the highest levels, as they did when aerodynamic javelins increased the rewards for technique as opposed to strength; when the new designs were excluded in 1986, for reasons we shall see, the power throwers moved back on top.[14]

But technology can turn from friend to an enemy of interest. We have already seen the problems that computers create when they help make people more efficient. Everybody competes better, and the game becomes increasingly harder to win. In his essay "Losing the Edge," explaining why baseball has no more .400 hitters, Stephen Jay Gould argues that decades of improved technique have reduced the gap between the strongest and the weakest professional players. Of course, technology has not driven all of this improvement. Just about any activity, developed carefully and systematically, can be improved as experience accumulates. Well before they applied new technologies, players gradually started to discover "optimal methods of positioning, fielding, pitching, and batting," reducing thereby the extremes of performance. But while technology did not begin the search for optimal performance, it has greatly extended it. Motion pictures and video, more recently digitized for computer analysis, have allowed frame-by-frame study of the techniques that separate masters from average players. Dynamometers and electric measuring instruments have helped raise the performance of baseball pitchers to the limits of safety. (As we have seen, though, some pitchers of the past, like Bob Feller and Walter Johnson, could exceed present-day performance even without electronic analysis.) Endurance and strength training on special equipment can improve scores more than conventional practice alone. Computers with databases of athletes and plays can optimize tactics and strategy. The scientific shaping of athletic accomplishment, like the convergence of the aerodynamic profiles of sedans, pushes all athletes toward the limits of what they can achieve.[15]

Is there anything wrong with everyone getting better? Gould himself insists that the recent systematic pursuit of the best technique has improved the game, making it even "more balanced and beautiful." If knowledge is good in itself, then the study of optimal performance should yield benefits. Who would deliberately forgo the most effective play to keep a game more interesting? But technology does not just refine. It intensifies. We have already seen that whether in the hospital, in the office, or on the highway, more advanced technology increases rather than reduces expense. It doesn't save the labor of training; it makes practice more rather than less necessary. As knowledge spreads, as the optimum is better understood, as athletes are recognized and recruited at younger ages, several things happen. Training escalates in cost, if only because it has to start earlier. On this point, at least, the former command economies of the East and the market societies of the West converged.

At every age, more effort and cost are needed to reach the highest level in any sport, whether as an amateur or as a professional. An Olympic-caliber figure-skating career costs as much per year as a Harvard education. But it lasts at least ten years, rather than four, and few candidates ever win the Olympic medal that offers the only hope of recovering the costs of training. Differences among the top performers seem smaller and smaller. In August 1991, the American runner Carl Lewis won the 100-meter sprint by 0.002 second, establishing a new world record of 9.86 seconds. A new hybrid imaging system allowed ranking Lewis and the runners-up, even though the first six runners were all within the smallest gradation of a traditional stopwatch, 0.11 second. The luge event in the Winter Olympics is timed in thousandths of a second; in Lillehammer in 1994, the German men's luge champion Georg Hackl defeated the Austrian 1992 winner Markus Prock by 0.004 second. When style reaches an optimum, when participants or viewers believe there are no more surprises, when technical strength begins to overwhelm personality, then the intensification of performance may begin to work against a sport's popularity.[16]

Just as professional football's kickers developed incredible consistency, they were punished with revised conversion rules; by becoming too good at what they did, they were endangering the interest of the game. They were too efficient. As retrained soccer players from overseas, they also represented a level of specialization (and globalization) with which many fans were uncomfortable. Still, they were applying the logic of rigorous selection and training—training that depended on all kinds of technological analysis and support. In most other human activities, organizations that hone some decisive skill and develop a cadre of specialists in it will be praised and rewarded. In professional sports they may actually be set back. Officials fought one revenge effect of improvement—diminished interest—by changing the rules. In fact, the national biases of subjective judging in Olympic events like figure skating and gymnastics seem to heighten interest in these sports.

Sometimes a rule change is in the interest of more rather than less action. In men's college baseball, the admission of aluminum bats, with their greater hitting power, has encouraged an offensive style of play that most coaches believe adds interest to the game. In 1993 the National Collegiate Athletic Association (NCAA) went a step further, revising the rules for women's baseball to permit a livelier, optic yellow ball with a polyurethane rather than a cork center. It was not a change in technique by players or coaches but a social and media transformation that stimulated the new rule; the less responsive ball and low scores of the traditional game were standing in the way of attractive cable television offers.

Changes that would offend fans in some cultures are positive attractions in others. While Japanese professional teams, like their American counterparts, use only wooden bats, aluminum alloy has long been the amateur favorite. For all the genius of Japan as a woodworking nation, for all its skill with natural materials, its high school players and spectators love the ping of a ball against a metallic bat and don't seem to share the sentimental attachment of older Americans to wood. Some early designs, lacking American-style sound insulation, produced sounds exceeding ninety-six decibels, a long-term hearing risk. Some new designs can suppress the sound, but Japanese teams see no need. In fact, every culture has ideas about what makes a game interesting or dull, and technological change can either enhance or threaten interest. When it seems to diminish it, no matter how much performance otherwise benefits, opinion holds the innovation to have revenge effects.[17]

The continuing evolution of the javelin shows how ambiguous technological progress in sports can be—and how the material and form of equipment, the recruitment and training of athletes, the dimensions of playing space, and even the safety of athletes in other sports all affect one another.

The fiberglass pole may be the most dramatic example of how a technology can transform a sport. Fiberglass, first used in sports for deep-sea fishing rods, was approved for pole vaulting in 1962. At first it appeared only as a way for the vaulters of the day to set new height records by storing more kinetic energy in the pole. But in fact it did much more. Ultimately it transformed pole vaulting into a more intense, riskier, but above all more interesting sport. It also removed vaulting from its presumed origins as a hunting maneuver in wet terrain and brought it into the twentieth-century realm of pure performance. Foam rubber cushioning, as writers about the sport point out, was as important as the pole itself; nobody could fall repeatedly into a pit of sawdust from eighteen feet or higher without serious injury. But the soft return has a revenge effect of its own. If an athlete misses the cushion, the consequences will be serious, and the fiberglass pole responds temperamentally.[18]

For many athletes the new poles must have been doubtful improvements. The biomechanics specialist Peter M. McGinnis commented: "It's scary. If you can't get enough bend in the pole, if you're not right on line at takeoff, if there's a little horizontal velocity to one side or the other, you might not land in the pit." Old-time vaulters from the stiff-pole era shake their heads about how unpredictable jumps have become. But there is no doubt that the pole vault has gained interest because of its new complexity and danger, proving (to borrow from computer jargon) that one person's bug can be another's feature. A successful vault is one of the tensest, most thrilling, most intricate maneuvers in sport. Far more than the bamboo-pole vault, it is almost made for high-speed photography and slow-motion video. A vaulter frozen at the beginning of an ascent, when stored kinetic energy is being released, seems to reveal the passion for maximum performance as do few other athletes. The Russian Sergei Bubka emerged as the new type of champion, clearing twenty feet in 1991: a combination of speed, strength, skill, and daring.[19]

The javelin throw (a Coubertin revival) shares with the more celebrated but much younger pole vault a frequent side effect of a change in sports technology: a changing of the guard in favor of athletes with a different mix of skills. As poles were growing springier and vaulters more powerful, javelins were moving in another direction. Designers were building them with profiles which gave them an aerodynamic lift that would have astonished the ancient Greeks who introduced the sport—and who were shooting at fixed targets rather than for distance. An optimum javelin throw demands a strong lift with the tip of the spear pointing upward until relatively late in its trajectory, when it tilts downward to earth. Flat and tail-first returns don't count, a rule not always easy to judge.

Because the new javelins were especially sensitive to initial conditions—and in this they did resemble the fiberglass poles—the biggest and strongest athletes did not necessarily remain champions. Control and judgment began to count as much as hurling power. The javelin throw was becoming much more interesting, but there was a revenge effect in that it was posing new risks for spectators. A standard track field is a hundred yards long, and javelin throwers may perform surrounded by runners. By the early 1980s, champions like

Tom Petranoff of the United States and Uwe Hohn of the German Democratic Republic were exceeding 100 meters. The world record then established, which still stands, was 104.8 meters. While the instability of the fiberglass pole has as yet endangered only the vaulter, the unpredictability of a javelin's flight began to worry judges and spectators alike. Beyond new records, many throws did not end with a neat entry of the tip into the ground; the spear often fell flat and continued to travel—slithering over the grass, as one official put it. Javelins could also make unexpected lateral moves. Other track athletes began to give javelin throwers a wide berth. At the 1984 Olympics in Los Angeles, a Norwegian thrower's javelin nearly landed among a group of judges preoccupied with a racing event at the other end of the field. It is no surprise that changes soon came.[20]

In 1984 the International Amateur Athletic Federation (IAAF) issued a new set of rules that moved the javelin's center of mass toward the front without permitting any advance in the center of pressure that gives the javelin its lift. This decision appeared to wipe out thirty years of careful design evolution and aerodynamic study, but it added an element of interest of its own. The advantage shifted again and the power hurlers soon came back into their own. Even more interesting was the reaction of the engineers who helped design the aerodynamic javelins. Far from abandoning their goals, they reformulated them. They began, and are continuing, to nudge javelin performance back to the old and possibly dangerous levels by designing to the letter of the new specifications. If they succeed, officials may have to revise the rules yet again. All this might sound futile, but it really has a positive side: the problem and controversy interests engineers, journalists, and athletes, and continues to offer some hope of resuming the quest for new world records.

Tennis and the Revenge of Technological Revolution

By the standards of professional sports, tennis officials once were casual about equipment. They strictly policed the dimensions and conditions of courts and nets and the specifications of balls, but well into the 1970s they left racket design and dimensions to the imagination of athletes and manufacturers. As late as 1977, the president of the U.S. Tennis Association could declare, "You can play with a tomato can on a broomstick, if you think you can win with it." This freedom, far from encouraging a profusion of fanciful designs, was permitting a slowly evolving uniformity. Equipment was not so important as skill—tennis professionals with frying pans can achieve excellent results against lesser players with rackets—but this was not the whole reason for conservatism. A similar advantage for skill as opposed to equipment exists in most sports and has not necessarily stopped invention.

What seemed to set the tennis racket apart was the properties of the materials that went into it. Cutting ash and beech into strips, laminating them, and more recently reinforcing them with a variety of plastics improves racket strength, but these techniques could not overcome an apparent natural size limit of about seventy square inches. A larger wooden or even aluminum racket face would tend to break with the force of the hardest shots, or reduce the speed of play because of weight. (Heavier rackets add little speed to balls, because a slower swing cancels out most of the advantage of the added mass, while lighter and faster-moving rackets do add speed significantly.) Steel rackets had been tried as early as the 1920s—a Birmingham manufacturer introduced a model using piano wire—but it was Jimmy Connors, who won in the late 1960s with the steel Wilson T2000, and Pancho Gonzalez, who used an aluminum Spalding Smasher at Wimbledon in 1969, who signaled the approach of an age of more rapid evolution of rackets.[21]

The real revolution in materials, however, did not begin until nearly ten years later, and it started at the bottom: with rackets designed to make the game easier for less skilled players. Howard Head, an engineer who made millions developing and producing laminated skis, saw that many amateur tennis players were frustrated by their inability to hit the ball consistently with conventional rackets. The early Wilson and Spalding metal rackets were of little help to the majority of players. Head realized that the absence of official specifications created a unique opportunity. A patent issued in 1974 for his aluminum model (marketed as the Prince in 1976) gave him a legal monopoly on oversized rackets. The original Prince has a surface of 130 square inches, nearly twice the area of conventional models. While sports physicists and engineers recognize three different plausible definitions of a "sweet spot," a zone of maximum efficiency in hitting a ball with any object, there was no doubt that the Prince had a significantly larger one than conventional models. (Because fewer shots twisted or vibrated players' arms, some believe larger rackets have reduced the incidence of tennis elbow, though this is hard to determine. In the early 1990s, half of all amateur frequent players still were reporting symptoms eventually. Midsized racket heads do reduce vibration and twisting, but the largest ones may twist more, and stiffer rackets are poor absorbers of shock.)[22]

Equipment that will forgive errors has never mattered much to professionals in any sport; nobody reaches top-level play without consistency. Those who first adopted the Prince racket were well-off, middle-aged, competitive but ordinary players—like Head himself. In fact, if the larger sweet spot had been its only benefit, bigger rackets might have suffered from a kind of prosthetic stigma in the face of traditional macho designs. But oversize construction has another, unexpected benefit that appeals to professionals. The new rackets—both the Prince and later models made of fiberglass, boron, graphite, and Kevlar in various combinations of materials—are both lighter and stiffer than traditional models. They permit velocities up to 30 percent greater than the old designs. And this improvement in performance had serious consequences for the sport.

(The market for more forgiving equipment can be anything but gentle. A slightly oversized, slower-moving tennis ball, the Wilson Rally, flopped among its intended market of older and less skilled amateurs in the early 1980s. Its weight conformed to regulations, but it felt heavier than standard balls when it hit the strings. Wilson soon had to withdraw it.)

Within only five years of the Prince's introduction, larger rackets had taken over tournament play. The move began with Pam Shriver and helped make women's tennis in the 1970s and 1980s one of the few female sports that could compete in media attention and cash with their male counterparts. But the effect on the men's game was much more complex. For individual stars, there was no alternative. Some leading male professionals were determined to show that they could win tournaments with wood—but they failed. John McEnroe was the last to use a wooden racket at Wimbledon in 1982; when Björn Borg tried one at the Monte Carlo Open in 1991, he lost twelve of seventeen games to a Spanish player ranked fifty-second but playing with a graphite fiber model. By the early 1990s, wooden rackets had become a niche product, available mainly from a single manufacturer in Cambridge, England.[23]

The triumph of metal and composite rackets, combined with the entry of stronger and better-conditioned young players, transformed the men's professional game. By the 1990s the sometimes monotonous serve-and-volley game was a thing of the past, with only a few of its specialists left on the tour. On the other hand, the new rackets multiplied the advantage of a powerful serve, especially on a fast surface like grass. Serves clocked at over 100 miles an hour became routine, and a number of top players have even been able to surpass 120 miles an hour. These results are all the more impressive because most top professionals are not yet using the most radical designs, exceptionally stiff wide-bodied rackets that they feel don't allow enough topspin. A growing number of serves are aces that no player could return, and more and more games have become serving contests. In the 1994 men's Wimbledon tournament, Pete Sampras defeated Goran Ivanisevic with a magnificent display of technique, but his 125-mile-per-hour serves bored many fans. The longest rally was just eight strokes, and the correspondent for the *Guardian,* David Irvine, appealed for action "to save the grass-court game from self-destructing."[24]

As of the mid-1990s, every proposed solution to the revenge effects of larger rackets in men's professional play appears to have unintended consequences. Higher nets or less lively balls in tournament play would affect not only the service but all other shots. Different court dimensions for professionals and amateurs would confuse training and make thousands of courts unusable at least part of the time. Requiring players to have both feet on the ground while serving would rob professionals of the benefits of countless hours of practice—possibly giving an advantage to some competitors better adapted physically to the new rules. New restrictions on rackets would not only raise questions about the usability of older models but invite U.S. antitrust action by manufacturers who might consider themselves penalized. And converting Wimbledon from grass to clay might affront tennis traditionalists more than any new racket design ever could.

The irony of the new rackets goes even further: they are not as profitable for the manufacturers as they once were. The large racket, for all the benefits it may give the average player, did not do very much for the tennis boom of the 1970s. According to the records of the Tennis Industry Association (TIA), the number of tennis players had already peaked in 1974, two years before introduction of the Prince in 1976. Participation remained stagnant, then began a sharp drop through the early eighties, bottoming out at ten to eleven million adults by the middle of the decade. This is not entirely surprising; a higher-performance product often needs a broad base of consumers eager to upgrade. What is unexpected is that participation continued to decline so sharply despite greater ease of learning and play. The TIA attributes the slump of the 1980s to the rise of aerobics and health clubs, but it still is not clear why these should have competed so successfully to the detriment of tennis but not of other outdoor sports. Could one reason be the higher price of the new rackets? Less affluent players might simply have rejected the prospect of a new $150 investment just to remain competitive. This cost would not, of course, deter a serious player but might give casual ones second thoughts. And some otherwise satisfied with their old rackets might have found the sweet spot unacceptably small, especially once their opponents began to play with large-head models.[25]

Just as the first boom in tennis ended before technological innovation, a recovery of participation began around 1985, three years before the introduction of wide-bodied rackets in 1988, with thinner but deeper frames that added stiffness—once more at a higher price point of $200 to $250. There was no doubt that these rackets made learning easier for beginners and gave serious players stronger shots.

Compared with wood they had fully twice the hitting area and were often twice as stiff, yet weighed 35 to 40 percent less. In the early 1990s the industry was expecting to regain something of the popularity it had reached at its peak.[26]

Once again, though, technology failed to save tennis. Instead of continuing to rebound, the sport was foundering by the mid-1990s, despite but also partly because of its success in innovation. The number of players continued its slow recovery from the trough of 1985, reaching 25 million by 1993, yet the sale of tennis balls—a measure of activity—dropped significantly between 1990 and 1993. Manufacturers and retailers were quick to blame inadequate marketing, but the game's explosion in the 1970s appeared to owe little to marketing campaigns, and even companies as adept as Nike have not been consistently successful. Whatever the reason, racket manufacturers began to slash prices in the mid-1990s and stores cut back on their space for tennis equipment. Meanwhile the higher quality of the new equipment seemed to work against the industry. The *New York Times* reported that the new metal rackets were lasting far longer than wooden models and needed less frequent restringing. This has not stopped the introduction of still more powerful rackets, but these show little prospect of bringing back the boom of the 1970s.[27]

Tennis shows how unpredictable technological change can be in any sport. For two decades, equipment improved for the average player as for the professional, yet participation never approached the peak of the wood-racket era at the end of the 1970s. The added power of male professionals did not seem to increase the game's appeal to spectators; if anything, the intensification of the game began to bore them.

Golf and the Advantages of Rationing Progress

Like tennis, golf is a sport that demands scarce resources—costly suburban or resort land, and especially time. While middle-American enthusiasts queue up overnight for inexpensive municipal links, the affluent pay tens of thousands of dollars for club memberships that reflect the high land and labor costs of maintaining a good course. In Japan, membership in some of the leading clubs has cost well over $500,000, and stood at twice current levels just before the golf boom ended in 1990. (Currency fluctuations make precise conversions difficult.) And in most climates a thriving golf course demands vast quantities of both water and chemicals. A single course may absorb as much water per day as hundreds of households, and up to seven or eight times the weight of pesticides per acre as nearby farmland. It takes surprisingly intensive chemical applications to produce serene bucolic landscapes—a far cry from the windswept seaside links of St. Andrews Old Course in Scotland where the game began.[28]

Golf's critics see in the superb landscape of a well-kept course not just an expense-account Eden but a green monster, devouring cropland and wildlife habitat. And for both golfers and employees, a serious concern is pesticide exposure. Greens require massive doses of pesticides and herbicides to offset the stress of close cropping. (Grounds crews once made liberal use of arsenic and other doubtful chemicals to achieve verdant greens.) Golf-course superintendents are said to have suspiciously high cancer rates. But not even the alleged higher health risk has so far interfered with the game itself, although it is changing practices in industrial countries and stiffening resistance to the spread of new courses in Asia and elsewhere in the high-growth developing world.

Golf is remarkable, in fact, as an experiment in managing technological change—possibly the most successful case in American life of the suppression of innovations that would make life easier. A friend, a mechanical engineer with some success in tennis racket design, began thinking of golf club improvements some years ago but was warned by a colleague that the U.S. Golf Association (USGA) would never accept them (unlike the U.S. Tennis Association, which had admitted his new models without serious objection).

Over the long term, golf has been one of the sports most open to new technology. The golf ball has experienced at least three technological revolutions. The first was in the mid-nineteenth century, when gutta-percha, a form of hard rubber, replaced the traditional boiled-feather core of golf balls. The second was late in the nineteenth century, when balls were made of three layers of rubbery materials (a rubber core surrounded by tightly wound strings, covered with gutta-percha or the product of another tropical tree, balata). By the late 1960s, manufacturers were introducing yet a third design: a synthetic cover of the DuPont thermoplastic surlyn over a plastic core. The two-piece surlyn ball travels farther after a typical amateur stroke and does not cut as easily; the three-piece balata-covered ball, however, still offers more spin and superior control.

Despite the USGA's reputation for technological conservatism, its policy toward ball design has been remarkably pragmatic. Officials, professionals, and amateurs all realize that if everybody could hit substantially and consistently longer drives, the layout of the average course would have to grow. The oldest and richest courses, with club memberships to match, might become technologically obsolete with no room for expansion. The solution has been to permit great variation in design and materials while regulating performance. Balls must be at least 1.68 inches in diameter, but they may be as large as will fit into the 4.25-inch hole. They must weigh at most 1.62 ounces, though they may be lighter. When struck with a mechanical driver based on the swing of the champion Byron Nelson, they may not exceed 296.8 yards of flight.[29]

While these rules for balls sound fussy and even reactionary, they actually are more liberal than they seem. Golf is one of the few competitive sports to let each player choose a ball from several possible combinations of materials, with two styles of covering that have distinctive playing characteristics. While the dimensions of the ball are more or less fixed, the geometry of the dimples that reduce turbulence and increase distance are subject to almost infinite experiment—provided, of course, they don't perform *too* well. Amateurs and professionals voluntarily obey the same standards, though players with higher handicaps nearly always choose surlyn-covered balls for greater distance and pros overwhelmingly stay with balata covering for its superior feel and control. For the amateur golfer, unlike the tennis player, there is always the hope that the design of a new ball will improve his or her game a notch. Yet there is also the risk that any radical improvement might destroy the sport's challenge. The USGA rules avoid the potential revenge effects of excessive improvement by setting absolute limits to performance. But they also keep the game from becoming static and dull by permitting and even encouraging small changes that approach limits slowly, decade by decade. Restrictions also have the fortunate side effect of promoting consistency rather than maximum possible performance as a goal, so that the difference between the fastest and the slowest ball in a batch has shrunk dramatically with better quality control. More predictable performance may be more satisfying to players than better performance.[30]

The 120-plus pages of *The Rules of Golf* are a paradox. Despite their scrutiny of equipment, the USGA and its Scottish partner, the Royal and Ancient Golf Club of St. Andrews, are astonishingly liberal about the design of courses. The hole must be a standard 4.25 inches (108mm) in diameter and at least 4 inches (100mm) deep, but otherwise definitions are minimal. A bunker consists of sand "or the like"; a water hazard is "any sea, lake, pond, river, ditch . . . and anything of a similar nature." Golf course design ranges from the indulgent to the stringent and unforgiving, though in choosing venues for the U.S. Open, the USGA clearly prefers the challenging variety. Where tennis authorities regulate court and net dimensions rigorously and equipment more casually, golf officials take the opposite approach. The result is that each course has dimensions and a personality of its own, increasing the game's interest.[31]

Clubs show the surprising benefits of limiting improvement, of inefficient means in Suits's sense, even more than balls do. Like balls, clubs have changed several times in the last hundred years, first with the introduction of steel shafts and heads in the late nineteenth century, then with the development of specialized and graded clubs—woods, putters, and especially wedges—in the early twentieth century. Golf grew impressively, in fact, when some of the new designs of irons and clubs made it easier for players to hack their way out of sand traps. Had the rules been frozen, perpetuating the specifications governing play at the beginning of the twentieth century, golf would have remained an eccentric sport for the rich. As it is, it manages to retain just a touch of that odd character. A professor of logic at Oxford, after a round with the future golf writer Horace G. Hutchinson, evaluated the sport as follows: "putting little balls in little holes with instruments ill adapted to the purpose."[32]

Sports scientists have actually quantified this mismatch. A legal clubhead is far too light to absorb the maximum possible energy from a golfer's swing—ideally it should weigh several kilograms—yet it also is much too heavy to impart to the ball the energy that would give it the highest possible velocity—ideally only a twentieth of a kilogram. In other words, like almost every other engineering solution, the design of clubs represents a compromise. It is also open to radical improvement, but with a revenge effect: a changed game that would either lose its challenge or require the costly redesign of courses. As Frank Thomas, the technical director of the USGA, has argued, making the game easier and thereby improving everyone's score requires no new technology. (Of course, hooks and slices frustrate most casual players more than missed putts do, but putter designs seem to be the favored objects of golf inventors, and radical new putters are always arriving for examination at USGA headquarters.) Doubling the diameter of the hole would be enough. Innovation has to preserve the sense of challenge while permitting small and gradual improvement. The interest in the game has to be managed.[33]

Thomas points out that golf is not competitive in the same way that most other sports are. Golfers, unlike tennis players, compete more against themselves than with others. And ability is as much mental as physical. Hundreds of touring professionals practice their shots with stunning consistency. Within this elite, individual response to a particular course, to the weather of the moment, and of course to competition itself contribute more to the outcome of a game than differences of technique alone. At all levels, golf depends far more on coordination of mind and body, on refinement, and on concentration than on strength or stamina. Champions can retain their standing well into middle age. What players lose first is not driving but putting. A psychological block called the yips, with no known neurological basis, unaccountably ruins even easy shots. It is probably because of the yips that so many radical putter designs appear. Thomas keeps in his office one design probably born of someone's frustration: it is not really a club, but more like a portable pendulum, and of course is barred by *The Rules of Golf.* (The long putter, held against the sternum and despised by many golfers, is still legal.)[34]

What makes a putter or any other new design work, Thomas and other golf researchers believe, is not the intended mechanical effect at all but a placebo effect. The golfer's unconscious mind knows how to swing. It is the conscious mind, with its anxieties, that throws the player off. What new technology does is to liberate the real golfer by disarming consciousness and letting underlying knowledge of the game take over. Invention does the work of meditation, and players credit manufacturers rather than themselves—for a while. As Thomas also points out, after a golfer is habituated to a new club, consciousness interferes again and begins to spoil things. Players sometimes go back to their attics years later, dust off an old set of clubs, and find they have the same magical powers that the new ones did when first bought. But of course these powers do not last, either. Thus golf has two contradictory myths: first, that the USGA is holding back innovation that could improve the sport, and second, that today's equipment is so good that skill is disappearing from the game. (The second complaint was appearing in golf magazines even before the First World War; one writer warned in 1907 that a coming age of three-hundred-yard drives would demand longer courses.)[35]

In 1994 the magazine *Golf Digest* tested the latest clubs and balls against twenty-five-year-old models supplied by the manufacturers. The old-style balls were fresh, made to original specifications by the Titleist company. They found that after tens of millions of dollars spent in developing and marketing new equipment, there was little difference in performance for professionals. A combination of new balls and new clubs produced hardly any additional distance from the tee, mainly because the balls carried about sixteen yards more in the air but rolled about thirteen yards less once they hit the ground. Modern drivers with oversize metal-alloy heads and graphite shafts were no more forgiving of off-center hits than 1970s laminated-wood drivers with steel shafts. On balance, there were slight benefits under certain conditions, but no revolution. Thomas's statistics confirm this. The average winning score on the U.S. Professional Golf Association Pro Tour has been cut by only one stroke per round per twenty-one years, and most of this improvement seems to have occurred in putting. Better recruitment, conditioning, and training rather than new clubs or balls are probably behind the gain in performance.[36]

Golf shows a paradox of technological conservatism: it has grown because its governing body has protected it from too much improvement. Politically conservative golfers who probably blame federal and state regulators for holding up new technology do not seem to object to the benevolent private despotism of the USGA. Of course, manufacturers take the USGA to court on antitrust charges from time to time. But somehow no lawsuit has ever resulted in a change that has transformed the game itself, partly because regulation is also in the best interests of the big manufacturers, the clubs, and the professionals.

Of course, it is possible that more men and women would take up golf if it were easier. It is said that three-quarters of the two million men and women who try it every year give up. But the present golfing population is known to be loyal club members and regular equipment buyers. Each year $5 billion worth of balls and clubs are sold in the United States. An invasion of neophytes armed with, say, pendulum-style putters might be a bonanza for the company that introduced them, though the example of tennis suggests that improved equipment may not help a sport's popularity. But if changes in the game drove out many regular club players, the new equipment could have a big financial revenge effect.

Part of the USGA's success in managing change—and thus interest—is in the attitude of the club players. Thomas has written, "[G]olfers have an intuitive understanding of a need for rules which will protect the traditions of the game and preserve the challenge it offers. This is the invisible bond between golfers and the rules-making bodies. . . . It is understood by the administrators of the game, as it is by the participants, that a golfer's *needs* and *wants* differ fairly dramatically at times."[37]

In setting rules for equipment the technical staff of the USGA are acutely aware of the unintended consequences of specifications. Balls are limited not only by minimum size, maximum weight, and maximum coefficient of restitution; there is also an Overall Distance Limit (ODL) even for otherwise conforming balls. Manufacturers still have room for improvement; new machinery can make balls approaching the limit closer than ever before and with greater consistency. Yet no brand can get a definitive advantage over any other. One exception is telling. In 1977, a company called Polara Enterprises introduced a ball with an asymmetrical dimple claimed to correct hooks and slices. The USGA argued that this design would change the game significantly, and it altered the specification to outlaw it. It took ten years of litigation and nearly $3 million in legal expenses and cash payments to Polara (which charged collusion with larger manufacturers), but the ball was withdrawn. The duffers who presumably would have benefited from the ball mounted no movement to support Polara. They didn't really want the game made less frustrating, and they evidently preferred the reign of private rulemakers to the marketplace. (It isn't clear that the Polara would have carried the market in any case; one golf journalist who later tested the ball reported that "the patented 'gyroscopic' effect makes hitting the Polara like hitting a can of Del Monte green beans.") And there was no rule against other new configurations of dimples. Even as the Polara case was moving through the courts, a dimple war broke out among the major manufacturers. They raised the ante from the standard 330 or so dimples per ball to 384, 392, and even 492. They reconfigured their dimples from the familiar "Atti" pattern into dodecahedron and icosahedron arrangements. All these changes conformed to *The Rules of Golf,* and all failed to affect play significantly.[38]

Clubs, unlike balls, need no prior approval and testing for admission to play. But since they can also be excluded, prudent manufacturers submit them for evaluation. For clubs, the rules become deliberately subjective. The head must be "plain in shape." This, for example, really means that it must look like a golf clubhead. Defining plainness precisely would have the revenge effect of promoting a search for loopholes. Likewise the requirement of a "straight" shaft prevents a manufacturer from incorporating a slight offset that still meets official tolerances. (A really tough quantitative specification, on the other hand, would raise the cost of clubs by forcing manufacturers to adhere to close tolerances, hindering the game's expansion.) The USGA initially fought the design of the Ping Eye 2 irons introduced by the Norwegian-born engineer Karsten Solheim in 1988; their closely spaced square grooves did not look like a radical change but imparted 20 percent more spin. Fortunately for both Solheim and the USGA, tests showed that in actual play the square grooves did not really lower players' scores. The USGA settled with Karsten, admitting the new irons but securing an agreement that the grooves would be more widely spaced in future models. It was in the Professional Golf Association (PGA) that opposition was strongest, led by Jack Nicklaus and other top professionals. They claimed the new design would reduce the skill level needed to play the game. In April 1993 the PGA also settled with Solheim, permitting the Pings, after calculating the ruinous cost of even an ultimately successful defense. (The U.S. district judge who was to have heard the case reportedly was a Ping user.) In 1994 the pro tour was back to following the rulings of the USGA.[39]

Many club golfers and golf writers acknowledge that new equipment may not change the professional game very considerably; after all, professionals in any sport played with a racket or club can work with small sweet spots. But surely, they argue, technology works differently for the average player, saving shots that would be ruined by off-center hits, at the very least adding a few good drives per round. The extra size of metal woods and the perimeter weighting of investment-cast irons would seem to give average players an extra chance, just as the oversize rackets did for tennis players. Golfers must believe so or they would not pay $1,500 to $2,000 or more for sets of top-brand clubs like Karsten, Callaway, and the offerings of Wilson, Dunlop, and MacGregor.

New designs would not have become as popular as they have without warm word-of-mouth. Amateurs feel more comfortable with this equipment. Yet as with professionals, the performance benefits of new designs are elusive. The USGA reported in 1993 that the average handicap had remained unchanged at a little over 16 ever since 1980—all through golf's equipment revolution. A typical rated golfer, one who has had his performance computed according to a complex formula, still takes about one stroke more per hole than the norm for an expert. These players are, of course, an active minority of 4 million out of 25 million who play some golf, but they are probably also the most likely to spend the money for the new clubs.[40]

It is hard to imagine technological change working out so well. Golf authorities preserve the integrity of the rules while allowing measured change. Players think they are getting better but really aren't: if they were, they would alarm USGA officials and would probably be unhappy that their results depended on technology and not on themselves. Manufacturers submit to the rulings of the USGA but gain tens of millions in new sales and are protected in their turn against competitors hawking radical designs not "plain in shape." Golfers are free to form and join clubs that admit disapproved equipment, but they haven't. Even garage inventors don't seem discouraged. Technological conservatism may stop most of them in their tracks, but it probably raises the payoff for those who persevere. Karsten Solheim has a reported personal fortune of more than $400 million; Callaway Golf announced profits of over $19 million in 1993 and was rated one of the twenty or so fastest-growing American corporations.[41]

Had participation in golf declined over the last twenty years, journalists would have no trouble finding explanations, as they did for tennis. Too hard to learn. Too slow. Long waits for facilities. Competition from aerobics and other fitness sports. They would point out how, even with all the modifications that have made courses more suitable for spectators and television, the pace of the game has little media appeal. And they would point to the USGA's opposition to significant change in equipment design.

Of course, nobody writes such things because golf has flourished. While the early-1990s recession hurt many second-tier private golf clubs, neither the famous old clubs nor public courses suffered any loss of demand. To the contrary, total rounds of golf increased and the number of golfers grew by 20 percent over five years from 1987 to 1992, even though the game remained as difficult and as costly as it had been for some time.[42]

Tennis remains easier to learn, less costly to play, better as aerobic exercise, and not necessarily more hazardous than golf. (Tennis elbow is common but also easily treated. The more serious risk for both sports is back injury, in one arising from rapid changes of direction, in the other from torsion. The top golfer Fred Couples told a reporter in 1994 that a "back's not made to do what we do," adding that his physical therapists told him that "the only backs that are worse [belong to] people in the rodeo.") Its revolution in equipment was if anything more radical than golf's; laminated wooden clubs, unlike laminated wooden racket frames, are still available. Yet tennis as a sport seems to be (in the mid-1990s) in decline again while golf holds and continues to increase its following. Could it be that part of golf's secret is its continuing difficulty? Or could it be that by allowing freer competition to build the "best" racket, tennis authorities have inadvertently put their game at a competitive disadvantage? An essay by the great geneticist J. B. S. Haldane on the idea of fitness points out that in evolution, a species can become so well equipped as a result of competition among its members that—once densities are higher—the whole species is handicapped in its environment. The peacock's tail and the fighting apparatus of various beetle species suggested to Haldane that competition within other species might have taken their own ornaments or defensive structures to an extent of development that doomed the species.[43]

The importance of interest and style in sport, as opposed to raw performance, is a lesson in technological development. Allowing a slow, measured increase of intensity keeps up interest and participation better than either freezing the rules or letting new inventions melt them away. Sports make clear what social scientists of technology have been urging for years, that what appear to be technological questions are really political ones, that outcomes depend even more on rules than on devices or physical structures. And golf in particular shows that where people agree on the nature of the activity, limits on technology can be good for manufacturers, professionals, and lay people.

12

Another Look Back, and a Look Ahead

"Doing Better and Feeling Worse." This phrase from a 1970s symposium on health care is more apt than ever, and not only in medicine. We seem to worry more than our ancestors, surrounded though they were by exploding steamboat boilers, raging epidemics, crashing trains, panicked crowds, and flaming theaters. Perhaps this is because the safer life imposes an ever-increasing burden of attention. Not just in the dilemmas of medicine but in the management of natural hazards, in the control of organisms, in the running of offices, and even in the playing of games there are, not necessarily more severe, but more subtle and intractable, problems to deal with.

To investigate why disasters should lead to improvement, and improvement should paradoxically foster discontent, it might help to look at three areas of technology we have not considered before: timekeeping, navigation, and motorization. The automobile first presented an acute problem—collisions—but its success reduced that difficulty while adding to it another, less easily soluble one—congestion. And the recent history of motoring also suggests a paradox of safety, that the better-made and less dangerous motor vehicles become, the greater are the burdens on the operator. The prognosis for revenge effects is hopeful: we will probably keep them under control. By replacing brute force with finesse, concentration with variety, and heavy traditional materials with lighter ones, we are already starting to overcome the thinking and habits that led to many revenge effects. Technology, too, is evolving and responding. The one thing we will not be able to do is avoid the endless rituals of vigilance.

In one example after another, revenge has turned out to be the flip side of intensity. The velocity of twentieth-century transportation and warfare produces trauma on an unprecedented scale, which in turn calls for equally intensive care; but the end result may be chronic brain damage that is beyond medical treatment. Intensive antibiotic therapy has removed the horror of some of the nineteenth century's most feared infections, yet it has also promoted the spread of even more virulent bacteria. Massive shielding of beaches from the energy of waves has deflected their intensity to other shores or robbed these beaches of replenishing sand. Smoke jumpers have suppressed small forest fires but have thereby helped build reservoirs of flammable materials in the understory for more intense ones. Towering smokestacks have propelled particulates at great velocity higher into the atmosphere than ever before—to the dismay of residents over an ever wider radius. Intensive chicken-pig-duck agriculture in China has rushed new influenza virus strains into production, for distribution internationally by the increasingly dense and speedy world network of commercial aircraft. Accelerating processor speed has multiplied computer operations without necessarily reducing costs to programmers, system managers, and end users. Rigid molded ski boots have helped prevent ankle and tibia fractures at the cost of anterior cruciate ligament injuries. And what are so-called pests but intensified life forms? Most of these animals and plants are unusually robust, prolific, and adaptive. The animals are mobile and the plants spread rapidly. Fire ants, Africanized bees, starlings, melaleuca go about their business single-mindedly. Even the dreamy-looking eucalyptus is capable of burning intensely to propagate itself—taking entire neighborhoods with it. And when intensity is a genuine protection against catastrophe, it may fail to address and even complicates persistent low-level problems.

We have learned the limits of intensiveness. What next? In the near time, intensification is still working. Human health and longevity have improved in most places and by most measures. As we have seen, people may feel sicker today because they are more likely to survive with some limitation or chronic illness. But they really are better off. It is hard to disagree with optimists like Leonard Sagan and Aaron Wildavsky when they point to the benefits of growth. Fortunately, every prediction of global famine and misery has failed—so far.

The second argument for optimism is humanity's success in digging deeper and looking harder for old resources and substituting new ones. In the crucible of technological change, shortages produce surpluses and crises yield alternatives. When the biologist Paul Ehrlich lost a bet with the economist Julian Simon on future prices of a bundle of commodities selected by Ehrlich—they dropped between 1980 and 1990, costing Ehrlich $576.06—the transaction seemed to bear out Simon's argument that inexhaustible human ingenuity would find a way around apparent shortages. Market forces appear to impose conservation and encourage discovery more efficiently than legislation generally can. We have seen how the feared hardwood shortage of the early twentieth century never happened, much to the dismay of Jack London and other hopeful eucalyptus growers. Of course, this analysis has revenge effects of its own for market economics: if constraint helps make us so much more clever, why should the state not prod the infinitely creative human mind with more taxes and restrictions? Heavy taxes on fossil fuels should, by the same logic, do wonders for conservation and alternative energy sources.[1]

When it comes to interpreting the last hundred years, the optimists have the upper hand. The future is another matter. Optimists counter projections of global warming, rising sea levels, population growth, and soil depletion with scenarios of gradual adjustment and adaptation. If the crisis of life in the oceans is the problem, then fish farming is the answer. A true optimist sees a silver lining even in the destruction of rain forests and wilderness: there may be much less acreage, but more and more people will be able to travel and see it. In terms of this strange anthropocentric, utilitarian calculus there will actually be more *available* forest and wilderness. As for soil depletion, genetic engineering and new methods of cultivation will presumably let us cope; the world can probably support a population of ten billion or more. (In 1994 it stood at 5.6 billion.) Optimists and pessimists disagree not so much on what is attainable, but on how long it will *be* attainable. What the first group welcomes as a successful adaptation the second belittles as a stopgap. Optimists and pessimists curiously agree that crisis is good for us, but for different reasons. Pessimists welcome emergency as a violent cure for profligacy. Optimists welcome it as an injection of innovatory stimulus.

The Ambiguity of Disaster

One reason for optimism is that disaster is paradoxically creative. It legitimizes and promotes changes in rules—changes that may be resisted as long as the levels of casualties remain "acceptable" prior to a disaster that leads to change. More important, disasters mobilize the kind of ingenuity that technological optimists believe exists in unlimited supply. Of course, new disasters may themselves be unintended consequences of prior solutions. It is uncertain whether, at least in developed countries, the incidence of new catastrophes is gradually declining or not. Should disasters be considered as waves that remain constant in amplitude, damped, or amplified? The unanswerable question about technological revenge effects is whether we are really learning. Even tragedies like Chernobyl and Bhopal are ambiguous as forewarnings. Are they just the most recent in an ongoing series that will strike again in Western Europe and North America, where matters are far less secure than their leaders admit? Or will they spark environmental consciousness and vigilance in the former Soviet bloc and the developing world? It is too soon to say, but there is excellent evidence that great disasters do have long-term reverse revenge effects.

The first great modern stimulus from disaster may have been the defeat of the Spanish Armada in 1588. The economic historian David Landes has speculated that this greatest setback in the history of Spain was what led its king Philip III to offer a perpetual pension of 6,000 ducats to "the discoverer of the longitude" when he ascended the throne ten years later. (Landes is not sure, however, what method would have kept the surviving ships from their fate on the rocks of Ireland and the Orkney Islands.) In France the Duc d'Orléans made a comparable offer. From Galileo to Newton, most of the giants of the scientific revolution of the late sixteenth and seventeenth centuries, with or without prizes in mind, joined the search. None of these thinkers produced a practical astronomical system, yet the shipwrecks and prizes did have other substantial benefits. The sociologist Robert K. Merton has suggested how many advances in mathematics, astronomy, mechanics, and magnetism could be traced to the vast losses that Spain and other maritime powers had suffered.[2]

It took a further disaster to complete the paradoxical work: the wreck of three ships from the fleet of Admiral Sir Clowdesley Shovell in 1707 on the Scilly Isles off the west coast of England, killing almost two thousand sailors. (The admiral reportedly struggled ashore, only to be murdered for his magnificent ring.) Today we know that bad geography, charts, and compasses, and poor navigation, complicated by fog and unpredictable currents, were mainly to blame. To contemporaries, though, the lesson was a new urgency in the search for a way to determine longitude at sea. Of course, a valid method would in turn make possible more accurate printed aids to navigation. The question of longitude was not immediately supported officially; only seven years later, in 1714, was an Act of Parliament passed, offering up to £20,000—at least $1 million in today's purchasing power—for a method of determining longitude on an oceangoing vessel.[3]

Entrepreneurs and cranks had been at work on solutions ever since the wreck, proposing lines of ships somehow "anchored" in mid-ocean, telepathic goats, and even dogs communicating through a "sympathetic powder" said by its promoters to relay sensations from an animal on land to one at sea after having been sprinkled on both. But the prize, after more than another decade had passed, attracted the attention of the gifted clockmaker John Harrison, who built a chronometer that met the specifications of the act. The steps and the time it took him to refine his timepiece (along with the fact that he did not secure payment of his claims until 1773, when he was eighty) are not the point here. What matters is that the magnitude of the Scilly Isles wreck eventually justified the great reward offered.

The earlier prizes contributed indirectly to the Act of Parliament. It was Newton, who had long worked on the problem, whose recommendation was essential for the act's passage. Only in hopes of the new prize did Harrison and other leading craftsmen abandon their normal clientele for a largely speculative project that had frustrated the scientific elite of Europe for decades. The search for longitude may represent the first great public high-technology program. In its costs and benefits it became one of the most successful. Anything like it would almost certainly have been long delayed in the absence of a spectacular new disaster.[4]

It took another two hundred years for a single marine disaster to have an international impact comparable to that of the Scilly Isles wreck. This was the sinking of the *Titanic,* pride of the White Star Line, on April 14, 1912. The ship's tragic end was memorialized not only as an enormous loss of life and property—over fifteen hundred passengers and crew perished, including the captain—but also as a cautionary tale. Some of its perceived lessons were social, the image of the frivolous rich fiddling as the world was about to burn, or even escaping in lifeboats as the poor drowned in steerage. Even the failure of other ships to respond to its distress calls has been blamed on the priority given by radio operators to the social cables of their first-class passengers. But in the long run, the dangers of technological pride rather than class conflict seemed to be the message of this disaster. Even more than the loss of the three English ships two centuries earlier, the sinking of the *Titanic* immediately became what risk analysts now call a *signal event*—one that reveals an ominous and previously underestimated kind of danger.[5]

The problem was not mainly in the operation of the ship's systems, useless though some of the lifeboat mechanisms turned out to be. Even though White Star officials never claimed the ship was actually unsinkable, the captain and crew acted with inappropriate confidence, steaming at high speed through waters notorious for sea ice.

After the *Titanic* hit the iceberg, the same confidence in the ship's safeguards delayed, with tragic consequences, the implementation of rescue procedures that could have reduced casualties immeasurably. (Her officers doubtless had faith in the owners' stringent design specifications, but marine archaeologists now believe that the vessel's steel plates did not meet these standards.) Belief in the safety of the ship became the greatest single hazard to the survival of its passengers, greater than the icebergs themselves. In fact, crews of other nearby vessels that might have rescued passengers believed the *Titanic*'s distress flares could only mean some celebration, not an emergency.

Less known is how important the *Titanic* disaster was in solving what had been a serious problem for international navigation: the prevalence of sea ice in the ocean lanes of the world's most active and lucrative route, the North Atlantic. The wreck had precedents: in the 1880s over fifty passenger ships reported sea ice damage in and around the Grand Banks off the Newfoundland coast where the *Titanic* later went down; fourteen of them had sunk. It was the loss of the *Titanic* that led not only to new regulations requiring lifeboat space for all passengers and crew, but to a series of international conferences on the Safety of Life at Sea (SOLAS) beginning in 1913. The International Ice Patrol, established in 1913, now uses aerial surveillance, satellite images, and radio-equipped oceanographic drifter buoys. The biggest bergs even have their own radio transmitters. Ships possess advanced radar systems. It would require extraordinary negligence for a captain to let an iceberg sink a ship.[6]

At least for passengers embarking in the United States, an ocean cruise now appears extraordinarily safe. From 1970 to 1989, only two of over thirty million passengers died in accidents involving cruise ships operating out of the United States, despite a number of collisions and fires. Each generation of ships meets higher standards. SOLAS now specifies a maximum thirty-minute evacuation time for cruise ships. Only one ship has ever sunk after hitting an iceberg since the *Titanic*, and that was in 1943, when the Ice Patrol was discontinued during the Second World War.[7]

Both tragedies and their consequences illustrate the engineer and historian Henry Petroski's point that a great disaster is often the best stimulus for new engineering ideas. Two things have changed, though, since the early eighteenth century. The growth of engineering as a profession has made a new type of error possible, as Petroski has also shown: overconfidence in the safety of a new design, the defects of which too often remain hidden until some new disaster occurs. But there is also a second type of error: failure to observe the repeated rituals that safe operation of advanced technology entails. The higher potential speed of steamships required (and requires) more rather than less care. The larger number of passengers and crew required (and requires) more careful drills and inspection of equipment. It is still difficult for a prospective passenger to tell how well trained a crew may be to handle an emergency. We know some technology has a built-in demand for care, a maintenance compulsion. But there is always a hidden catch of technological improvement: the need for enhanced vigilance that we have already seen in medicine, in environmental modification, in the translocation of plants and animals, in electronic systems, and even in some aspects of athletics.[8]

At this point in the history of technology we can draw a fundamental lesson from an unexpected source, the law of negligence. In a number of important articles, the legal scholar Mark Grady correlates better and safer technology with the number of lawsuits for malpractice and personal injury. During the centuries when bleeding, purging, and mercury compounds (as we have seen) hastened the deaths of many of the West's elite, legal action against the physicians who pursued these remedies was rare. The public did not hold doctors in awe; neither did they really expect heroic remedies to work. In fact, it was precisely because they doubted the scientific basis of contemporary treatments that a malpractice suit had little point.

According to Grady, "the first negligence explosion occurred during the 1875–1905 period. In that time of industrial revolution, claims increased by fully 800%, and the negligence rule did not change significantly. When machines abound, negligence claims increase. Put differently, a doctor who forgot to perform a modern fetal health procedure could not have been liable in 1960, before the procedure was invented." On this view, a dialysis machine reduces the risk of kidney failure in nature but adds a new risk: that physicians and technicians operating the machines under their supervision may fail to make safe connections, test the hemodialytic solution, or observe all the other precautions of good practice. Anyone who has watched the pilot and copilot of a common two-engine commuter aircraft carry out their extensive preflight procedures, flipping through pages in a printed notebook as they read their scripts, has been struck by the number of precautions that a long-accepted and well-developed technology imposes.[9]

By the standards of its day, the *Titanic* was a ship relatively high in what Grady calls "durable precautions," the safety hardware that popular opinion supposed made it unsinkable. It is true that size, luxury, and speed had higher priority than safety in her design—but she had the latest in communications and damage-containment equipment. Grady's analysis suggests, though, that the very presence of these measures increased the importance of "nondurable precautions"—all the things an officer or crew member must remember to do—in keeping the ship afloat. The flow of messages on the ship's radio demanded constant attention: did a given message warrant immediate transmission to the bridge? Once the captain was aware of it, did it necessitate a change of speed or course? And with lifeboats come other questions. Have they been inspected regularly? Does each crew member know his or her part in supervising a possible abandonment? If a major marine loss occurs, it is the way an emergency plan is carried out, not physical safeguards alone, that will determine whether or not it becomes a disaster for human life.

Here is where the difference between early and industrial technology becomes telling. The captain of a seventeenth-century oceangoing ship needed excellent navigation skills, and the management of cargo, ballast, and rigging were already arts for specialists. Some captains and pilots of Renaissance and early modern Europe had superb intuition which let them achieve amazing feats of "dead reckoning": the estimation of position from relatively crude measurements of last position, direction, and speed. A gifted mariner could go beyond the limits of the technology of the day. Yet because of the difficulty of measuring longitude, compounded by the other defects of instruments, disaster could happen to the best of seafarers. That is why Sir

Clowdesley Shovell still got an overbearing tomb by Grinling Gibbons in Westminster Abbey after his catastrophic end. (On the other hand, Joseph Addison ridiculed it as "the figure of a beau, dressed in a long periwig, and reposing himself upon velvet cushions under a canopy of state," and deplored that it commemorated only his demise and not his victories.) The better and the safer technology becomes, the more we presume human error when something goes seriously wrong. If it is not the error of the captain or crew, it is one of the engineers or designers of equipment, or of executives and their maintenance policies.[10]

THE AUTOMOBILE AND REVENGE EFFECTS

Intensity—disaster—precaution—vigilance: the cycle appears on land as well as at sea. The rise of motoring shows this more clearly than the transformation of sailing, but in a different way. As we saw in Chapter 1, nineteenth-century railroad accidents were the first of a new type of technological disaster unknown in the eighteenth century. Historians of technology have long pointed out the importance of indignation over early railroad tragedies in developing the first complex control systems in American business, not to mention safety hardware like signals and air brakes. But there is an equally interesting side to the intensification of transportation by the railroad: the rise of automobile transport. Casualties from car accidents occur as a steady series of small disasters, not the few-but-great wrecks involving trains and steamships. The automobile invited chronic catastrophes. Indignation built more slowly.[11]

The growing capacity of the nation's railroad network had an unforeseen consequence that few scholars have noted—chaos in the horse-drawn city. Nearly every passenger journey or freight shipment began and ended with a horse-drawn vehicle or a horse, at least until cable cars and electric trolleys spread late in the century. Even the physical size of horses increased throughout the nineteenth century, to move the heavier loads and serve the larger populations of European and American cities. By the 1880s, massive Percherons were a familiar sight on American streets. Teamstering already was a crucial trade, and the number of horses for every teamster was growing. Local delivery by horse could cost nearly as much as hundreds of miles by rail. Today's Budweiser Clydesdales, a magnificent public relations asset, are the heritage of yesterday's logistical nightmares.[12]

Herds of horses multiplied. Even after cable and electric power had begun to replace horse traction for streetcars, horses were everywhere. The Fiss, Doerr, and Carroll horse auction mart on East 24th Street in New York drew up to a thousand buyers and boasted its own seven-story, block-long stable. New York City's horses alone produced over 300 million pounds of manure annually; stables accumulated tens of thousands of cubic feet for months at a time. In fact, as we have seen, one imported pest, the English sparrow, thrived on the bounty of grain in horse droppings. Repeated horse epidemics—technically epizootics—paralyzed commerce and interfered with firefighting. Despite limitation of their workdays to four hours, horses died after only a few years of service, usually in the middle of the street, up to fifteen thousand a year in New York. Dust from powdered horse manure helped spread tuberculosis and tetanus. As railroads grew safer, the horse-drawn city became more dangerous.[13]

Less remembered today than the sanitary problems caused by horses were the safety hazards they posed. Horses and horse-drawn vehicles were dangerous, killing more riders, passengers, and pedestrians than is generally appreciated. Horses panicked. In frequent urban traffic snarls, they bit and kicked some who crossed their path. Horse-related accidents were an important part of surgical practice in Victorian England and no doubt in North America as well. In the 1890s in New York, per capita deaths from wagons and carriage accidents nearly doubled. By the end of the century they stood at nearly six per hundred thousand of population. Added to the five or so streetcar deaths, the rate of about 110 per million is close to the rates of motor vehicle deaths in many industrial countries in the 1980s. On the eve of motorization, the urban world was not such a gentle place.[14]

The automobile was an answer to disease and danger. In fact, private internal-combustion transportation was almost utopian. The congested tenements of the center city spread dirt and disease. Dispersing people into the green suburbs was a favorite theme of city reformers. Progressive mayors supported the extension of horsecars and then trolleys. But at least on city stretches, these had an unpleasant intensity of their own. In 1912 the *Los Angeles Record* found their air "a pestilence . . . heavy with disease and the emanations from many bodies. . . . A bishop embraced a stout grandmother, a tender girl touched limbs with a city sport. . . ." And hard-pressed straphangers objected to allegedly high fares, reckless drivers, and rude conductors.[15]

Automobiles may have begun as rich people's toys, but thanks largely to Henry Ford, they soon came to represent independence *from* the rich: from railroad interests, traction (streetcar) companies, center-city landlords. By the 1950s and the 1960s, the automotive industry had come to represent big business at its most arrogant, but motorization won because it rallied so many small businesses. Diffuse interests were its political strength. Motoring did not benefit only car manufacturers and petroleum producers and refiners. It enriched tens of thousands of small businesses: trucking companies, suburban developers, construction contractors, dealers and parts retailers, service station operators. Of course, as Clay McShane and other urban historians before him have documented, road improvement was not really populist, or uniformly popular. It did change the nature of the street, but to the disadvantage of residents. The roadway ceased to be a gathering place and became a thoroughfare. Many neighborhoods resisted asphalt paving, and children even stoned passing cars. Still, motoring showed the political advantages of spreading benefits to many small and medium-sized interests.[16]

In spite of clear damage to urban greenery and space, using roads to help disperse people in private suburban houses remained not just a popular but a politically correct idea for a long time, and not only in America. Franklin D. Roosevelt thought that spreading population would lower the cost of government and directly reduce the expense of urban services. One radical planner, Carol Aronovici, wrote in 1932: "Let the old cities perish so that we may have great and beautiful cities." Aronovici called for "a thorough emancipation of the suburban communities from the metropolis" that was threatening to "suck their very physical existence into the body politic of decayed and corrupted political organization." (More than sixty years later, these same towns—now aging demographically and economically—are beginning to make common cause with the old central cities against the flight of businesses and residents to the sprawling outer suburbs.)[17]

At virtually the same time a school of Soviet planners called the "disurbanists" were dreaming of dispersing their own overcrowded urban masses into new settlements amid the fields and forests by building new road networks. A distinguished visitor, the French architect Le Corbusier, summed up the mood in his book *La Ville Radieuse* (1930):

> People were encouraged to entertain an idle dream: "The cities will be part of the country; I shall live 30 miles away from my office under a pine tree; my secretary will live 30 miles away from it too, in the other direction, under another pine tree. We shall both have our own car. We shall use up tires, wear out road surfaces and gears, consume oil and gasoline. All of which will necessitate a great deal of work . . . enough for all. . . ."[18]

It is almost as though postwar American suburbia was the realized fantasy of Soviet planners. Or, more accurately, the victory of motorization was an unintended consequence of an international decentralizing mood. As Kenneth Jackson has pointed out, even the *Bulletin of the Atomic Scientists* embraced dispersion of cities in a 1951 special issue, "Defense Through Decentralization." It promoted satellite cities and low-density suburbs in which former urbanites could be housed more safely for the duration of the Cold War.[19]

Automobiles and road systems promoted an old technological utopia, the community of private villas. Automobiles also have an immense advantage over railroads and trolleys: they make it possible to go directly from one outlying point to another. America never had an integrated national or even regional transportation network as

European countries did. Its trains and even some of its urban transport systems were run by competing corporations. Nostalgic admirers of railroad transportation forget how many trips required completing two sides of a triangle, sometimes with hours of waiting between them. K. H. Schaeffer and Elliot Sclar, transportation analysts, exposed these shortcomings trenchantly in *Access for All.* A trip of fourteen miles from the small town of New Washington, Ohio, to its county seat could take all day by rail, even when train travel was at its peak. And New Washington's two depots were a half mile apart.[20]

Usually, motorization bought space rather than time. Ivan Illich wrote in 1974: "The typical American spends over 1,600 hours a year (or thirty hours a week or four hours a day including Sundays) in his car. This includes the time spent behind the wheel, moving or stopped, the hours of work needed to pay for it and for gasoline, tires, tolls, insurance, fines, and taxes. . . . For this American it takes 1,600 hours to cover a year total of 6,000 miles, four miles per hour. This is just as fast as a pedestrian and slower than a bicycle."[21]

In fact, the greatest surprise of motoring was the speed at which traffic clogged the roads, including freeways and other limited-access highways built to relieve congestion. When the Washington Beltway was dedicated in 1964, the governor of Maryland, who cut the ribbon on its last segment, called it "a road of opportunity." The federal highway administrator compared it to a wedding ring. The *Washington Post* declared that "the stenographer in Suitland will be able to get to the Pentagon without finding the day ruined almost before it begins." Twenty-two years later, another *Post* correspondent reported: "The dream turned to nightmare. The Great Belt tightened to the point where right now it resembles nothing less than a noose around the communal neck. . . . We could die on the Beltway and rot until vultures pick clean our bones." London's counterpart, the M25, had already exceeded its projected traffic for the year 2001 by the late 1980s, only three years after completion. Surprisingly, even states like Kentucky, Missouri, Nebraska, South Carolina, Tennessee, and Texas classify more than half their interstate highway mileage as congested. And mature suburbs of large cities have become so traffic-choked that the American Automobile Association itself has moved its headquarters from Fairfax County, Virginia, to Florida.[22]

There are social reasons for this recongestion: not just two-commuter families but the multiple motorized errands that suburban living demands. Saturday afternoons may be the most crowded times of all. Traffic engineers, applied mathematicians, and economists have also discovered that expanding old routes and adding new ones may actually increase travel time. An enlarged bridge will redirect traffic that had been taking a longer route around it, but unless it is substantially larger, it will be just as slow. New highways also may increase total travel time for all travelers when they draw traffic from alternative rail systems. And the ultimate recongesting effect is called Braess's Paradox, in honor of a pioneering investigator of the subject. Where each of two congested routes has a bottleneck, adding what appears to be a shortcut between them may actually increase travel time for everyone. The reason: the new, "direct" road actually funnels traffic through *both* the old bottlenecks. Thanks to quirks of driver psychology, even common operations like merging traffic can produce equally counterproductive results. Because motorists tend to close up spaces to discourage entering cars from cutting in front of them, especially when these attempt to enter from other roads, mysterious traffic jams can appear a mile or more from the actual merge. Because spaces are tight, a driver decelerating slightly at the head of a clump can unwittingly induce one following motorist after another to brake a bit harder. And when congestion reaches a certain maximum roadway capacity, the flow of cars falls so sharply that traffic engineers recognize (but still can't fully explain) a "breakdown." What appears rational to an individual driver becomes irrational for the motoring population and for society. Recongesting turns out to be a form of recomplicating, of creating a machine of parts coupled in poorly understood ways.[23]

What is interesting technologically about this new congestion is its unexpectedly positive side. It has helped make driving safer than anyone thought it would ever be. Congestion may be a chronic negative side effect of mobility, but safety is a positive outcome of congestion. There is a school of thought that denies that driving or anything else can ever be made safer. This is called risk homeostasis. The phrase means simply that people unconsciously seek a certain level of hazard. They compensate for "dangerous" conditions by driving more cautiously—and offset safety measures by taking more risks. The geographer John G. U. Adams looked into the accident rate of England's "adventure playgrounds," loosely supervised assortments of high wooden ladders and platforms that offer "opportunities to test skills appropriate to chimpanzees." They are visibly more dangerous than "fixed equipment" playgrounds with their smooth surfaces and rubberized matting. Yet insurance companies quote lower liability rates for the adventure playgrounds, and the secretary of the National Playing Fields Association has written that "the accident rate is lower than in orthodox playgrounds since hooliganism which results from boredom is absent." Adams and others (mainly social scientists) have argued conversely that seat belts, by making drivers feel more secure, actually cause more pedestrian casualties even as they reduce motorist injuries.[24]

Few traffic engineers accept risk homeostasis as a principle, or the seat belt as an instance of it. In fact, as Leonard Evans, a physicist and safety researcher, argues, some safety measures save more lives than we might have predicted, but others may actually increase casualties. The fifty-five-mile-per-hour speed limit reduced deaths more than anyone had expected. Seat belts met expectations. Studded tires, improved acceleration, and antilock braking systems (ABS) have a moderate benefit, though there is some evidence that ABS-equipped drivers may have as many crashes as those not similarly equipped, or perhaps even more. New traffic signals seem to have a neutral effect. So do strict inspections. And surprisingly, zebra stripes and flashing lights at crossings actually increase pedestrian injuries significantly. (That does not mean they are useless. As another leading traffic specialist, Frank Haight, has put it, the benefit of some safety measures is fair access rather than safety. They give pedestrians not absolute protection from reckless motorists but the welcome ability to cross roads that would otherwise be almost impassable.) Changing any single piece of hardware, or any law, may or may not have the desired effect.[25]

It isn't only safer equipment, then, that has brought down the rate of deaths per million passenger-miles. In fact, cause and effect might be reversed. Only when drivers start giving up speed and price to protection do manufacturers start selling safer cars. And that seems to depend on the amount of driving. The British mathematician and traffic engineer R. J. Smeed had that most remarkable gift, the ability to point out an obvious pattern that others had missed. In 1949, Smeed began to plot the relationship between fatalities per vehicle and vehicles per capita. What he found then, and what he and others have noticed ever since, is that more driving makes fatal accidents less likely per mile driven.[26]

In the late 1960s, for example, the nations with the highest fatality rates were developing countries with few private automobiles per capita. Even today within Europe, the riskiest countries are those on the periphery, like Portugal, where automobile ownership is still twenty years or more behind England or Germany. John Adams, while dissenting from Smeed's conclusions about the reasons for greater safety, found that later data supported Smeed's original 1949 paper.[27]

Smeed's observations point to a very complex process: a set of technological, legal, and social changes that more general driving brings. Countries with few roads, wide-open spaces, and few vehicles may be dangerous to motorists' health. A colleague once recalled from her childhood in Iran that on long stretches of country road, chauffeurs raced toward each other in the center, playing a local variation of chicken. Visibility was excellent, but there were no lane-dividing stripes. One driver *nearly* always turned off; the point was to wait long enough to maintain one's honor. What in the United States are adolescent rites may elsewhere be the serious contests of middle-aged men.

Early motorization's mix of human, animal, and motor power can be equally fatal. In India in the early 1980s, there were seventy-five road deaths a day, half as many as in the United States, which had forty times as many vehicles. Twenty times more people were killed in accidents than in floods. In 1989, more than a thousand died on the Grand Trunk Road from New Delhi to Calcutta alone.

Uniquely Indian, or Third World? Not at all. Early in the century, New York also had a mix of animal-drawn vehicles, automobiles, streetcars, bicycles, and pedestrians, and saw casualties double during the earliest driving boom.[28]

Congestion leads to demands for limited-access roads that in turn promote safer high-speed driving. U.S. national statistics also suggest that the most dangerous roads are straight, two-lane desert highways, with the worst being the notorious U.S. 66 near Gallup, New Mexico. One study of motor-vehicle crash mortality found a hundredfold variation; in Esmeralda County, Nevada, the death rate was 558 per 100,000, while in Manhattan it was 2.5.[29] In hilly country with old roads and many new drivers, the results are similar. Whereas the United States had 248 deaths per million vehicles in 1989, Britain 248, and the Netherlands 236, Portugal had 1,163. The Portuguese-based writer Robert D. Kaplan has written of drivers on the twisting and crumbling Sintra-Lisbon highway: "Instead of going slow, they race along . . . passing on curves at night, with the ease and tranquillity of a blind person reading Braille."[30] Yet these same people are impeccably courteous pedestrians and have passed stiff written tests requiring a three-month course. The gentle Malaysians, mostly teetotaling Muslims, also reflect the spirit of early motorization. Malaysian drivers "love to pass on blind curves or approaching hills," wrote one visiting American. "They routinely ride up on each other's tails, going 50, 60, 70 miles an hour, then impertinently flash their headlights."[31]

The traffic congestion of highly motorized countries poses a chronic rather than an acute health menace. Road safety statistics do not reflect the health consequences of vehicle emissions. A car that covers ten miles in thirty minutes of rush-hour crawl produces three and a half times the hydrocarbons of one that takes eleven minutes at off-peak hours. Idling engines produce three hundred times as much carbon monoxide as those running freely. While automotive emissions were reduced by 76 to 96 percent from 1967 to 1990, the number of cities with hazardous ground-level ozone increased to over one hundred by the late 1980s. Estimates of the health and agricultural damage done by carbon monoxide and smog range from $5 billion to

$16 billion per year. All of these are serious chronic consequences, but they don't alter the fact that riding in a motor vehicle has become far safer.[32]

There is an unexpected discipline in the apparently more dangerous congested road. Interstates and other limited-access highways would not be feasible without a minimum traffic volume. Density forces slower and more uniform speeds. It also makes possible greater police supervision, better rescue services, and easier access to emergency treatment facilities. The safest part of the New Jersey Turnpike is the crowded metropolitan portion north of New Brunswick; the more rural South Jersey section has twice its accident rate. And much of the reason is that congestion compels vigilance. The chairman of the Turnpike Authority explained: "[I]n the north . . . there is so much going on, you're pumping adrenaline just to stay on top of it. We're keeping you alert up here. Down there you're dozing." In fact, as Albert O. Hirschman has pointed out, proper driving is actually easier in the city than in the country. "The city traffic requires greater technical mastery, but this increase in the difficulty of driving is outweighed by the fact that intense traffic helps [the driver] in the task of focusing his attention."[33]

In spite of countless incidents of violence on the highway, in spite of all our experiences to the contrary, mature motorization seems to engender (relatively) more courteous and disciplined behavior, "collective learning" in Leonard Evans's phrase, or, as the *Washington Post*'s Malcolm Gladwell puts it, "driving under the influence of society."[34]

A spokesman from the Insurance Institute for Highway Safety reports that while safety-related advertising once appeared to harm sales by substituting fear for fantasy, "safety is only second to quality and well ahead of price in the consumer's mind."[35] We are far from the free spirits of early motoring, of Booth Tarkington's George Amberson Miniver, of Kenneth Grahame's Mr. Toad, "the terror, the traffic-queller, the Lord of the lone trail, before whom all must give way or be smitten into nothingness and everlasting night."[36] Or, as the columnist Richard Cohen has written, "Jay Gatsby never dreamed of gridlock."[37]

CONSERVATION OF CATASTROPHE?

Marine navigation and motoring alike seem to argue for optimism, for the idea that intensification can be tamed, in fact that disasters are self-correcting. Society learns. Progress, that long-despised concept, comes in by the back door. The point is not that disasters continue, but that on balance and by most measures, people continue to be better off. Unfortunately for technological optimism, things are not quite so simple.

The *Titanic*'s sinking has been moralized so much that we have to remember the incident would have turned out much differently had her plates not fractured. No one had tested (and possibly no one could have tested) metal for the kind of brittle fracture her hull experienced. Even if the crew had been able to evacuate every passenger safely, the loss of the ship would have been one of the greatest material disasters of peacetime marine history.

The disturbing fact about the accident is that we can never be completely confident of the behavior of any new material as part of a complex system. Splinters from fiber-optic cables, to take just one case, can pose serious health risks for telephone workers (and especially for self-taught laypeople) who have to cut and splice them. Yet it is rare to see this problem mentioned in most discussions of networking.

Software adds another dimension to complexity. We have seen in Chapter 9 how high the risk of fatal bugs in life-critical systems can be. Malfunction in software control of processes is also less likely to produce the warning signals familiar in the mechanical world—heat, noise, color change, vibration. A system crash may be much more sudden. It is harder to achieve what engineers call "graceful degradation."

The historian of science Michael S. Mahoney has observed that computers do not eliminate artisans but reintroduce them in the new guise of programmers. Recomplication has made software so bulky that only teams of programmers can write it, yet talented programmers are individualists who do not usually work efficiently as part of a team. This affects not only operating systems and applications software for desktop computers, but the code that runs everything from aircraft navigation to automotive fuel injection and medical equipment. As John Shore, a software engineer, has pointed out, vigilance works well for mechanical systems; high-rise elevators need constant maintenance, but they rarely injure people. Software requires maintenance, too, but this makes it less rather than more reliable. Every feature that is added and every bug that is fixed adds the possibility of some new and unexpected interaction between parts of the program. A small change to solve a minor problem may create a larger one. The technical writer Lauren Wiener has noted that the repeated paralysis of local and regional telephone systems in 1991 resulted from only a few changed lines in the millions of lines of code that drove call-routing computers. A meaningful test of the revisions would have taken thirteen weeks.[38]

Catastrophic risk will stay with us because more rather than less of life is likely to depend on complex software. Intelligent vehicle-highway systems (IVHS) may someday squeeze more capacity out of existing limited-access roads. Individual vehicles under electronic control would join formations called platoons. These convoys could be spaced more tightly than today's normal traffic. And they could control some of our daily highway nightmares, such as the tailgater, the lane jumper, and the sleepless trucker. But if software or communication or even a lead vehicle's tire failed, the results could be catastrophic. If we add the dependence of government, banking, and commerce on global electronic networks that in turn depend on software, a revival of catastrophic errors cannot be ruled out. (And the critics of IVHS insist that electronically controlled roads will soon be recongested anyway.)[39]

Even more serious than hidden risk may be displaced risk. The safety of one technology has a way of creating danger in another. Our current successes may be preparing us for failures where we least expect them. We have seen how good hygiene left the well-scrubbed children of the middle and upper classes more susceptible to polio than the dirty kids of the poor. We have also noted the suggestion of Mirko D. Grmek that success in suppressing bacterial infection indirectly promoted the rise of AIDS and other new viral infections by leaving a niche for virulent pathogens.

If hidden risk is the concern of the liberal, distrustful of corporate assurances of safety through technology, displaced risk is the objection of the conservative to regulation. And conservative skepticism is directed less often at technologies themselves than at attempts to limit, regulate, or impose them. Requiring parents to place their infants in (paid-for) child carrier seats on airlines instead of carrying them on their laps may lead more families to drive instead of fly. Since aircraft are safer than highways, the argument runs, the rule may injure more infants than would have been hurt in the air. Pesticide-free fruit and vegetables at high prices may be more harmful to public health, by reducing consumption by the poor, than cheap produce with pesticide residues would have been. Taking this line of thought to an extreme, one British researcher has even found that male physicians who quit smoking tend to offset their health gains with higher rates of alcoholism, accidents, and suicide. (Not surprisingly, tobacco industry sources supported this study.)[40]

Like hidden risks, displaced risks appear impossible to rule out of any proposed change. The natural and social worlds interact in too many poorly understood ways. Risk analysts call these unexpected effects Type III errors. (A Type I error is an unnecessary preventive step, like evacuating a coastline when storm warnings turn out to be a false alarm, or delaying the approval of a lifesaving drug. A Type II error is a decidedly harmful action like releasing a drug that turns out to have lethal side effects.) When strict directives on meat radiation after the Chernobyl meltdown of 1986 destroyed the Lapp reindeer-meat economy, as a recent report of the Royal Society pointed out, the unexpectedness of the result made it a Type III rather than a Type I error. Many market-oriented risk analysts like Aaron Wildavsky urge resilience and gradual responses to unforeseen consequences as they occur, rather than attempts to calculate and balance all possible results. The report of the Royal Society points to clearly organized schools of "anticipationism" and "resilientism." Resilience often turns out to be an excellent policy, provided the phenomena cooperate and appear distinctly and gradually on the horizon.[41]

In the real world, few trends emerge without ambiguity, beyond a reasonable doubt, before precious time is lost. It is now more than 150 years since the eccentric French utopian socialist Charles Fourier predicted that the increasing cultivation of the earth would bring about higher temperatures and eventually a melting of the polar icecap. While Fourier's scientific credentials were dubious—he thought the northern seas would become "a sort of lemonade" and humanity would move about on "antilions" and get their fish from "antisharks"—he was on to something. In fact, at about the same time, another Frenchman, coincidentally also named Fourier (the physicist Jean-Baptiste), discovered that the earth's atmosphere maintains the planet's warmth by trapping heat. As early as a hundred years ago, the Swedish geochemist Svante Arrhenius speculated on a possible increase of up to 6 degrees C. in air temperature if industrial carbon dioxide emissions continued to grow. Yet even now, the science we need most gives us not the precision we want but a set of possible tempos and consequences. We want numbers, but instead our best models give us ranges. We want a truth that will apply to the whole globe, or at least to our own continent, and face the likelihood of patchy local change. We want an idealized eighteenth-century celestial mechanics to rule our world, but we find only probabilistic models.[42]

We can't even count on conditions continuing to drift slowly. As Stephen Jay Gould and others have often reminded us, steep rather than gradual natural change is the norm, and it is extremely hard to predict the future state of a complex system even without the added imponderables of human culture and behavior. Well before climate became an issue, human culture (including technology) set off bizarre chains of cause and effect. The fashion for feathers and entire dead birds on women's hats in the late nineteenth century devastated whole species; but it also drew women and men into bird preservation movements that outlived the fad. The early automobile spread its own nemesis, the puncture weed with its tire-killing spiked seedpods. Decades after safer and puncture-resistant tubeless tires appeared, this technology unexpectedly abetted another pest: the Asian tiger mosquito, a vector for dengue fever, which traveled the Pacific in recycled tires and now enjoys an extended breeding season in water that collects in tire dumps. We have already seen how cleaning up European harbors probably helped spread tenacious zebra mussels to North America. Yet motorization also helped reduce the population of European sparrows.

Anyone correctly predicting these sequences well in advance would have seemed a crank or an alarmist. In fact, most of the greatest changes of the twentieth century simply did not occur to the nineteenth-century imagination. Air war and weapons of mass destruction were outstanding exceptions, and even these were logical extensions of pre-1900 sieges and bombardments. Otherwise the human ability to envision something truly new, good *or* bad, is surprisingly limited. Late-twentieth-century personal computers are radically different not just from nineteenth-century analytical engines and mechanical calculators but even from those (far slower) behemoths, the postwar data processors of the von Neumann era. High European mortality in tropical Asia and Africa did not prepare the Western mind for the emergence of AIDS and other "new" viruses—nor for the influenza epidemic of 1918.

Extrapolation doesn't work, because neither nature nor human society is guaranteed to act reasonably. Some things like computer processor power and data storage get better and cheaper more quickly than the optimists expected; on the other hand, the tasks that they are supposed to perform, like machine translation, turn out to be more difficult than most people had thought. What is almost a constant, though, is that the real benefits usually are not the ones that we expected, and the real perils are not those we feared. What prevail are sets of loosely calculable factors and ranges of outcomes, with no accepted procedure for choosing among them. And since we have seen that it is impossible to rid any computer models of bugs, we have no assurance that reality will not be well beyond our projected range in either direction. Instead of the malice of the isolated object, we face ever more complicated possible linkages among systems of objects.

It is impossible, then, to prove that large-scale disasters will not reassert themselves in North America and the rest of the developed world, that we will not intensify not only our chronic problems but our acute ones. William H. McNeill has a telling phrase for this possibility: the Conservation of Catastrophe. Just as engineers will continue to explore the bounds of a "safe" bridge design, test pilots will "push the envelope," regional planners will overrate the capacity of roads to evacuate a hurricane zone, and engineers will disregard all they have learned about O-rings. We can even find analogies in the realm of finance: the New Deal's precautions against the bank failures of the Depression created institutions that helped promote the wave of savings-and-loan bankruptcies of the 1980s. International electronic networks for communication and commerce make new kinds of disasters possible, as localized malfunctions now have unprecedented opportunities for spreading. If the postal carriers of one city start hoarding or discarding mail, it is a major problem but no immediate threat to the system's integrity. If a network node were to go wrong in some unforeseen way, worldwide systems could fail before the cause was even identified.

The real question is not whether new disasters will occur. Of course they will. It is whether we gain or lose ground as a result. It is whether our apparent success is part of a long-term and irreversible improvement in the human condition or a deceptive respite in a grim and open-ended Malthusian pressure of human numbers and demands against natural limits. It is whether revenge effects are getting worse or milder. I think, but cannot prove, that in the long run they are going to be good for us. And I would like to suggest why.

Retreating from Intensity

Revenge effects reached their peak in the hundred years between the 1860s and the 1960s, during the very acme of technological optimism. Clobbering nature into submission united North Americans and Europeans, Communists and Republicans. Explosives, heavy machinery, agriculture, and transportation seemed at last to be fulfilling the injunction of Genesis 1:28 to "fill the earth and subdue it."

Soviet citizens named their children for Henry Ford and his tractors. Contemporaries thought they were living at the beginning of an era of open-ended change; but it is also clear that few of them reckoned with the tendency of nature to strike back. Although (as the historian Douglas Weiner has documented) Friedrich Engels himself wrote of how nature "avenges" humanity against exploitation, the Eastern Bloc kept subjugating its part of the planet until the bitter end.[43]

The real meaning of Communism's collapse had less to do, in fact, with collectivism than with a fixation on intensity that continued through the Gorbachev years. Officially the regime campaigned to conserve materials. But it also set output goals by weight, not performance. Industrial quotas, meted out in metric tons, were filled with heavy stuff—sometimes incredibly sturdy, more often simply bad. The alleged Soviet boast of producing the world's largest microchips may be apocryphal, but Marshall I. Goldman, an economist who visited the USSR often, noticed an exceptional proportion of office typewriters with unnecessary extra-long carriages.[44]

The Soviet fixation on goals by gross weight and volume was only an extreme case of the pathology of intensity: the single-minded overextension of a good thing. We should keep in mind that the West went through even more serious crises of intensification. Potatoes, a great benefit for the European popular diet, were genetically vulnerable when grown from a single strain and used as a primary source of nutrition by the very poor. Yet terrible as the Irish potato famine of the 1840s was, nothing like it has recurred. The crash of the French raw silk industry in the 1850s, so important for Louis Pasteur's career, also showed how dangerous it could be for so many families to link their economic fate to a single organism.

It is curious how many resource-rich nations and regions have faltered because they relied too strongly on exploiting only one or two sources of natural wealth. The Mississippi delta, the deserted mining towns of the Rockies, and the desolate coal patches of the Pennsylvania anthracite country all have their counterparts overseas: Sicily, the Ukraine, and Argentina as former world breadbaskets, Romania and Azerbaijan as fabled energy reserves, Zaire and Siberia as gold vaults, the Ruhr as ironworks. The nature of the resource does not seem to matter. Nor do colonialism or foreign rule, though absentee ownership may. It was wealth that became an enemy of a vital diversity. On the other side, resource-poor islands and formerly isolated regions like Switzerland, Japan, Taiwan, and Singapore have become the twentieth century's economic stars.[45]

Of course, it is too optimistic to say that we have overcome the perils of intensity. We have already seen how "rationalized" forestry in England and Scotland has helped turn the familiar ground squirrel of North America into a significant woodland pest. The science writer Matt Ridley has described how even in Tory England, state-promoted conversion of "unproductive" downland to wheat fields and ancient forests to conifer plantations had endangered butterflies and other native wildlife and plants. In Spain and Portugal, the ancient *dehesas* of mixed cork oak and holm oak in a setting of grain and grasslands have also been threatened by clearance for Euro-subsidized crops. Elsewhere, clear-cut forestry and overfishing continue. The greatest risk of any new natural technology, especially a genetic one, is not a superpest. It is an apparently harmless organism or chemical that begins as a stunning success and displaces alternatives in the marketplace. Making anything so hardy and productive is like announcing a huge prize for the first naturally selected pests and parasites. Sooner or later there will be a big winner.[46]

All this hardly means that science or technology has overintensified life, or that traditional agriculture was always environmentally benign. In the Mediterranean and elsewhere, preindustrial agriculture could devastate as well as foster diversity; it is hard to imagine any biologically engineered organism as catastrophic in the wild as the otherwise useful and endearing goat. And technologies can follow a number of alternative paths, depending on the assumptions and interests of those who develop and support them. Technologies can help preserve old genetic resources, evaluate new crops, reduce the quantity of pesticide and herbicide applications, consume less water. In other words, they can diversify and *de*-intensify. This implies a new balance between public and market-driven research, since (as the geneticist Richard C. Lewontin and others have shown) commercial research necessarily neglects natural, nonpatentable varieties of organisms that would be in the public domain after the first sale.

In agriculture, the retreat from intensity means forgoing applications of heavy fertilizer in favor of planting complementary crops in the same fields, increasing both productivity and resilience. In medicine, the retreat from intensity demands a shift away from the heavy reliance on a handful of antibiotics. In business computing, it implies a heavy dose of skepticism about the functional value of "more powerful" new releases of both hardware and software. It also suggests doubts as to whether higher workloads and longer days always yield more profit; sometimes it even calls for deliberately slowing or interrupting the pace of work. In sports, it provokes a harder look at whether stiffer and more powerful equipment necessarily makes for a better game. The retreat from intensification does not necessarily require giving it up; it does mean subjecting it to much greater scrutiny.

It isn't enough, of course, to modify intensity. Reducing revenge effects demands substituting brains for stuff. And the record of human ingenuity in making brainpower do the work of energy and raw materials is impressive. Balloon-frame houses, the invention of anonymous carpenters on the nearly treeless prairies of the nineteenth-century American Midwest, became famous for their durability as well as their economy. In our own time the cheapest electronic computers available today from any discount store can calculate many times faster than the room- and building-size arrays of relays and vacuum tubes of the industry's pioneer days. Steel is lighter and stronger, yet certain plastics are lighter and stronger than steel. Automobiles now weigh less and use less gasoline per mile. A CD weighs a fraction of an LP, and a CD player is lighter and more compact than a conventional turntable. New mathematical algorithms allow the same information to be stored on smaller disks—or more information on the same size disk.

The engineer Robert Herman, the technology analyst Jesse H. Ausubel, and their associates argue that technological change exerts powerful forces both for increasing and for reducing the amount of energy and other resources used. Electronic storage can reduce the consumption of paper, but as we have seen, it can also multiply it.

Lighter goods may heighten rather than diminish the need for materials if they are marketed or treated as throwaways rather than durables. (Thick, returnable glass bottles may, for example, demand less intense use of energy and other resources than even recyclable aluminum.) In fact, as Herman and Ausubel have suggested, lighter and more efficient automobiles promote resource-consuming if dispersed suburban living and thus materialization. Nuclear power generation begins with low-weight raw materials but ends with vast contaminated structures that probably can never be reused.[47]

What appears to be a technological question—how much of anything we really need—is in the end a social one. It is the size and appearance of a yard or a lawn or a house, the taste for (or repudiation of) meat, and so forth. Often what is most crucial, and most uncertain, is not invention and discovery but taste and preference. The open question, raised during the upheavals of the 1970s and then forgotten during the boom of the 1980s, is whether cultural change can lead to new preferences that will in turn relieve humanity's pressure on the earth's resources. Human culture, not some inherent will of the machine, has created most revenge effects. Without the taste for silk, there would have been no gypsy moths in North America. Without the preference for detached housing, there would still be congestion, perhaps, but more economical congestion. Without the love of oceanside living, shore erosion yes, but no social disruption.

Even more promising than diversification and dematerialization is an attitude that has not yet found its rightful name. It is the substitution of cunning for the frontal attack, and it is not new. It began with immunization against smallpox—as we have seen, a folk practice long before Edward Jenner introduced it to medicine—and continued with the vaccines of the late nineteenth and twentieth centuries.

Finesse means abandoning frontal attacks for solutions that rely on the same kind of latent properties that led to revenge effects in the first place. Sometimes it means ceasing to suppress a symptom. In medicine, finesse suggests closer attention to the evolutionary background of human health and illness, to the positive part that fever plays, for example, in fighting infection. At other times, finesse means living with and even domesticating a problem organism. As we have seen, researchers like Stanley Falkow and Paul Ewald have suggested a kind of evolutionary compromise with what are now lethal bacteria and viruses, turning them into common but harmless companions. In the office, finesse means producing more by taking more frequent breaks and conveying more information by, for example, limiting rather than multiplying color schemes. In construction, finesse means allowing skyscrapers to sway slightly in the wind instead of bracing them to resist it. On the road, finesse means a calmer approach to driving, improving the speed and economy of all drivers by slowing them down at times when impulse would prompt accelerating. It can mean moving more traffic by metering access to some roads and even closing off others. (Some German analysts have written of the "softening," *Besänftigung*, of traffic.) Diversification, dematerialization, and finesse are far from a rejection of science. To the contrary, it is science that points us away from crude reductionism and counterproductive brute force toward technologies that improve human life. But the improvement has a cost.

As the Red Queen said in *Through the Looking-Glass*, we are no longer in the "slow sort of country" where running gets one somewhere: "Now, *here*, you see, it takes all the running you can do, to keep in the same place. If you want to get somewhere else, you must run at least twice as fast as that!" And in fact the alternatives to the intensified, revenge-prone modes of earlier technology seem to take nearly all the running we can do. Even the optimistic report of the Council for Agricultural Science and Technology (CAST) makes clear that most of our agricultural research goes to "maintenance," that is, to keeping the gains we have made: dealing with deteriorating water quality and increasing costs, and offsetting "biological surprises like the appearance of more virulent pests." The same could probably be said of many medical efforts. Similarly, the power of personal computer hardware seems driven by the need to compensate for the way that more elaborate interfaces and features slow the fundamentals of performance.[48]

Technological optimism means in practice the ability to recognize bad surprises early enough to do something about them. And that demands constant monitoring of the globe, for everything from changes in mean temperatures and particulates to traffic in bacteria and viruses. It also requires a second level of vigilance at increasingly porous national borders against the world exchange of problems. But vigilance does not end there. It is everywhere. It is in the random alertness tests that have replaced the "dead man's pedal" for train operators. It is in the rituals of computer backup, the legally mandated testing of everything from elevators to home smoke alarms, routine X-ray screening, securing and loading new computer-virus definitions. It is in the inspection of arriving travelers for products that might harbor pests. Even our alertness in crossing the street, second nature to urbanites now, was generally unnecessary before the eighteenth century. Sometimes vigilance is more of a reassuring ritual than a practical precaution, but with any luck it works. Revenge effects mean in the end that we will move ahead but must always look back just because reality is indeed gaining on us.

For Further Reading

There are few books and articles only and explicitly on unintended consequences, but countless ones with a strong element of irony or paradox. This list is selective and personal and does not attempt to repeat even all the main sources in the notes to the chapters. Some of the books listed here could not be discussed explicitly in the text, sometimes because the issues they raise would have demanded whole chapters.

Robert K. Merton first called the attention of social scientists to the importance of what I have called revenge effects in his paper "The Unanticipated Consequences of Purposive Social Action," *American Sociological Review*, vol. 1, no. 6 (December 1936), 894–904, and continued it in his paper "The Self-Fulfilling Prophecy," in Aaron Rosenblatt and Thomas F. Gieryn, eds., *Social Research and the Practicing Professions* (Cambridge, Mass.: Abt Books, 1982), 248–67. Charles Perrow, *Normal Accidents: Living with High-Risk Technologies* (New York: Basic Books, 1984), is the most influential contemporary analysis of the perils of complex systems. Aaron Wildavsky, *Searching for Safety* (New Brunswick, N.J.: Transaction Books, 1988), uses unintended consequences to argue for a strategy of "resilience." Ulrich Beck, *Risk Society: Towards a New Modernity* (London: Sage, 1992), and Niklas Luhmann, *Risk: A Sociological*

Theory (New York: A. de Gruyter, 1993), are two important recent German contributions. The fall 1990 issue of *Daedalus,* on risk, is a useful collection of papers.

Albert H. Teich, ed., *Technology and the Future,* 5th edn. (New York: St. Martin's, 1990), has a number of important contributions. Charles Piller, *The Fail-Safe Society: Community Defiance and the End of American Technological Optimism* (New York: Basic Books, 1991), and Chellis Glendinning, *When Technology Wounds: The Human Consequences of Progress* (New York: Morrow, 1990), are able defenses of technological skepticism. Of many excellent books in technology studies, Langdon Winner, *The Whale and the Reactor: A Search for Limits in an Age of High Technology* (Chicago: University of Chicago Press, 1986), Howard P. Segal, *Technological Utopianism in American Culture* (Chicago: University of Chicago Press, 1985), and Howard P. Segal, *Future Imperfect: The Mixed Blessings of Technology in America* (Amherst: University of Massachusetts Press, 1994), are worth special mention.

Of books by scientists and engineers on risk and failure, two of the best are Henry Petroski, *To Engineer Is Human: The Role of Failure in Successful Design* (New York: St. Martin's, 1985), and H. W. Lewis, *Technological Risk* (New York: Norton, 1990). J. G. U. Adams, *Risk and Freedom: The Record of Road Safety Regulation* (London: Transport Publishing Projects, 1985), and Gerald J. S. Wilde, *Target Risk* (Toronto: PDE Publications, 1994), are the most provocative statements of "risk homeostasis," the theory that safety measures lead human beings to compensate by taking greater risks. Leonard Evans, *Traffic Safety and the Driver* (New York: Van Nostrand Reinhold, 1991), is a masterly summary of research on highway risk, opposed (as are the great majority of highway safety specialists) to the Adams/Wilde version of risk homeostasis.

The best source of news about, and general-interest interpretation of, technological risks may be the British weekly magazine *New Scientist.* It also offers substantial historical insight. Mick Hamer's "Lessons from a Disastrous Past" (vol. 128, no. 1748 [22 December 1990], 72–74), suggests why the potential for disaster was so high a hundred years ago.

While this book could not deal extensively with social institutions, several studies of unintended social consequences are worth reading for their range and insight: Fred Hirsch, *Social Limits to Growth* (Cambridge, Mass.: Harvard University Press, 1976); Albert O. Hirschman, *The Rhetoric of Reaction: Perversity, Futility, Jeopardy* (Cambridge, Mass.: Harvard University Press, 1991); and Robert H. Frank, *Choosing the Right Pond: Human Behavior and the Quest for Status* (New York: Oxford University Press, 1985). I have largely left housework out of this book because Ruth Schwartz Cowan treats it so well in *More Work for Mother: The Ironies of Household Technology from the Open Hearth to the Microwave* (New York: Basic Books, 1983).

On the unintended consequences of medicine, the most destructively brilliant book remains Ivan Illich, *Medical Nemesis: The Expropriation of Health* (New York: Pantheon, 1976). Arthur J. Barsky, *Worried Sick: Our Troubled Quest for Wellness* (Boston: Little, Brown, 1988), Thomas McKeown, *The Role of Medicine: Dream, Mirage or Nemesis* (Oxford: Basil Blackwell, 1979), and Leonard A. Sagan, *The Health of Nations* (New York: Basic Books, 1987), are all thoughtful responses by physicians of very different backgrounds and temperaments to the issues Illich raises. Randolph M. Nesse and George C. Williams, *Why We Get Sick: The New Science of Darwinian Medicine* (New York: Times Books, 1994), explores the evolutionary background of medical dilemmas and presents a hopeful message while suggesting how and why nature has come to be booby-trapped.

On emerging health hazards, some of the most important recent works have been Stephen S. Morse, ed., *Emerging Viruses* (New York: Oxford University Press, 1993), Joshua Lederberg, ed., *Emerging Infections: Microbial Threats to Health in the United States* (Washington, D.C.: National Academy Press, 1992), and Laurie Garrett, *The Coming Plague: Newly Emerging Diseases in a World Out of Balance* (New York: Farrar, Straus & Giroux, 1994). Richard Preston, *The Hot Zone* (New York: Random House, 1994), is a riveting popular account of a potential catastrophe. Of hundreds of books on AIDS, Mirko D. Grmek, *History of AIDS: Emergence and*

Origin of a Modern Pandemic (Princeton, N.J.: Princeton University Press, 1990), is the fundamental historical work. Paul W. Ewald, *Evolution of Infectious Disease* (New York: Oxford University Press, 1994), offers an important theoretical perspective on dealing with HIV. Elizabeth Fee and Daniel M. Fox, eds., *AIDS: The Burdens of History* (Berkeley: University of California Press, 1988), and Elizabeth Fee and Daniel M. Fox, eds., *AIDS: The Making of a Chronic Disease* (Berkeley: University of California Press, 1992), are two important collections on AIDS and society. See also Daniel M. Fox, *Power and Illness: The Failure and Future of American Health Policy* (Berkeley: University of California Press, 1993).

On paradoxes of the environment, John McPhee, *The Control of Nature* (New York: Farrar, Straus & Giroux, 1989), is superb. Ian Burton, Robert W. Kates, and Gilbert F. White, *The Environment as Hazard,* 2nd edn. (New York: Guilford Press, 1993), is a sophisticated overview. Anders Wijkman and Lloyd Timberlake, *Natural Disasters: Acts of God or Acts of Man?* (London: Earthscan, 1984), treats unintended consequences of development. Edward Bryan, *Natural Hazards* (Cambridge: Cambridge University Press, 1991), is the most comprehensive survey. On special topics, Wallace Kaufman and Orrin Pilkey, *The Beaches Are Moving* (Garden City, N.Y.: Anchor Press/Doubleday, 1970), and Stephen J. Pyne, *Fire in America: A Cultural History of Wildland and Rural Fire* (Princeton, N.J.: Princeton University Press, 1982), deserve special mention.

Thousands of books and papers exist on pests, but only a small number on unintended introductions, and on the consequences of trying to suppress them. Two collective works are excellent starting points: U.S. Congress, Office of Technology Assessment, *Harmful Non-Indigenous Species in the United States,* OTA-F-565 (Washington, D.C.: U.S. Government Printing Office, September 1993), and Bill N. McKnight, ed., *Biological Pollution: The Control and Impact of Invasive Exotic Species* (Indianapolis: Indiana Academy of Science, 1993). David Ehrenfeld, *The Arrogance of Humanism* (New York: Oxford University Press, 1981), is an impassioned argument against the anthropocentrism underlying the concept "pest." As Illich attacks progressive as well as authoritarian medicine, Ehrenfeld questions not only the exploitation of nature but the self-deception underlying enlightened "management" of resources.

On computer issues, Rob Kling, *Computerization and Controversy,* 2nd edn. (San Diego: Academic Press, 1995), is the most comprehensive and diverse collection of articles, with some original contributions. Joseph Weizenbaum, *Computer Power and Human Reason* (San Francisco: W. H. Freeman, 1976), is still on target after twenty years. On the computerized office, Shoshana Zuboff, *In the Age of the Smart Machine* (New York: Basic Books, 1984), and Juliet B. Schor, *The Overworked American: The Unexpected Decline of Leisure* (New York: Basic Books, 1991), illustrate the sources of our ambiguity; Daniel Crevier, *AI: The Tumultuous History of the Search for Intelligence* (New York: Basic Books, 1993), blends sober realism with unquenchable optimism, as does Thomas K. Landauer, *The Trouble with Computers* (Cambridge, Mass.: MIT Press, 1995). And on the folklore of computer revenge effects, there is Karla Jennings, *The Devouring Fungus: Tales of the Computer Age* (New York: Norton, 1990).

No history of the technology of sports exists. Probably the most influential interpretation of sports history is Allen Guttmann, *From Ritual to Record: The Nature of Modern Sports* (New York: Columbia University Press, 1978). Peter J. Brancazio, *Sport Science: Physical Laws and Optimum Performance* (New York: Simon & Schuster, 1984), is an excellent overview and reference, with some valuable points on the effects of technological change on sports. John Jerome, *The Sweet Spot in Time* (New York: Summit Books, 1980), is a fascinating and insightful survey of technology and performance but lacks references. Eric W. Schrier and William F. Allman, *Newton at the Bat* (New York: Charles Scribner's Sons, 1984), is a lively set of articles.

Notes

Preface

1. Paul Valéry, "Unpredictability," in *History and Politics,* trans. Denise Folliot and Jackson Matthews (New York: Pantheon, 1962), 71.

2. See *Today Then: America's Best Minds Look 100 Years into the Future on the Occasion of the 1893 World's Columbian Exposition,* compiled and introduced by Dave Walter (Helena, Mont.: American & World Geographic Pub., 1992).

3. See John von Neumann, "Can We Survive Technology?," in *The Fabulous Future: America in 1980* (New York: Dutton, 1956), 36–37.

1 Ever Since Frankenstein

1. Rod Serling, "A Thing About Machines," in *More Stories from the Twilight Zone* (New York: Bantam Books, 1961), 51–71.

2. Hilaire Belloc, "Lord Finchley," in *Hilaire Belloc's Cautionary Verses* (New York: Alfred A. Knopf, 1941), 268–69.

3. Michael Matz, "A Lemon-Law Loophole Gets New Scrutiny," *Philadelphia Inquirer,* 8 February 1992, A1.

4. David L. Cohn, *The Good Old Days* (New York: Arno Press, 1976), 153.

5. Erik Sandberg-Diment, "A Computer Comes In from the Cold," *New York Times,* 21 April 1987, C4.

6. See Bruce Watson, "For a While, the Luddites Had a Smashing Success," *Smithsonian,* April 1993, 140–54; Bob Sipchen, "Stop the World: Neo-Luddites Fear Steamroller of Technology Threatens Humanity," *Los Angeles Times,* 25 February 1992, E1; Chellis Glendenning, "Notes Toward a Neo-Luddite Manifesto," *Utne Reader,* March–April, 50–53.

7. Paul Jennings, "Report on Resistentialism," in Dwight Macdonald, ed., *Parodies: An Anthology from Chaucer to Beerbohm—and After* (New York: Modern Library, 1965), 394–95.

8. "Vehicle 'Lock-Outs' Pose Threat to Personal Safety," *AAA Spotlight,* July–August 1993; Jeff Gammage, "Cracking Down on False Alarms," *Philadelphia Inquirer,* 1 July 1993, A1; Robert Schulman, "Burglar, Fire, Flood—Name It, You Can Guard Against It," *New York Times,* 10 January 1993, Business 10.

9. William Stolzenburg, "Bad Move for Tortoises," *Nature Conservancy Magazine,* vol. 43, no. 3 (May/June 1993), 7.

10. See Theodore A. Postol, "Lessons of the Gulf War Experience with Patriot," *International Security,* vol. 16, no. 3 (Winter 1991–1992), 119–70, esp. 139–51; also Eliot Marshall, "Patriot's Scud Busting Record Is Challenged," *Science,* vol. 252, no. 5006 (3 May 1991), 640–41.

11. Peter G. Ryan and Coleen L. Moloney, "Marine Litter Keeps Increasing," *Nature,* 7 January 1993, 23.

12. *Hamlet,* III.iv.205–6.

13. Gerhard Heilfurth, *Der Bergbau und seine Kultur: Eine Welt zwischen Dunkel und Licht* (Zurich: Atlantis Verlag, 1981), 222.

14. See Chris Baldock, "Taming the Monster," *Times Literary Supplement,* 27 July–2 August 1990, 801, reviewing Steven Early Forry, *Hideous Prodigies: Dramatizations of "Frankenstein" from the Nineteenth Century to the Present* (Philadelphia: University of Pennsylvania Press, 1990).

15. Mary Wollstonecraft Shelley, *Frankenstein, or the Modern Prometheus,* ed. James Rieger (New York: Pocket Books, 1976), 50–51.

16. See Anne K. Mellor, "*Frankenstein:* A Feminist Critique of Science," in George Levine, ed., *One Culture: Essays in Science and Literature* (Madison: University of Wisconsin Press, 1987), 287–312; Laura Kranzler, "Frankenstein and the Technological Future," *Foundation,* no. 44 (Winter 1988–89), 42–49; and Susan Squier, " 'Frankenstein' Has Message for Feminists," *New York Times,* 19 April 1992, sec. 4, 10 (letter).

17. Shelley, *Frankenstein,* 59.

18. Ibid., 54, 266–67.

19. See Edit Fél and Tamás Hofer, *Geräte der Átányer Bauern* (Copenhagen: Royal Danish Academy of Sciences, 1974), 291–304, for a fascinating discussion of peasant tools as extensions of the body. Robert C. Davis, *Shipbuilders of the Venetian Arsenal: Workers and Workplace in the Preindustrial City* (Baltimore: Johns Hopkins University Press, 1991), suggests that the thousands of arsenal workers were organized in a more industrial way than historians had thought. But even here there is little evidence that they saw a ship or the work of shipbuilding as a system.

20. James R. Blackaby, "How the Workbench Changed the Nature of Work," *American Heritage of Invention and Technology,* Fall 1986, 26–30.

21. Paul Hoffman, "Six-Legged Saboteurs," *Discover,* May 1989, 81–83.

22. Edwin Gabler, *The American Telegrapher: A Social History, 1860–1900* (New Brunswick, N.J.: Rutgers University Press, 1988), 45, 48.

23. This account follows the *Boston Globe* article of Brad Pokorny, "In Boston, a Weakening Foundation," reprinted in the *Philadelphia Inquirer,* 28 November 1985.

24. Charles Perrow, *Normal Accidents: Living with High-Risk Technologies* (New York: Basic Books, 1984).

25. Clinton V. Oster, Jr., John S. Strong, and C. Kurt Zorn, *Why Airplanes Crash: Aviation Safety in a Changing World* (New York: Oxford University Press, 1992), 136.

26. Walter Muir Whitehill, *Boston: A Topographical History,* 2nd edn. (Cambridge: Harvard University Press, 1968), 164; Thomas P. Hughes, *American Genesis: A Century of Invention and Technological Enthusiasm* (New York: Viking, 1989), 300–1.

27. Bob Beyers, "Future Shock a 'Popular Misconception,' McCarthy Says," *Stanford Campus Report,* 26 January 1983, 11. Even by 1994, microcomputers were still not meeting what McCarthy thought were the conditions of a true revolution: changing systems of political representation and providing true mass access to government documents, for example.

28. Daniel J. Boorstin, *Hidden History: Exploring Our Secret Past* (New York: Harper & Row, 1987), 252.

29. Hughes, *American Genesis,* 446–50.

30. "Gremlin," *Oxford English Dictionary,* supp. vol. 1, 1294.

31. See Dianna Waggoner's excellent reporting in "Murphy's Law Really Works . . ." *People Weekly,* 31 January 1983, 81–82; Hugh Kenner, "Things Do Go Wrong; Does That Mean Nothing Works?" *Byte,* 1 January 1990, 416; Arthur Block, *Murphy's Law* (Los Angeles: Price/Stern/Sloan, 1979), especially George E. Nichols's letter on 4–5.

32. Lee Dye, "The Man Who Proved Seat Belts Can Save Lives," *Philadelphia Inquirer,* 1 February 1987, 24A.

33. Greedy landlords built these *insulae* as cheaply as possible, and Roman leasing practices actually could make collapse profitable. The emperors' occasional efforts to limit the size of these buildings were feeble, and construction standards beneath elegant facades remained dismal. Since the richest Romans lived near the *insulae,* and sometimes on their palatial ground floors, debris and fires endangered all classes. Why was Roman building not reformed? The pressure of population and the limits of government were probably the main reasons. Disaster does not ensure improvement. See Jérôme Carcopino, *Daily Life in Ancient Rome,* trans. E. O. Lorimer (New Haven: Yale University Press, 1940), 22–44.

34. Stuart Flexner with Doris Flexner, *The Pessimist's Guide to History* (New York: Avon Books, 1992), 179–80.

35. A. Bryant, *Natural Hazards* (Cambridge: Cambridge University Press, 1992), 68. Fred Pearce, "Back to the Days of Deadly Smogs," *New Scientist,* vol. 136, no. 1850 (8 December 1992), 24–28.

36. Flexner, *Pessimist's Guide to History,* 361–63, 352, 354–55.

37. Oster et al., *Why Airplanes Crash,* 23.

38. Spencer Weart, "From the Nuclear Frying Pan into the Global Fire," *Bulletin of the Atomic Scientists,* vol. 48, no. 5 (June 1992), 19–27.

39. Marilyn Chase, "X-Rays Shed Light on Diagnosis, Cure—But Know the Risks," *Wall Street Journal,* 17 July 1995, B1; Steve Bates, "Shards of Truck Brake Drums Are Common Highway Hazard," *Washington Post,* 30 July 1993, B3; Steve Bates, "Death Out of Nowhere," *Washington Post,* 29 July 1993, A1; Steve Bates, "Tracing a Fatal Crash," *Washington Post,* 31 July 1993, A1.

40. Will Steger and Jon Bowermaster, *Saving the Earth* (New York: Alfred A. Knopf, 1990), 27–37; Michael Weisskopf, "CFCs: Rise and Fall of Chemical 'Miracle,'" *Washington Post,* 10 April 1988, A1; Aaron Wildavsky, *Searching for Safety* (New Brunswick, N.J.: Rutgers University Press, 1988), 196–97.

41. Nita A. Davidson and Nick D. Stone, "Imported Fire Ants," in Donald L. Dahlsten and Richard Garcia, eds., *Eradication of Exotic Pests: Analysis with Case Histories* (New Haven: Yale University Press, 1989), 208 [196–217]; Chris Offutt, "Troubles Rise as the Water Drops," *New York Times,* 1 September 1993, A19; Wildavsky, *Searching for Safety,* 198–99.

42. Larry Thompson, "The Atomic Bomb: Scientists Reassess the Long-Term Impact of Radiation," *Washington Post,* 14 April 1990, Z12. For

a full scientific discussion, including variations in additional risk for specific cancers, see Yushiko Shimizu, William J. Schull, and Hiroo Kato, "Cancer Risk Among Atomic Bomb Survivors," *Journal of the American Medical Association,* vol. 264, no. 5 (1 August 1990), 601–9. An excellent recent overview is Ken Ringle, "A Fallout over Numbers," *Washington Post,* 5 August 1995, A16.

2 Medicine: Conquest of the Catastrophic

1. Arthur J. Barsky, *Worried Sick: Our Troubled Quest for Wellness* (Boston: Little, Brown, 1988), 3–20; "Americans Healthier, but Not Feeling as Well," *Los Angeles Times,* 22 February 1988, part 2, 5; Ivan Illich, *Medical Nemesis: The Expropriation of Health* (New York: Pantheon, 1976), 40–41; Aaron Wildavsky, "Doing Better and Feeling Worse: The Political Pathology of Health Policy," *Daedalus,* vol. 106, no. 1 (Winter 1977), 105–25.

2. Amartya Sen, "Economics of Life and Death," *Scientific American,* vol. 268, no. 5 (May 1993), 40–47.

3. Thomas McKeown, *The Role of Medicine: Dream, Mirage, or Nemesis?* (Oxford: Basil Blackwell, 1979).

4. John B. McKinlay and Sonja M. McKinlay, "The Questionable Contribution of Medical Measures to the Decline of Mortality in the United States in the Twentieth Century," *Milbank Memorial Fund Quarterly, Health and Society,* vol. 55, no. 3 (Summer 1977), 414.

5. Boyce Rensberger, "Evidence of Ill Health Erodes Legend of Ancient Colony's Well-Being," *Washington Post,* 16 November 1992, A3; Daniel Pool, *What Jane Austen Ate and Charles Dickens Knew* (New York: Simon & Schuster, 1993), 203–6.

6. Leonard A. Sagan, *The Health of Nations* (New York: Basic Books, 1987), 42–53.

7. Donald S. Kenkel, "Health Behavior, Health Knowledge, and Schooling," *Journal of Political Economy,* vol. 99, no. 2 (1991), 287–305.

8. Tony Horwitz, "To Die For: Lethal Cuisine Takes High Toll in Glasgow, West's Sickest City," *Wall Street Journal,* 22 September 1992, A1.

9. Roy Porter and Dorothy Porter, *In Sickness and in Health: The British Experience 1650–1850* (London: Fourth Estate, 1988), 106; Edward L. Shorter, *Bedside Manners: The Troubled History of Doctors and Patients* (New York: Simon & Schuster, 1985), 33; John O'Shea, *Was Mozart Poisoned? Medical Investigations into the Lives of the Great Composers* (New York: St. Mar-

tin's Press, 1991), 186; Erwin H. Ackerknecht, *A Short History of Medicine,* rev. edn. (Baltimore: Johns Hopkins University Press, 1982), 150.

10. Porter and Porter, *In Sickness and in Health,* 265–69; O'Shea, *Was Mozart Poisoned?* 65–88. "Side effect" is a relatively recent phrase. It first appeared in a book title with Louis Lewin's *Side Effects of Medication* (*Die Nebenwirkungen der Arzneimittel*) in 1881. The *Oxford English Dictionary* didn't list it or the alternative "adverse effect" until an entry in a supplement citing a 1939 reference book, Wright and Montag's *Materia Medica* (*Pharmacology and Therapeutics,* vol. 10, 112); the same supplement's first reference to "iatrogenic" (medically caused) illness dates from 1923. See Erwin H. Ackerknecht, "Zur Geschichte der iatrogenen Krankheiten," *Gesnerus,* vol. 27, no. 1/2 (1970), 57–63.

11. Stanley Joel Reiser, *Medicine and the Reign of Technology* (Cambridge: Cambridge University Press, 1981), 41.

12. Ibid., 77.

13. Ibid., 36–37, 61–62.

14. Lewis Thomas, *The Lives of a Cell* (New York: Bantam Books, 1991), 35–42.

15. John Maddox, "Adventures with a Vulnerable Knee," *Nature,* vol. 361, no. 6407 (7 January 1993), 13.

16. Charles E. Rosenberg, "The Therapeutic Revolution: Medicine, Meaning, and Social Change in Nineteenth-Century America," in Morris J. Vogel and Charles E. Rosenberg, eds., *The Therapeutic Revolution: Essays in the Social History of American Medicine* (Philadelphia: University of Pennsylvania Press, 1979), 7–8 on the rationality of the old therapeutics.

17. Early quack remedies reflected the systems of established medicine. Professors at the University of Pennsylvania Medical School did not hesitate to prescribe and endorse one of the early hits, the bookbinder William Swaim's Panacea. See James Harvey Young, *The Toadstool Millionaires* (Princeton, N.J.: Princeton University Press, 1961), 58–62. On twentieth-century developments, see Young's *The Medical Messiahs* (Princeton, N.J.: Princeton University Press, 1967).

18. Gina Kolata, "Wariness Is Replacing Trust Between Healer and Patient," *New York Times,* 28 February 1990, A1-D15.

19. Richard Holmes, "Woeful Crimson," *Times Literary Supplement,* 22 January 1993, 5.

20. William H. McNeill, *The Pursuit of Power: Technology, Armed Force, and Society Since A.D. 1000* (Chicago: University of Chicago Press, 1982), 223–61.

21. Richard A. Gabriel and Karen S. Metz, *A History of Military Medicine,* 2 vols. (New York: Greenwood Press, 1992), vol. 2, 169–70.

22. Ibid., 182, 185; Richard H. Shryock, "A Medical Perspective on the Civil War," *American Quarterly,* vol. 14, no. 2, part 1, 164.

23. Gabriel and Metz, *Military Medicine,* 2:258.

24. Roger O. Egeberg, "Caring for the Wounded in Wartime," *Journal of the American Medical Association,* vol. 264, no. 17 (7 November 1990), 2263–64; Donald L. Custis, "Military Medicine from World War II to Vietnam," *Journal of the American Medical Association,* vol. 264, no. 17 (7 November 1990), 2259–62; Gabriel and Metz, *Military Medicine,* 260.

25. Charles E. Rosenberg, *The Care of Strangers: The Rise of America's Hospital System* (New York: Basic Books, 1987), 101.

26. Ronald Kotulak, "Scientists Offer Hope for Reversing Paralysis," *Chicago Tribune,* sec. 1, 1.

27. Marylou Tousignant, "CIA Shooting Victim Pulled Back from 'Precipice' of Death," *Washington Post,* 28 January 1993, B1–B4.

28. Colin McEvedy, "The Bubonic Plague," *Scientific American,* February 1988, 118–24.

29. William H. McNeill, *Plagues and Peoples* (Garden City, N.Y.: Anchor Press/Doubleday, 1976), 154.

30. "Pneumonic Plague—Arizona, 1992," *Journal of the American Medical Association,* vol. 268, no. 16 (28 October 1992), 2146–47.

31. On quacks and pills: Porter and Porter, *In Sickness and in Health,* 105.

32. Lloyd M. Krieger, "A Folder Full of X-Rays and Computer Printouts, Not a 20-Year-Old Squirming in Pain," *Los Angeles Times,* 24 February 1988.

33. Troyen A. Brennan et al., "Incidence of Adverse Events and Negligence in Hospitalized Patients," *New England Journal of Medicine,* vol. 324, no. 6 (7 February 1991), 370–76; Lucian L. Leape et al., "The Nature of Adverse Events in Hospitalized Patients," *New England Journal of Medicine,* vol. 324, no. 6 (7 February 1991), 377–84.

34. Cristine Russell, "Human Error," *Washington Post,* 18 February 1992, Health, 7.

35. "Cheaper Surgery, but Higher Costs," *New York Newsday,* 26 September 1993, 75. See also Elisabeth Rosenthal, "Questions Raised on New Technique for Appendectomy," *New York Times,* 14 September 1993, C3; Lawrence K. Altman, "Standard Training in Laparoscopy Found Inadequate," *New York Times,* 14 December 1993, C3; Antonio P. Legorreta et al., "Increased Cholecystectomy Rate After the Introduction of Laparoscopic

Cholecystectomy," *Journal of the American Medical Association,* vol. 270, no. 12 (22 September 1993), 1429–32.

36. Lawrence K. Altman, "Surgical Injuries Lead to New Rule," *New York Times,* 14 June 1992, A1; Lawrence K. Altman, "When Patient's Life Is Price of Learning New Kind of Surgery," *New York Times,* 23 June 1992, C3.

37. David Naylor et al., "Pulmonary Artery Catheterization: Can There Be an Integrated Strategy for Guidelines Development and Research Promotion?" *Journal of the American Medical Association,* vol. 269, no. 18 (12 May 1993), 2407–11; Eugene D. Robin and Robert F. McCauley, "Risk-Benefit Analysis in Cardiovascular Disease," in Amar S. Kapoor and Bramah N. Singh, eds., *Prognosis and Risk Assessment in Cardiovascular Disease* (New York: Churchill Livingstone, 1993), 17–25.

38. Janny Scott, "Hospitals Can Make You Sick," *Los Angeles Times,* 28 July 1992, A1; "Hand-Washing Habits Studied in Hospitals," *Washington Post,* Health sec., Z5; Donald Goldmann and Elaine Larson, "Handwashing and Nosocomial Infections," *New England Journal of Medicine,* vol. 327, no. 2 (9 July 1992), 120–22; Ulla Bettge, "Claim That Hospital Patients Are Dying Because of Unhygienic Practices," *German Tribune,* 2 September 1990, 13.

3 Medicine: Revenge of the Chronic

1. Daniel M. Fox, "AIDS and the American Health Polity," in Elizabeth Fee and Daniel M. Fox, eds., *AIDS: The Burdens of History* (Berkeley: University of California Press, 1988), 319–20; René Dubos, *Mirage of Health: Utopias, Progress, and Biological Change* (Garden City, N.Y.: Anchor, 1959), 137; Abdel R. Omran, "The Epidemiologic Transition: A Theory of the Epidemiology of Population Change," *Milbank Memorial Fund Quarterly,* vol. 49, no. 4 (October 1971), 509–38.

2. Ernest M. Gruenberg, "The Failures of Success," *Milbank Memorial Fund Quarterly: Health and Society,* vol. 55, no. 1 (Winter 1977), 3–24.

3. See Alvan R. Feinstein, "Scientific Standards in Epidemiologic Studies of the Menace of Daily Life," *Science,* vol. 242, no. 4883 (2 December 1988), 1261; Paul Weindling, "From Infectious to Chronic Diseases: Changing Patterns of Sickness in the Nineteenth and Twentieth Centuries," in Andrew Wear, ed., *Medicine in Society: Historical Essays* (Cambridge: Cambridge University Press, 1992), 306.

4. See the excellent treatment in David Rosner and Gerald Markowitz, *Deadly Dust: Silicosis and the Politics of Occupational Disease in Twentieth-Century America* (Princeton: Princeton University Press, 1991), 4–48.

5. Ibid., 201. The same official pointed out that while the public reads about "mine tragedies as being explosions, fires, and other disasters," silicosis actually killed more men.

6. On nineteenth-century medical views of women and neurasthenia, see Carroll Smith-Rosenberg and Charles Rosenberg, "The Female Animal: Medical and Biological Views of Woman and Her Role in Nineteenth-Century America," *Journal of American History,* vol. 60, no. 2 (September 1973), 332–56; on the uses of invalidism in Victorian culture, see Miriam Bailin, *The Sickroom in Victorian Fiction* (Cambridge: Cambridge University Press, 1994); on Lyme disease as a "new" chronic illness and the complex negotiations between physicians and patients on its meaning, see Robert A. Aronowitz, "Lyme Disease: The Social Construction of a New Disease and Its Social Consequences," *Milbank Quarterly,* vol. 69, no. 1 (Spring 1991), 79–112.

7. Peter J. Bowler, *Evolution: The History of an Idea,* rev. edn. (Berkeley: University of California Press, 1989), 175; John Winslow, *Darwin's Victorian Malady: Evidence for Its Medically Induced Origins* (Philadelphia: American Philosophical Society, 1971); Ralph Colp, Jr., *To Be an Invalid: The Illness of Charles Darwin* (Chicago: University of Chicago Press, 1977).

8. Anselm L. Strauss et al., *Chronic Illness and the Quality of Life,* 2nd edn. (St. Louis: C. V. Mosby, 1984), 1–18; see also Daniel M. Fox, *Power and Illness: The Failure and Future of American Health Policy* (Berkeley: University of California Press, 1993).

9. David M. Eisenberg, "Unconventional Medicine in the United States," *New England Journal of Medicine,* vol. 328, no. 4 (28 January 1993), 246–52.

10. Rosie Mestel, "Diabetics Protected by Strict Regime," *New Scientist,* vol. 138, no. 1878 (19 June 1993), 7.

11. Wolfgang Schivelbusch, *The Railway Journey: The Industrialization of Time and Space in the Nineteenth Century* (Berkeley: University of California Press, 1986), 139.

12. Roger J. Spiller, "Shell Shock," *American Heritage,* May–June 1990), 77, 84–85.

13. Ghislaine Boulanger, "Post-Traumatic Stress Disorder: An Old Problem with a New Name," in Stephen M. Sonnenberg, Arthur S. Blank, Jr., and John A. Talbott, eds., *The Trauma of War: Stress and Recovery in Viet Nam Veterans* (Washington: American Psychiatric Press, 1985), 20.

14. Robert J. White and Matt J. Likavec, "The Diagnosis and Initial Management of Head Injury," *New England Journal of Medicine,* vol. 327, no. 21 (19 November 1992), 1507–11; Peter Pae, " 'Silent Epidemic' Raises

Voices; Va. Man Urges Resources for the Head-Injured," *Washington Post,* 25 July 1988, D1; D. Neil Brooks, "The Head-Injured Family," *Journal of Clinical and Experimental Neuropsychology,* vol. 13, no. 1 (1991), 169.

15. Brooks, "Head-Injured Family," 165–88.

16. Sid Moody, "Phantom of Poliomyelitis Still Haunting Ex-Patients," *New Brunswick Home News,* 12 September 1988, A8; Theodore L. Munsat, "Poliomyelitis: New Problems with an Old Disease," *New England Journal of Medicine,* vol. 324, no. 17 (25 April 1991), 1206–7.

17. Jane E. Brody, "Treating Ills in Childhood Cancer Survivors," *New York Times,* 3 February 1993, C13; Joan Kirchner, "Tracking Childhood Cancer Survivors," *Washington Post,* 23 March 1993, Health, 11.

18. James C. Riley, *Sickness, Recovery and Death: A History and Forecast of Ill Health* (Iowa City: University of Iowa Press, 1989), 48–49, 122.

19. Ibid., 120, 242–43.

20. Lynn Rogers, *Dirt and Disease: Polio Before FDR* (New Brunswick, N.J.: Rutgers University Press, 1992), 165–66; Dubos, *Mirage of Health,* 164–65; Lynn Payer, *Medicine and Culture: Varieties of Treatment in the United States, England, West Germany, and France* (New York: Henry Holt, 1988), 69–70.

21. Steven Austad, "On the Nature of Aging," *Natural History,* February 1992, 34.

22. Robert M. Sapolsky and Caleb E. Finch, "On Growing Old," *The Sciences,* vol. 31, no. 2 (March–April 1991), 30–38.

23. H. Stewart, "A Mandate for State Action," presented at the Association of State and Territorial Health Officers, Washington, D.C., 4 December 1967, cited in Laurie Garrett, *The Coming Plague* (New York: Farrar, Straus & Giroux, 1994), 33, 624n9.

24. Stuart B. Levy, *The Antibiotic Paradox: How Miracle Drugs Are Destroying the Miracle* (New York: Plenum, 1992), 109–14.

25. Mike Toner, "The Best of Drugs, the Worst of Drugs," *Atlanta Journal-Constitution,* 5 September 1992, E1; Ann Gibbons, "Exploring New Strategies to Fight Drug-Resistant Microbes," *Science,* vol. 257, no. 5073 (21 August 1992), 1036–38.

26. Mike Toner, "Antibiotics, Salmonella a Deadly Combination," *Atlanta Journal-Constitution,* 29 August 1992, E1; Mitchell L. Cohen, "Epidemiology of Drug Resistance: Implications for a Post-Antimicrobial Era," *Science,* vol. 257, no. 5073 (21 August 1992), 1050–55.

27. Phyllida Brown, "The Return of the Big Killer," *New Scientist,* vol. 135, no. 1842 (10 October 1992), 30–37; Mario C. Raviglione, Dixie E. Snider,

Jr., and Arata Kochi, "Global Epidemiology of Tuberculosis," *Journal of the American Medical Association,* vol. 273, no. 3 (18 January 1995), 220–26.

28. Raviglione et al., "Global Epidemiology," 224.

29. Roy M. Anderson and Robert M. May, "Understanding the AIDS Pandemic," *Scientific American,* vol. 266, no. 5 (May 1992), 58–59.

30. Mirko D. Grmek, *History of AIDS: Emergence and Origin of a Modern Pandemic* (Princeton, N.J.: Princeton University Press, 1990), 156–70, 180–81; the word derives from Greek roots meaning "disease" and "common" or "general." See Grmek's extended discussion in *Diseases in the Ancient Greek World,* trans. Mireille Muellner and Leonard Muellner (Baltimore: The Johns Hopkins University Press), 2–4 and 88–92.

31. Sunetra Gupta and Roy Anderson, "Sex, AIDS, and Mathematics," *New Scientist,* vol. 136, no. 1838 (12 September 1992), 34; U.S. Department of Commerce, Bureau of the Census, *Statistical Abstract of the United States 1993,* 96, 134.

32. Elizabeth Fee and Daniel M. Fox, "Introduction: The Contemporary Historiography of AIDS," in Elizabeth Fee and Daniel M. Fox, eds., *AIDS: The Making of a Chronic Disease* (Berkeley: University of California Press, 1992), 4–5; Michael Waldholz, "New Discoveries Dim Drug Makers' Hopes for Quick AIDS Cure," *Wall Street Journal,* 26 May 1992, A1.

33. Anderson and May, "Understanding the AIDS Pandemic," 66.

34. Kenneth E. Warner and John Slade, "Low Tar, High Toll," *American Journal of Public Health,* vol. 82, no. 1 (January 1992), 17–18. The consequences of "safer" cigarettes are among the best-documented revenge effects. See Annamma Augustine, Randall E. Harris, and Ernst L. Wynder, "Compensation as a Risk Factor for Lung Cancer in Smokers Who Switch from Nonfilter to Filter Cigarettes," *American Journal of Public Health,* vol. 79, no. 2 (February 1989), 188–91; Lynn T. Kozlowski, Marilyn A. Pope, and Joann E. Lux, "Prevalence of the Misuse of Ultra-Low-Tar Cigarettes by Blocking Filter Vents," *American Journal of Public Health,* vol. 78, no. 6 (June 1988), 694–95; David J. Maron and Stephen P. Fortmann, "Nicotine Yield and Measures of Cigarette Smoke Exposure in a Large Population: Are Lower-Yield Cigarettes Safer?" *American Journal of Public Health,* vol. 77, no. 5 (May 1987), 546–49; Neal L. Benowitz et al., "Smokers of Low-Yield Cigarettes Do Not Consume Less Nicotine," *New England Journal of Medicine,* vol. 309, no. 3 (21 July 1983), 139–42; Helen Saul, "Chancing Your Arm on Nicotine Patches," *New Scientist,* vol. 137, no. 1860 (13 February 1993), 12–13; David Margolick, "Ex-Kent Smoker Blames Filter of Past for Illness," *New York Times,* 30 August 1991, B20. The industry once attempted to argue that there

are risks in quitting. A statistician at the Tobacco Advisory Council in London suggested that doctors who gave up smoking in the 1950s and 1960s may have suffered higher mortality rates from stress-related hazards, especially suicide and accidents, offsetting lower rates of cancer and heart disease. And weight gain, with related health risks, often does follow cessation. See P. N. Lee, "Has the Mortality of Male Doctors Improved with the Reductions in Their Cigarette Smoking?" *British Medical Journal,* 15 December 1979, 1538–40. In the mid-1990s there is limited evidence that cigarette smokers have lower rates of Parkinson's disease and Alzheimer's disease than nonsmokers. For the controversy surrounding research on this topic, see Ian Mundell, "Peering Through the Smoke Screen," *New Scientist,* vol. 140, no. 1894 (9 October 1993), 14–15, and related editorial, "Challengers of Fashion," 3.

35. Natalie Angier, "Theory Hints Sunscreens Raise Melanoma Risks," *New York Times,* 9 October 1990; Cedric F. Garland, Frank C. Garland, and Edward D. Gorham, "Rising Trends in Melanoma: A Hypothesis Concerning Sunscreen Effectiveness," unpublished manuscript (1991).

36. Kelly D. Brownell, "Personal Responsibility and Control Over Our Bodies: When Expectation Exceeds Reality," *Health Psychology,* vol. 10, no. 5 (1991), 303–10; Kelly D. Brownell, "Dieting and the Search for the Perfect Body: Where Physiology and Culture Collide," *Behavior Therapy,* vol. 22 (1991), 1–12; Jean Mayer and Jeanne Goldberg, "Is the Whole World on a Diet? Not Quite, but the Figures Are Hefty," *Philadelphia Inquirer,* 27 September 1992, K7; Claude Bouchard, "Is Weight Fluctuation a Risk Factor?" *New England Journal of Medicine,* vol. 324, no. 26 (27 June 1991), 1887–88, citing "Obesity and Eating Disorders" in National Research Council, *Diet and Health: Implications for Reducing Chronic Disease Risk* (Washington, D.C.: National Academy Press, 1989), 563–92.

37. Brownell, "Personal Responsibility and Control over Our Bodies," 307; Janet Polivy and C. Peter Herman, "Dieting as a Problem in Behavioral Medicine," *Advances in Behavioral Medicine,* vol. 1 (1985), 1–37.

38. Gina Kolata, "The Burdens of Being Overweight: Mistreatment and Misconceptions," *New York Times,* 22 November 1992, A1.

39. Roy J. Martin, B. Douglas White, and Martin G. Hulsey, "The Regulation of Body Weight," *American Scientist,* vol. 79, no. 6 (November–December 1991), 528–41.

40. Peter Herman, personal communication, 30 January 1992.

41. Janet Polivy and Peter Herman, "Dieting and Binging: A Causal Analysis," *American Psychologist,* vol. 40, no. 2 (1985), 193–201; C. Peter Herman and Janet Polivy, "Studies of Eating in Normal Dieters," in B. T.

Walsh, ed., *Eating Behavior in Eating Disorders* (Washington, D.C.: American Psychiatric Association Press, 1988), 97–111; William J. Cromie, "Study: One in Five Female Undergraduates Has Eating Problem," *Harvard Gazette,* vol. 88, no. 34 (7 May 1993), 3, 10.

42. Trish Hall, "One Who Filled Out on the Low-Fat Fill-Yourself-Up Diet," *New York Times,* 8 January 1992, C6.

43. Trish Hall, "Diet Pills Return as Long-Term Medication, Not Just Diet Aids," *New York Times,* 14 October 1992, C6.

44. Rick Weiss, "Viruses: The Next Plague?" *Washington Post,* 8 October 1989, C1; Marlene Cimons, "Nature's Tiny Killers Are Back," *Los Angeles Times,* 8 April 1993, A1.

45. Paul W. Ewald, "The Evolution of Virulence," *Scientific American,* vol. 268, no. 4 (April 1993), 86–93; Paul W. Ewald, "Transmission Modes and the Evolution of Virulence," *Human Nature,* vol. 2, no. 1 (1991), 1–30; and Paul W. Ewald, *Evolution of Infectious Disease* (New York: Oxford University Press, 1994).

46. Terence Monmaney, "Marshall's Hunch," *New Yorker,* 20 September 1993, 64–72; Feinstein, "Scientific Standards in Epidemiologic Studies of the Menace of Daily Life," *Science,* vol. 242, no. 4883, 1257–63, including the quotation from Lewis Thomas; C. E. Woteki and P. R. Thomas, eds., *Eat for Life: The Food and Nutrition Board's Guide to Reducing Your Risk of Chronic Disease* (Washington, D.C.: National Academy Press, 1992).

4 Environmental Disasters: Natural and Human-Made

1. Arthur J. Barsky, *Worried Sick: Our Troubled Quest for Wellness* (Boston: Little, Brown, 1986), 171–72; Andrew Coburn and Robin Spence, *Earthquake Protection* (New York: Wiley, 1992), 319–20.

2. Ian Burton, Robert W. Kates, and Gilbert F. White, *The Environment as Hazard,* 2nd edn. (New York: Guilford, 1993), 10–16.

3. Anders Wijkman and Lloyd Timberlake, *Natural Disasters: Acts of God or Acts of Man?* (London: Earthscan, 1984), 26–29.

4. Burton, Kates, and White, *Environment as Hazard,* 2–5; United States National Academy of Sciences, Advisory Committee on the International Decade for Natural Hazard Reduction, *Confronting Natural Disasters* (Washington, D.C.: National Academy Press, 1987), xii.

5. Shari Rudavsky, "Technology Brings Forecasters' Eyes Closer to Eye of the Storm," *Washington Post,* 31 August 1992, A3; Peter Applebome,

"Hurricane Rips Louisiana Coast Before Dying Out," *New York Times,* 27 August 1992, A1; Joseph B. Treaster, "Troops Begin Work in Storm-Hit Area, but Misery Mounts," *New York Times,* 31 August 1992, A1; Peter Kerr, "Insurers' Florida Tab: $7.3 Billion," *New York Times,* 2 September 1992, D1; Sheryl Stolberg, "Battlefield Medicine Lifts Up a Weary South Florida," *Los Angeles Times,* 19 September 1992, A1; Burton, Kates, and White, *Environment as Hazard,* 15–18.

6. "Is the United States Headed for Hurricane Disaster?," *Bulletin of the American Meteorological Society,* vol. 67, no. 5 (May 1986), 537–38; Peter Applebome, "Storm Cycles and Coastal Growth Could Make Disaster a Way of Life," *New York Times,* 30 August 1992, sec. 4, 1. See John R. Stilgoe, *Metropolitan Corridor: Railroads and the American Scene* (New Haven: Yale University Press, 1985), for this and other unintended consequences of the American railroad network.

7. "Hurricane Disaster," 537.

8. William M. Bulkeley, "Software Can Speed Hurricane Response," *Wall Street Journal,* 22 September 1992, B6.

9. Donald Worster, *Dust Bowl: The Southern Plains in the 1930s* (New York: Oxford University Press, 1979), 198–209; Sears citation, 200.

10. Richard A. Warrick, "Drought in the U.S. Great Plains: Shifting Social Consequences," in K. Hewitt, ed., *Interpretations of Calamity: From the Viewpoint of Human Ecology* (Boston: Allen & Unwin, 1983), 67–82; Bryant, *Natural Hazards,* 113–14; Anders Wijkman and Lloyd Timberlake, *Natural Disasters: Acts of God or Acts of Man?* (London: Earthscan, 1984), 45–46.

11. Bruce A. Bolt, *Earthquakes,* 3rd edn. (New York: Freeman, 1993), 276–77; Bryant, *Natural Hazards,* 198.

12. Bryant, *Natural Hazards,* 198–99. The best-known recent panel, reporting in November 1992, placed the probability of quake of at least magnitude 7 as 5 to 12 percent each year, or 47 percent within five years. See Kenneth Reich, "Scientists Hike Probability of Major Quake," *Los Angeles Times,* 1 December 1992, A1; Jane Gross, "Is 1993 the Year of the Big One?" *New York Times,* 3 January 1993, Opinion, 7; Kenneth Reich, "Quake Expert's Message Is the Same, but He Uses a Lot Less Magnitude," *Los Angeles Times,* 10 August 1992, A3; Bryant, *Natural Hazards,* 49–50.

13. Coburn and Spence, *Earthquake Protection,* 10; Bolt, *Earthquakes,* 194–97.

14. Wijkman and Timberlake, *Natural Disasters,* 87; Coburn and Spence, *Earthquake Protection,* 188–213, 215–52.

15. Chris Hedges, "Old Cairo: Ricketiness Killed Poor," *New York Times,* 14 October 1992, A24; Coburn and Spence, *Earthquake Protection,* 12; National Academy of Sciences, *Confronting Natural Disasters,* 12.

16. Robert L. Ketter, "Eastquake: Countdown to Catastrophe," *Washington Post,* 4 December 1988, L3.

17. Christine Spolar, "L.A. Shaken to Its Foundations," *Washington Post,* 8 February 1994, A1; "Experiences of Florida, San Francisco Suggest What L.A. Is In For," *Wall Street Journal,* 19 January 1994, A1; Jane Gross, "California Shows Signs of Recovery as Jobs Increase," *New York Times,* 11 April 1994, A1. The Northridge quake helped alert seismologists to a new type of fault that will be a challenge to map and to monitor. In the short run, residents have the unwelcome knowledge that risks are greater than expected; ultimately, greater knowledge will probably reduce the risks. On the risks of blind thrust faults: Keay Davidson, "Waiting for the Big One," *New Scientist,* vol. 141, no. 1918 (26 March 1994), 24–28.

18. Bryant, *Natural Hazards,* 164.

19. Dirk Johnson, "As Fires Rage in West, Damage Is Less Than Expected," *New York Times,* 8 August 1992; Tom Knudson, "Colorado Inferno Spawns Grief, Questions," *Sacramento Bee,* July 8, 1994, A1.

20. Norman L. Christensen et al., "Interpreting the Yellowstone Fires of 1988," *BioScience,* vol. 39, no. 10 (November 1989), 691.

21. Charles Little, "Smokey's Revenge," *American Forests,* vol. 99, no. 5–6 (May–June 1993), 24–25, 58–60; Bill Richards, "As Fires Sear the West, Forest Service Policies Come Under Scrutiny," *Wall Street Journal,* 6 October 1992, A1.

22. Stephen J. Pyne, personal communication, 11 June 1993.

23. Dan Morain, "Housing Puts Millions at Risk of Recurring Wildfires," *Los Angeles Times,* 11 November 1991, A1; Herbert E. McLean, "Five Hot Tips for Homeowners on the Edge," *American Forests,* vol. 99, no. 5–6 (May–June 1993), 30–31; Stephen J. Pyne, "The New American Fire," *New York Times,* 28 October 1991, A19; Stephen J. Pyne, "The Summer We Let Wild Fire Loose," *Natural History,* August 1989, 34–51; Stephen J. Pyne, "Letting Wild Fire Loose: The Fires of '88," *Montana,* vol. 39, no. 3 (1989), 76–79; Stephen J. Pyne, "Keeper of the Flame; Our Management of This Element Ranges from Poor to Criminal; We Suppress Natural Burning and Take Arson to Brutal Extremes," *Los Angeles Times,* 1 November 1993, B7.

24. Timothy Egan, "New Hazard in Fire Zones: Houses of Urban Refugees," *New York Times,* 16 September 1994, A1; Stephen J. Pyne, "Flame and Fortune," *Wildfire,* September 1994, 33–36.

25. Orrin H. Pilkey, Jr., and Robert Thieler, "Erosion of the United States Shoreline," in *Quaternary Coasts of the United States: Marine and Lacustrine Systems,* SEPM Special Publication No. 48 (1992), 1–7; Orrin H. Pilkey and William H. Neal, "Save Beaches, Not Buildings," *Issues in Science and Technology,* vol. 8, no. 3 (Spring 1992), 38.

26. Cory Dean, "A New Theory: A Beach Has a Right to Its Sand," *New York Times,* 29 November 1991, B9; Katherine E. Stone and Benjamin Kaufman, "Sand Rights: A Legal System to Protect the 'Shores of the Sea,'" *Shore & Beach,* vol. 56, no. 3 (July 1988), 8–14.

27. Orrin H. Pilkey, "Coastal Erosion," *Episodes,* vol. 14, no. 1 (March 1991), 50.

28. See Rutherford H. Platt, Timothy Beatley, and H. Crane Miller, "The Folly at Folly Beach and Other Failings of U.S. Coastal Erosion Policy," *Environment,* vol. 33, no. 9 (November 1991), 6–9, 25–32.

29. Anthony F. C. Wallace, *St. Clair: A Nineteenth-Century Coal Town's Experience with a Disaster-Prone Industry* (New York: Alfred A. Knopf, 1987), 249–58; *Statistical Abstract of the United States 1993,* 434, 703.

30. See Mary Procter and Bill Matuszeski, *Gritty Cities* (Philadelphia: Temple University Press, 1978), for a sympathetic but clear-eyed view of the Northeast's old industrial communities.

31. United States National Acid Precipitation Assessment Program, *Acidic Deposition: State of Science and Technology,* September 1991, is a summary of the four-volume NAPAP report.

32. Jeremy M. Hales, *Tall Stacks and the Environment,* Report EPA-450/3-76-007 (Research Triangle Park: Environmental Protection Agency, 1976), 13–15.

33. United States National Acid Precipitation Assessment Program, *1990 Integrated Assessment Report* (1991), 198–208; John F. Harris, "In Shenandoah Park, an Outspoken Shepherd," *Washington Post,* 11 November 1991, A10; W. Page, "Acid Rain Is Killing Streams," *Richmond Times-Dispatch,* 1 November 1992, C7.

34. Robert Ostmann, Jr., *Acid Rain: A Plague upon the Waters* (Minneapolis: Dillon, 1982), 76–77; Natalie Angier, "Debate on Buildings: To Scrub or Not," *New York Times,* 14 January 1992, C1; Janet Daley, "Guarding Heritage from the Masses," *The Times* (London), 21 June 1991. Even after installing an advanced new climate control system, the Vatican has been concerned about the effects of four million feet and two million voices each year. See Richard L. Wentworth, "Controversy Clings to Sistine Chapel Ceiling," *Christian Science Monitor,* 1 June 1994, Arts, 16.

35. Bruce A. Ackerman and William T. Hassler, *Clean Coal: Dirty Air* (New Haven, Yale University Press, 1981), 68.

36. Ostmann, *Acid Rain,* 75–77.

37. Peter G. Ryan and Coleen L. Moloney, "Marine Litter Keeps Increasing," *Nature,* vol. 361, no. 6407 (7 January 1993), 23; Michael Weisskopf, "Pollution from Plastics Ravaging Marine Life," *Washington Post,* 15 December 1986, A1.

38. See Roy Church, *The History of the British Coal Industry,* vol. 3: *Victorian Pre-eminence* (Oxford: Clarendon Press, 1986), 324–28, 582–87; Bill Paul, "Weak Link: High Strength Steel Is Implicated as Villain in Scores of Accidents," *Wall Street Journal,* 16 January 1984, Part 1, 6; Caleb Solomon and Daniel Machalaba, "Oil Tankers' Safety Is Assailed as Mishaps Average Four a Week," *Wall Street Journal,* 20 June 1990, A1.

39. Michael Cross and Mick Hamer, "How to Seal a Supertanker," *New Scientist,* vol. 133, no. 1812 (14 March 1992), 40–44; Donald Smith, "Obsolete Nautical Charts Blamed for Accidents," *Los Angeles Times,* 20 December 1992, A19; Solomon and Machalaba, "Oil Tankers' Safety," A4.

40. Shaunagh Kirby, "The Thick Black Line," *New Scientist,* vol. 137, no. 1858 (30 January 1993), 24–25; Eliot Marshall, "Valdez: The Predicted Oil Spill," *Science,* vol. 244, no. 4900 (7 April 1989), 20–21; Leslie Roberts, "Long, Slow Recovery Predicted for Alaska," *Science,* vol. 244, no. 4900 (7 April 1989), 22–24.

41. David Kennedy, interview, 17 April 1992; James E. Mielke, "Oil in the Ocean: The Short- and Long-Term Impacts of a Spill," U.S. Congressional Research Service Report 90–356 SPR, 24 July 1990; Jonathan P. Houghton et al., "Evaluation of the Condition of Intertidal and Shallow Subtidal Biota in Prince William Sound following the *Exxon Valdez* Oil Spill and Subsequent Shoreline Treatment," National Oceanographic and Atmospheric Administration Report HMRB 91–1 (March 1991), 4–13; Marguerite Holloway, "Soiled Shores," *Scientific American,* vol. 265, no. 4 (October 1991), 102–16.

42. Janet Raloff, "An Otter Tragedy," *Science News,* vol. 143, no. 13 (27 March 1993), 200–2.

43. Mielke, "Oil in the Ocean," 32–33.

44. Will Steger and Jon Bowermaster, *Saving the Earth: A Citizen's Guide to Environmental Action* (New York: Alfred A. Knopf, 1990), 173.

45. Sean Kelly, "Waste Oil Pits May Have Killed 500,000 Migratory Birds in '89," *Washington Post,* 6 April 1990, A17; "100 Spills, 1,000 Excuses" (Washington, D.C.: Wilderness Society, 1990), 1; Mary H. Cooper,

"Oil Spills," *CQ Researcher,* vol. 2, no. 2 (17 January 1992), 35; John J. Fried, "Leaking Oil Tanks: Who Protects Water Supply," *Philadelphia Inquirer,* 23 August 1992, D1.

46. Michael Allaby and Jim Lovelock, "Wood Stoves: The Trendy Pollutant," *New Scientist,* vol. 88, no. 1227 (13 November 1981), 420–22.

47. Matthew L. Wald, "Wood Stoves Facing Curbs as Polluters," *New York Times,* 30 November 1986, sec. 1, 1; Nelson Bryant, "Wood Fires Under Scrutiny," *New York Times,* 10 November 1988, Sports, 10; "The Fire for Wood as Fuel Is Cooling," *New York Times,* 17 November 1989, A16; Richard Stone, "Environmental Toxicants Under Scrutiny at Baltimore Meeting," *Science,* vol. 267, no. 5205 (24 March 1995), 1770–71.

48. Fred Bayles, "Hurricane Experts See Scenario for Horror on East Coast," *Minneapolis Star-Tribune,* 30 May 1993, 12A.

5 Promoting Pests

1. John Balzar, "A Deadly Plague of Stowaways," *Los Angeles Times,* 17 May 1993, A1.

2. James E. Childs, "And the Cat Shall Lie Down with the Rat," *Natural History,* June 1991, 16–19; John Terborgh, "Why Our Songbirds Are Vanishing," *Scientific American,* vol. 266, no. 5 (May 1992), 98–104.

3. Daniel Machalaba, "Lovers of Bluebirds Are Wringing in the Breeding Season," *Wall Street Journal,* 6 May 1992, A1; Michael Pollan, "Against Nativism," *New York Times Magazine,* 15 May 1994, 52; Sheryl Stolberg, "Time to Die for Killer Virus?," *Los Angeles Times,* 18 May 1993, A1.

4. L. Jones, "Creative Disruptions in American Agriculture, 1620–1820," *Journal of Agricultural History,* vol. 48, no. 4 (October 1974), 526–27.

5. Michael Moss, "The Trouble with Gribbles," *New York Newsday,* 14 February 1993, 3; Lindsey Gruson, "Cleaner Harbor Brings Back a Menacing Creature," *New York Times,* 27 June 1993, part 1, 1; Michael J. Ganas, Michael P. Hunnemann, and Danni R. Goulet, "Marine Borer Activity on the Rise in New York Harbor," *Public Works,* vol. 124, no. 1 (January 1993), 32ff.

6. Carrol B. Fleming, "Unwelcome Immigrants: Ballast-Water Stowaways," *Sea Frontiers,* vol. 37, no. 3 (May–June 1991), 22–29.

7. Michael L. Ludyanskiy, Derek McDonald, and David MacNeill, "Impact of the Zebra Mussel, a Bivalve Invader," *BioScience,* vol. 43, no. 8 (September 1993), 533–44; James T. Carlton and Jonathan B. Geller, "Ecological Roulette: The Global Transport of Nonindigenous Marine Organisms," *Science,* vol. 261, no. 5117 (2 July 1993), 78–82; David W. Garton et

al., "Biology of Recent Invertebrate Invading Species in the Great Lakes: The Spiny Water Flea, *Bythotrephes cederstroemi,* and the Zebra Mussel, *Dreissena polymorpha,"* in Bill McKnight, ed., *Biological Pollution: The Control and Impact of Invasive Exotic Species* (Indianapolis: Indiana Academy of Science, 1993), 63–84; Ben Barber, "Proliferating Zebra Mussels Cost U.S. Industry Billions of Dollars," *Christian Science Monitor,* 14 January 1994, 11.

8. James T. Carlton, "Dispersal Mechanisms of the Zebra Mussel (*Dreissena polymorpha*)," in Thomas F. Nalepa and Donald W. Schloesser, eds., *Zebra Mussels: Biology, Impacts, and Control* (Boca Raton, Fla.: Lewis Publishers, 1993), 677–97; Charles Bosworth, Jr., "Tiny Mussel, Mighty Menace: Slippery Invader Surfs Flood South in Surprising Numbers," *St. Louis Post-Dispatch,* 18 December 1993, 1C.

9. Ludyanskiy et al., "Impact of the Zebra Mussel"; Rogers Worthington, "Science May Have Struck upon a Task for Pesky Zebra Mussels," *Chicago Tribune,* 28 November 1994, 7.

10. Charles McCoy, "Sea Lions, Protected by Law, Are Thriving, but at Trout's Expense," *Wall Street Journal,* 3 April 1992, A1; Amal Kumar Naj, "Parasite Infections of Raw-Fish Eaters Show Recent Rise," *Wall Street Journal,* 3 November 1988, B4, citing James H. McKerrow et al., "Anisakiasis: Revenge of the Sushi Parasite," *New England Journal of Medicine,* vol. 319, no. 18 (3 November 1988), 1228–29.

11. Debora MacKenzie, "Green Detergents 'Foul Up' Italian Coastline," *New Scientist,* vol. 138, no. 1869 (17 April 1993), 7.

12. Carlton and Geller, "Ecological Roulette," 81.

13. Ruth Schwartz Cowan, *More Work for Mother: Ironies of Technology from the Open Hearth to the Microwave* (New York: Basic Books, 1983), 101.

14. Terri Shaw, "Can a Carpet Make You Sick?" *Washington Post,* 24 June 1993, Washington Home, 9; Thad Godish, *Sick Buildings: Definition, Diagnosis and Mitigation* (Boca Raton, Fla.: Lewis Publishers, 1995); Testimony of Victor J. Kimm, Acting Assistant Administrator for Prevention, Pesticides, and Toxic Substances, U.S. Environmental Protection Agency, Before the Subcommittee on Environment, Energy, and Natural Resources, Committee on Government Operations, United States House of Representatives, 11 June 1993 (transcript furnished by EPA).

15. Linda Gamlin, "The Big Sneeze," *New Scientist,* vol. 126, no. 1719 (2 June 1990), 37–41; "A Blast from the Recent Past," *Daily Telegraph,* 2 July 1990, 17; M. B. Emanuel, "Hay Fever, a Post Industrial Revolution Epidemic: A History of Its Growth During the 19th Century," *Clinical Allergy,*

vol. 18, no. 3 (May 1988), 295–304; Ronald Finn, "John Bostock, Hay Fever, and the Mechanism of Allergy," *Lancet,* vol. 340, no. 8833 (12 December 1992), 1453.

16. Margaret Studer, "Just Thinking About All of Them Has Folks Scratching Their Heads," *Wall Street Journal,* 17 May 1993, B1; Sean Ryan, "Foreign Insects Plague Humid Britain," *The Sunday Times* (London), 2 August 1992; Julia Llewellyn Smith, "The Flea Plague Is Coming," *The Times* (London), 21 April 1994; James Langton, "It's a Jungle in There," *Sunday Telegraph,* 26 September 1993, 15; Celia Haddon, "What Makes Fleas Flee?" *Daily Telegraph,* 24 July 1993, 11; Brendan McWilliams, ". . . And the Living Is Fleasy," *Irish Times,* 14 August 1992, Weather, 2; Jay Rayner, "How Clean Is Your House? Mites on the Sofa, Bugs in the Rugs," *Mail on Sunday,* 11 September 1994, 9.

17. R. MacFadden, Jr., and Ileen A. Gilbert, "Asthma," *New England Journal of Medicine,* vol. 327, no. 27 (31 December 1992), 1928–37; Richard Sporik et al., "Exposure to House-Dust Mite Allergen (*Der p* 1) and the Development of Asthma in Childhood," *New England Journal of Medicine,* vol. 323, no. 8 (23 August 1990), 502–7. Thomas A. E. Platts-Mills et al., "Role of Allergens in Asthma and Airway Hyperresponsiveness," in Michael A. Kaliner et al., eds., *Asthma: Its Pathology and Treatment* (New York: Marcel Dekker, 1991), 595–631; Paul Harvey and Robert May, "Matrimony, Mattresses, and Mites," *New Scientist,* vol. 125, no. 1706 (3 March 1990), 48–49.

18. Betty Beard, "Sick at Hearth: Host of Hidden Hazards May Be Lurking at Home," *Arizona Republic,* 26 June 1993, D1.

19. "For Millions, Allergy Means Roaches," *New York Times,* 6 September 1990, B15.

20. "Study Links Death Risk, Overuse of Asthma Drug," *Washington Post,* 20 February 1992, A3.

21. Art Thomason, "Monitors Help Noses Know What Ails Them," *Arizona Republic,* 2 March 1992, A1.

22. Donald Worster, *Rivers of Empire: Water, Aridity, and the Growth of the American West* (New York: Pantheon, 1985), 274–75; Mark R. Sneller, "Molds Have Messages, Secrets," *Arizona Daily Star,* 26 June 1991, 6FM.

23. Thomason, "Monitors Help Noses Know What Ails Them," A1.

24. Margery Rose-Clapp, "Pollen Catcher Knows When New Allergens Take Flight," *Arizona Republic,* 2 December 1988, 5N1; Dee Ralles, "Law Trims Tree Pollen in Tucson," *Arizona Republic,* 17 June 1989, A1; Pat Kossan, "Sneeze Free Pollenless Landscape on the Horizon," *Phoenix Gazette,* 20 December 1991, D1.

25. René Sanchez, "D.C. Council Unleashes a Debate on Wolf Dogs," *Washington Post,* 10 January 1992, B3; Maia Davis, "Hybrid Breed of Wolf, Dog Can Pack Nasty Disposition," *Los Angeles Times,* 29 June 1993, B1; William K. Stevens, "Terror of Deep Faces Harsher Predator," *New York Times,* 8 December 1992, C1.

26. Stephen Jay Gould, "How Does a Panda Fit?" in *An Urchin in the Storm* (New York: Norton, 1987), 19–25, discusses the animal's "primary, unsurmounted dilemma of trying to eat bamboo with a carnivore's digestive tract"; George Laycock, *The Alien Animals* (Garden City, N.Y.: Natural History Press, n.d.), 204; R. E. Kenward, "Bark-Stripping by Gray Squirrels in Britain and North America: Why Does the Damage Differ?" in R. J. Putman, ed., *Mammals as Pests* (London: Chapman & Hall, 1989), 144–54.

27. James Whorton, *Before Silent Spring: Pesticides and Public Health in Pre-DDT America* (Princeton, N.J.: Princeton University Press, 1974), 70–72, 90.

28. Kenneth Mellanby, "With Safeguards, DDT Should Still Be Used," *Wall Street Journal,* 12 September 1989, A26; Kenneth Mellanby, *The DDT Story* (Farnham, Surrey: British Crop Protection Council, 1992), 73–82. One obituary called Standen's 1943 book *Insect Invaders* "an indictment of insects as harmful to humans, beasts and crops from someone with a passion for insecticides that slay insects wholesale." See Wolfgang Saxon, "Anthony Standen Is Dead at 86; Chemist Who Deflated Pomposity," *New York Times,* 25 June 1993, B7. Constance Matthiessen, "The Day the Poison Stopped Working," *Mother Jones,* March–April 1992, 48–49; Whorton, *Before Silent Spring,* 248–55.

29. Harland Austin et al., "A Prospective Follow-up Study of Cancer Mortality in Relation to Serum DDT," *American Journal of Public Health,* vol. 79, no. 1 (January 1989), 43–46; Sharon Begley, "Silent Spring Revisited?" *Newsweek,* 14 July 1986, 72ff.

30. George P. Georghiou and Roni B. Mellon, "Pesticide Resistance in Time and Space," in George P. Georghiou and Tetsuo Saito, eds., *Pest Resistance to Pesticides* (New York: Plenum, 1983), 8.

31. Paul A. Colinvaux, *Introduction to Ecology* (New York: Wiley, 1973), 413–14; Mellanby, *DDT Story,* 64; Stephen Lacey, "Sex-Driven Males Killed—by Appointment," *Daily Telegraph,* 20 February 1993, 6; Jane Adler, "Fight Pesky Spider Mites with Some of Their Own," *Fresno Bee,* 23 July 1994, H4.

32. David Pimentel et al., "Environmental and Economic Costs of Pesticide Use," *BioScience,* vol. 42, no. 10 (November 1992), 753–54; on losses,

Robert M. May and Andrew P. Dobston, "Population Dynamics and the Rate of Evolution of Pesticide Resistance," in *Pesticide Resistance: Strategies and Tactics for Management* (Washington, D.C.: National Academy Press, 1986), 170–71.

33. Fred Gould, "The Evolutionary Potential of Crop Pests," *American Scientist,* vol. 79, no. 6 (November–December 1991), 500–2; "Insects Start Resisting Safe Type of Pesticide," *Wall Street Journal,* 10 December 1991, B1; Adrienne Cook, "A Plague of Beetles: Organic Treatments Yield Disappointing Results," *Washington Post,* 20 August 1992, Washington Home, 15.

34. "Tactics for Prevention and Management," in *Pesticide Resistance,* 313–26; Mike Toner, "A Bumper Crop of Fleas," *Atlanta Journal-Constitution,* 3 October 1992, A1.

35. Mike Toner, "Herbicide-Resistant Weeds a Serious Threat," *Atlanta Journal-Constitution,* 26 September 1992, E6.

36. Richard Conniff, "Fire Ants: Too Hot to Handle?" *Smithsonian,* vol. 21, no. 4 (July 1990), 48–57; Andrew C. Revkin, "March of the Fire Ants," *Discover,* vol. 10, no. 3 (March 1989), 70–76; Peter H. Lewis, "Mighty Fire Ants March Out of the South," *New York Times,* 24 July 1990, C1; Pete Daniel, "A Rogue Bureaucracy: The USDA Fire Ant Campaign of the Late 1950s," *Agricultural History,* vol. 64, no. 2 (Spring 1990), 99–104; M. R. Orr et al., "Flies Suppress Fire Ants," *Nature,* vol. 373, no. 6512 (26 January 1995), 292–93.

37. Mike Toner, "George Beats Boll Weevil, but Nature Evens Score," *Atlanta Journal-Constitution,* 6 September 1992, A14.

38. Emily Yoffe, "The Ants Come Marching In," *Washington Post,* 23 August 1988, Health, 8–11; Charles L. Mann, "Fire Ants Play Their Queens into a Threat to Biodiversity," *Science,* vol. 263, no. 5153 (18 March 1994), 1560–61.

39. Edwin Gabler, *The American Telegrapher: A Social History, 1860–1900* (New Brunswick, N.J.: Rutgers University Press, 1988), 48.

40. Keith Bradsher, "Electrifying News: Bug Zapper Sales Are Stagnating," *Los Angeles Times,* 30 May 1988, part 4, 5.

41. Roger S. Nasci, Credric W. Harris, and Cyresa K. Porter, "Failure of an Insect Electrocuting Device to Reduce Mosquito Biting," *Mosquito News,* vol. 43, no. 2 (June 1983), 180–84.

42. Alberto B. Broce, "Electrocuting and Electronic Insect Traps: Trapping Efficiency and Production of Airborne Particles," *International Journal of Environmental Health Research,* vol. 3, no. 1 (1993), 47–58; "Resistant Little Critters Prove It's Not Nice to Fool with Mother Nature," *Fort Worth Star-Telegram,* 22 May 1994, 6.

43. Pimentel et al., "Assessment of Environmental and Economic Impacts of Pesticide Use," in David Pimentel and Hugh Lehman, eds., *The Pesticide Question: Environment, Economics, and Ethics* (New York: Chapman & Hall, 1993), 54–56.

44. John Madeley, "Beyond the Pestkillers . . .," *New Scientist,* vol. 142, no. 1924 (7 May 1994), 24–27. On the other hand, use of herbicides is actually growing in spite of evidence that seed cleanliness, tillage, and flooding are more effective, and emergence of resistant strains is inevitable. See Bob Holmes, "A Natural Way with Weeds," *New Scientist,* vol. 142, no. 1924 (7 May 1994), 22–23.

6 Acclimatizing Pests: Animal

1. Warwick Anderson, "Climates of Opinion: Acclimatization in Nineteenth-Century France and England," *Victorian Studies,* vol. 35, no. 2 (Winter 1992), 134–57.

2. Daniel Headrick, *The Tentacles of Progress: Technology Transfer in the Age of Imperialism, 1850–1940* (New York: Oxford University Press, 1988), 20–21; Marcus Lee Hansen, *The Atlantic Migration, 1607–1860* (New York: Harper Torchbooks, 1961), 299–300; Kenneth Lemmon, *The Golden Age of Plant Hunters* (London: Phoenix House, 1968), 54, 182–85, 217–18 (quotation).

3. Isidore Geoffroy Saint-Hilaire, *Rapport Général sur les Questions Relatives à la Domestication et la Naturalisation des Animaux Utiles* (Paris: Imprimerie Nationale, 1849), 18–23, 39–40; Michael A. Osborne, *Nature, the Exotic, and the Science of French Colonialism* (Bloomington: Indiana University Press, 1994), 62–129.

4. Osborne, *Nature,* 145; Christopher Lever, *They Dined on Eland: The Story of the Acclimatisation Societies* (London: Quiller, 1992).

5. Andrew Balfour, "Problems of Acclimatisation," *Lancet,* vol. 205 (14 July 1923), 84–88. The author is grateful to Douglas C. Zinn for calling this article to his attention.

6. American Acclimatization Society, *Charter and By-Laws* (New York: George W. Averell, 1871), 5.

7. "Eugene Schieffelin," *Historical Families of America,* vol. 1 (1907), 356. The St. Nicholas Society has since moved to smaller quarters in the New York Genealogical and Biographical Society Building, and its director does not know the Schuyler portrait's present location.

8. See most recently Ted Gup, "100 Years of the Starling," *New York Times,* 1 September 1990, 19.

9. Felton Gibbons and Deborah Strom, *Neighbors to the Birds: A History of Birdwatching in America* (New York: Norton, 1988), 214–15; Chandler S. Robbins, "Introduction, Spread, and Present Abundance of the House Sparrow in North America," in S. Charles Kendeigh, ed., *A Symposium on the House Sparrow* (Passer Domesticus) *and European Tree Sparrow* (P. Montanus) *in North America,* (Anchorage, Ky.: American Ornithologists' Union, 1973), 3–9.

10. Frank M. Chapman, "The European Starling as an American Citizen," *Natural History,* vol. 25, no. 5 (September–October 1925), 480–85; Bill Lawren, "Starlings Are No Darlings," *National Wildlife,* vol. 28, no. 5 (April–May 1990), 24–27.

11. Alexander Wetmore, *Song and Garden Birds of North America* (Washington, D.C.: National Geographic Society, 1964), 238–41; Brina Kessel, "Distribution and Migration of the European Starling in North America," *Condor,* vol. 55, no. 2 (March–April 1953), 49–67.

12. Alfred Russel Wallace, "Acclimatization," *Encyclopaedia Britannica,* 11th edn., vol. 1, 114–21.

13. Pierre Clerget, *Les Industries de la Soie en France* (Paris: Librairie Armand Colin, 1925), 4–10. On the latest research on the Silk Road, see John Noble Wilford, "New Finds Suggest Even Earlier Trade on Fabled Silk Road," *New York Times,* 16 March 1993, C1.

14. Clerget, *Soie,* 4–10; Gerald Geison, "Louis Pasteur," *Dictionary of Scientific Biography,* vol. 10 (New York: Charles Scribner's Sons, 1974), 372–76.

15. *Sericulture* (New York: Silk Association of America, 1903), 1–9; L. P. Brockett, *The Silk Industry in America* (n.p., 1876), 9–14, 26–37; Richard S. Peigler, "Wild Silks of the World," *American Entomologist,* vol. 39, no. 3 (Fall 1993), 131–61.

16. Léopold Trouvelot, "The American Silk Worm," *American Naturalist,* vol. 1, no. 1 (March 1867), 30–38; vol. 1, no. 2 (April 1867), 85–94, and vol. 1, no. 3 (May 1867), 145–49; quotation on 85.

17. Edward H. Forbush and Charles H. Fernald, *The Gypsy Moth* (Boston: Wright & Potter, 1896), 273–84, 3–4.

18. Ibid., 234–35.

19. Ibid., 94–113.

20. Ibid., 117–45, 158–72.

21. W. Hitchcock, "Early History of the Gypsy Moth in Connecticut," in *25th Anniversary Memoirs, Connecticut Entomological Society* (New Haven: Connecticut Entomological Society, 1974), 87–97.

22. Tina Adler, "Squelching Gypsy Moths: What's Hot and What's Not

in the Arsenal Against Leaf Eaters," *Science News,* vol. 145, no. 12 (19 March 1994), 184ff.

23. Jack C. Schultz, "The Multimillion-Dollar Gypsy Moth Question," *Natural History,* June 1991, 40–45; Adler, "Squelching Gypsy Moths."

24. Mark Jerome Walters, "The Dubious War on Gypsy Moths," *Nature Conservancy,* vol. 43, no. 3 (May–June 1995), 8–9; Ann E. Hajek, "Enter the Fungal Factor," *Natural History,* June 1991, 42; Laurie Goodrich, "A Case for Doing Nothing," *New Jersey Audubon,* vol. 18, no. 1 (Spring 1992), 20–21.

25. This account of Baird's influence in American fisheries is based on Dean Conrad Allard, Jr., "Spencer Fullerton Baird and the U.S. Fish Commission: A Study in the History of American Science" (Ph.D. dissertation, George Washington University, 1967).

26. Commission of Fish and Fisheries, *Report of the Commissioner for 1878,* part VI, xlv–liii; J. T. Bowen, "A History of Fish Culture as Related to the Development of Fishery Programs," in Norman G. Benson, ed., *A Century of Fisheries in North America* (Washington, D.C.: American Fisheries Society, 1970), 71–94.

27. B. Moyle, "Fish Introductions into North America: Patterns and Ecological Impact," in Harold A. Mooney and James A. Drake, eds., *Ecology of Biological Invasions of North America and Hawaii* (New York: Springer-Verlag, 1986), 27.

28. Pete Thomas, "The Rodney Dangerfield of Fish," *Los Angeles Times,* 8 April 1992, C6; U.S. Commission of Fish and Fisheries, *Report of the Commissioner for 1877,* part V, 40–44; Leon J. Cole, "The Status of the Carp in America," *Transactions of the North American Fisheries Society,* vol. 34 (1905), 201–6.

29. Rudolph Hessel, "The Carp and Its Culture in Rivers and Lakes: And Its Introduction in America," in U.S. Commission of Fish and Fisheries, *Report of the Commissioner for 1875–1876,* part IV, 865–97.

30. Laycock, *Alien Animals,* 162–63.

31. Ibid.

32. "Fish Hunks," *Time,* 30 November 1992, 26; Christer Bromark and Jeffrey G. Miner, "Predator-Induced Phenotypical Change in Body Morphology in Crucian Carp," *Science,* vol. 258 (20 November 1992), 1348–50.

33. Laycock, *Alien Animals,* 165–68; Cole, "Status of Carp," 204; Rob Hotakainen, "Chemical Treatment of Lake Raises Fears," *Minneapolis Star-Tribune,* 19 June 1989, 1A.

34. Seth Norman, "Carp: Fish of Your Future," *Field and Stream,* July

1990, 22–24; Steve Grant, "Delving Deeper into Nuclear Plant Outflows," *Hartford Courant,* 26 January 1993, A1.

35. "Where Carp Are Concerned, It's No Holds Barred," *Milwaukee Journal,* 6 June 1993, C14; Phil Borchmann, "City OKs Chemical Fish Kill on McCollum Lake," *Chicago Tribune,* 21 June 1993, NW3; Dean Rebuffoni, "Tests Show Poison DNR Used in Lakes Has Toxic Chemicals," *Minneapolis Star-Tribune,* 21 August 1990, 1B; "DNR Decides Fish-Poisoning Project Near Mora Can Proceed," *Minneapolis Star-Tribune,* 5 August 1989, 3B; "Arkansas River Cleanup," *Washington Post,* 21 August 1988, A11.

36. Commercial carp fishing also bore the brunt of anglers' hostility toward freshwater commercial fishing in general. See Arnold W. Fritz, "Commercial Fishing for Carp," in Cooper, ed., *Carp in North America,* 17–30.

37. Sam Jameson, "Buddhist Rite Brings Bad Fortune to Fish," *Los Angeles Times,* 15 December 1992, H2.

38. The standard general-interest work is Mark L. Winston, *Killer Bees: The Africanized Honey Bee in the Americas* (Cambridge: Harvard University Press, 1992). For the early history the best account in English is Wallace White, "The Bees from Rio Claro," *New Yorker,* 16 September 1991, 36–60.

39. On behavior, see Thomas E. Rinderer, "Evolutionary Aspects of the Africanization of Honey-Bee Populations in the Americas," in Glen R. Needham et al., eds., *Africanized Honey Bees and Bee Mites* (Chichester: Ellis Horwood, 1988), 13–28; Warren E. Leary, "Trying Times for U.S. Honeybee," *New York Times,* 22 January 1991, C4; Glynn Mapes, "In U.S. Apiaries This Is the Buzz: Long Live the Queen," *Wall Street Journal,* 13 June 1991, A1.

40. Ronald B. Taylor, " 'Killer Bees': State Wins the First Round," *Los Angeles Times,* 3 December 1985, part 1, 3; Robert Cooke, "Calm Amid the Buzz over Killer Bees," *New York Newsday,* 6 November 1990, Discovery, 9; Tom Gorman, " 'Killer Bees' About to Join Facts of Life in Southland," *Los Angeles Times,* 13 March 1994, A1.

41. William Booth, "Invasion of the Surprisingly Hardy 'Killer Bees,' " *Washington Post,* 30 November 1990, A10.

42. Sue Hubbell, "Maybe the 'Killer' Bee Should Be Called the 'Bravo' Instead," *Smithsonian,* September 1991, 116ff; White, "Bees from Rio Claro," 60.

43. Lever, *They Dined on Eland,* 181–82, 129, 170–71; Douglas R. Weiner, *Models of Nature: Ecology, Conservation, and Cultural Revolution in Soviet Russia* (Bloomington: Indiana University Press, 1988), 194–223 and figs. 19–20; Georgie Anne Geyer, "The Dictator of the Cows," *Saturday Evening Post,* July 1991, 65.

44. Charles H. Hocutt, "Toward the Development of an Environmental Ethic for Exotic Fishes," in Walter R. Courtenay, Jr., and Jay R. Stauffer, Jr., *Distribution, Biology, and Management of Exotic Fishes* (Baltimore: Johns Hopkins University Press, 1984), 384 [374–86]; Gilbert C. Radonski and Robert G. Martin, "Fish Culture Is a Tool, Not a Panacea," in Richard H. Stroud, ed., *Fish Culture in Fisheries Management* (Bethesda, Md.: American Fisheries Society, 1986), 7–13; Jon R. Luoma, "Boon to Anglers Turns into a Disaster for Lakes and Streams," *New York Times,* 17 November 1992, C4.

45. Craig N. Spencer, B. Riley McClelland, and Jack A. Stanford, "Shrimp Stocking, Salmon Collapse, and Eagle Displacement," *BioScience,* vol. 41, no. 1 (January 1991) 14–21. See David Ehrenfeld's account in *Beginning Again: People and Nature in the New Millennium* (New York: Oxford University Press, 1993), 149–51.

46. Luoma, "Boon to Anglers," C4; on the proposed introduction of the Pacific or Japanese oyster to replace familiar but depleted Eastern oysters, see Beth Becker, "Botcher of the Bay or Economic Boom," *BioScience,* vol. 42, no. 10 (November 1992), 744–47.

47. Steve Grant, "Not All the Pests Come Uninvited," *Hartford Courant,* 31 May 1993, A1.

48. Robert Johnson, "A Homely Duck Stirs Up the Wrath of Tidy Suburbia," *Wall Street Journal,* 22 February 1989, A1.

49. Peter B. Moyle, "America's Carp," *Natural History,* vol. 93, no. 9 (September 1984), 42–51.

50. William H. McNeill, *Plagues and Peoples* (Garden City, N.Y.: Anchor Press/Doubleday, 1976), 21–22; Daniel Machalaba, "Bloodthirsty Flies Invade Adirondacks, Stirring Hostilities," *Wall Street Journal,* 23 June 1989, A1; Anthony G. Lawrence, letter, *Wall Street Journal,* 14 July 1989, A11.

51. Paul R. Ehrlich, "Which Animal Will Invade?" in Mooney and Drake, eds., *Ecology of Biological Invasions,* 79–95.

52. Grant, "Not All Pests Uninvited," A6; Walter R. Courtenay, Jr., "The Introduced Fish Problem and the Aquarium Fish Industry," *Journal of the World Aquaculture Society,* vol. 21, no. 3 (September 1990), 145–59.

53. Bruce E. Coblentz, "Invasive Ecological Dominants: Environments Boar-ed to Ears and Living on Burro-ed Time," in McKnight, *Biological Pollution,* 223–24; U.S. Office of Technology Assessment, *Non-Indigenous Species,* 267–85.

7 Acclimatizing Pests: Vegetable

1. On the versatility of plants, see Rick Weiss's survey "The Green Kingdom's Secrets of Movement," *Washington Post,* 6 June 1994, A3. On imports in the New World: Alfred W. Crosby, *Ecological Imperialism: The Biological Expansion of Europe, 900–1900* (Cambridge: Cambridge University Press, 1986), 146–70, 194. The classical account remains Charles S. Elton, *The Ecology of Invasions by Animals and Plants* (London: Chapman & Hall, 1958), though today's population biologists are less sure about the importance of disrupted habitats.

2. Holzner, "Concepts, Categories, and Characteristics of Weeds," in W. Holzner and M. Numata, eds., *Biology and Control of Weeds* (The Hague: Dr. W. Junk, 1982), 5; Dan Carroll, "Subduing Purple Loosestrife," *Conservationist,* vol. 49, no. 1 (August 1994), 6ff.; Michael Pollan, "Weeds Are Us," *New York Times Magazine,* 5 November 1989, 97ff; Greg Fewer, "Hands Off, That's No Weed," *New Scientist,* vol. 145, no. 1965 (18 February 1995), 47–48.

3. Fairchild cited in Elton, *Ecology of Invasions,* 51; Anne Raver, "A Tree Grows in Brooklyn, but Maybe Not for Long," *New York Times,* 11 January 1993, 21; Michael Martin Mills, "Woodman, Spare That Ailanthus," *Philadelphia Inquirer,* 7 August 1991, 13A; Clare Ansberry, "Those Aren't Weeds in Cleveland: They're Really Rare, Exotic Plants," *Wall Street Journal,* 4 September 1986, sec. 1, 29; Harold Faber, "Wild Orchids? In Schenectady? Aloha!" *New York Times,* 6 October 1991, sec. 1, 32. In Arizona, the orchids suffered when cattle were excluded from the area; their grazing helped the orchids by keeping down the grass.

4. David Pimentel, "Plant and Animal Invasions in Agriculture," in Harold A. Mooney and James A. Drake, eds., *Ecology of Biological Invasions of North America and Hawaii* (New York: Springer-Verlag, 1986), 151. W. T. Haller, "Disturbed Habitat Not Necessary for Invasion," in Florida Department of Environmental Protection, *An Assessment of Invasive Non-Indigenous Species in Florida's Public Lands,* Technical Report No. TSS-94–100 (1994), 17–18, provides examples of plants that spread rapidly even in habitats that have not been disturbed. Michael Pollan, "Against Nativism," *New York Times,* 15 May 1994, sec. 6, 52ff, invokes the fanatic botanical jingoism of some National Socialist authorities while ignoring the equally powerful symbolism of the European appropriation of Native American territory.

5. R. N. Mack, "Plant Invasion in the Intermountain West," in Mooney and Drake, *Ecology of Biological Invasions,* 198; John Stilgoe, *Metropolitan Corridor* (New Haven, Conn.: Yale University Press, 1983), 142–43;

James A. Young, "Tumbleweed," *Scientific American,* vol. 264, no. 3 (March 1991), 82–87.

6. On the history of federal plant introductions, see Fred Wilbur Powell, *The Bureau of Plant Industry* (Baltimore: Johns Hopkins Press, 1927); on the lawn, Kenneth T. Jackson, *Crabgrass Frontier* (New York: Oxford University Press, 1985), 54–61.

7. Charles Richards Dodge, "Hemp Culture," in U.S. Department of Agriculture, *Yearbook 1895,* 215–22; Richard N. Mack, "Catalog of Woes," *Natural History,* March 1990, 44–53; Richard N. Mack, "The Commercial Seed Trade: An Early Disperser of Weeds in the United States," *Economic Botany,* vol. 45, no. 2 (1991), 257–73; C. G. McWhorter, "Introduction and Spread of Johnsongrass in the United States," *Weed Science,* vol. 19, no. 5 (September 1971), 498–99; Michael Pollan, "How Pot Has Grown," *New York Times Magazine,* 19 February 1995, 31ff.

8. Mack, "Commercial Seed Trade," 270–71; Mack, "Plant Invasion," 197.

9. Ashleigh Brilliant, *The Great Car Craze* (Santa Barbara, Calif.: Woodbridge Press, 1989), 126–27.

10. Peter Friederici, "The Alien Saltcedar," *American Forests,* vol. 101, nos. 1–2 (January–February 1995), 44–47.

11. William Shurtleff and Akiko Aoyagi, *The Book of Kudzu* (Brookline, Mass.: Autumn Press, 1977), 8–12.

12. Doria R. Gordon and Kevin P. Thomas, "Introduction Pathways for Invasive Non-Indigenous Species," in Florida Department of Environmental Protection, *Assessment,* 32.

13. Shurtleff and Aoyagi, *Book of Kudzu,* 12–17; John J. Winberry and David M. Jones, "Rise and Decline of the 'Miracle Vine': Kudzu in the Southern Landscape," *Southeastern Geographer,* vol. 13, no. 2 (November 1973), 61–70; James H. Miller and Boyd Edwards, "Kudzu: Where Did It Come From? And How Can We Stop It?" *Southern Journal of Applied Forestry,* vol. 7, no. 4 (1983), 165–69; David Fairchild, *The World Was My Garden: Travels of a Plant Explorer* (New York: Charles Scribner's Sons, 1938), 328.

14. Winberry and Jones, "Rise and Decline," 67–68; Sy Montgomery, ". . . Agricultural Scientists Are Losing the Battle Against Invading Plants," *Los Angeles Times,* 30 May 1988, Metro, part 2, 4.

15. Shurtleff and Aoyagi, *Book of Kudzu,* 15–16; Gordon Baxter, "Computer Session Is Not All Vine and Dandy: Kudzu's Power Stronger Than Utility's," *Dallas Morning News,* 30 October 1988, 44A.

16. Shurtleff and Aoyagi, *Book of Kudzu,* 16–17.

17. Neil Santaniello, "Will Kudzu KO Everglades?" *Toronto Star,* 12 September 1992, D6.

18. James W. Amrine, Jr., and Terry A. Stasny, "Biocontrol of Multiflora Rose," in Bill N. McKnight, ed., *Biological Pollution: The Control and Impact of Invasive Exotic Species* (Indianapolis: Indiana Academy of Sciences, 1993), 9–21.

19. David K. Roberts, "Another Alien Species Invades," *St. Petersburg Times,* 30 January 1994, 1B.

20. G. Shilling and J. F. Gaffney, "Cogon Grass (*Imperata Cylindrica* (L.) Beauv.): Is It a Threat to Natural Areas?" in Florida Department of Environmental Protection, *Assessment,* 200–01.

21. Roberts, "Another Alien Species Invades," 1B.

22. J. Ewel, "Invasibility: Lessons from South Florida," in Mooney and Drake, eds., *Ecology of Biological Invasions,* 214–30; Daniel Simberloff, "Why Is Florida Being Invaded?" in Florida Department of Environmental Protection, *Assessment,* 7–9.

23. Denis A. Saunders, Richard J. Hobbs, and Chris R. Margules, "Biological Consequences of Ecosystem Fragmentation: A Review," *Conservation Biology,* vol. 5, no. 1 (March 1991), 18–33.

24. Don C. Schmitz et al., "Exotic Aquatic Plants in Florida: A Historical Perspective and Review of the Present Aquatic Plant Regulation Program," in Ted C. Center et al., eds., *Proceedings of the Symposium on Exotic Pest Plants* (Washington, D.C.: U.S. Department of the Interior, National Park Service, 1991), 303–26; Luz Villarreal, "Grand Exotics Can Turn Ugly for the Environment," *Orlando Sentinel,* 13 February 1992, I1.

25. National Park Service, *Exotic Weeds I: Kudzu, Saltcedar, and Brazilian Pepper,* report (16 November 1984), VIII–6 to VIII–15; Ingrid Olmsted and Steve Yates, "Florida's Pepper Problem," *Garden,* May–June 1984, 20–23.

26. Dean G. Barber, "The Expansion of Brazilian Pepper in Central Florida," in Florida Department of Environmental Protection, *Assessment,* 158; Olmsted and Yates, "Pepper Problem," 21–23.

27. Daniel F. Austin, "Exotic Plants and Their Effects in Southeastern Florida," *Environmental Conservation,* vol. 5, no. 1 (Spring 1978), 25–34; Eric Morgenthaler, "What's Florida to Do with an Explosion of Melaleuca Trees?" *Wall Street Journal,* 8 February 1993, A1; Don C. Schmitz et al., "The Ecological Impact of Non-indigenous Plants in Florida," in Florida Department of Environmental Protection, *Assessment,* 20–21.

28. Stephen J. Pyne, *Burning Bush: A Fire History of Australia* (New York: Henry Holt, 1991), 5–9.

29. Schmitz et al., "Ecological Impact," 19–23; Morgenthaler, "What's Florida to Do?" A1; Craig Diamond, Darrell Davis, and Don C. Schmitz, "Economic Impact Statement: The Addition of *Melaleuca quinquenervia* to the Florida Prohibited Aquatic Plant List," in Center et al., eds., *Exotic Plant Pests,* 87–110.

30. Jeffrey D. Schardt and Don C. Schmitz, *1990 Florida Aquatic Plant Survey* (Tallahassee: Florida Department of Natural Resources, 1991), 75; Morgenthaler, "What's Florida to Do?"; Don C. Schmitz, personal communication.

31. Yvonne Baskin, "Ecologists Dare to Ask: How Much Does Diversity Matter?" *Science,* vol. 264, no. 5156 (8 April 1994), 202–3.

32. Pyne, *Burning Bush,* 5–6, 15–25.

33. Zacharin, *Emigrant Eucalypts,* 30–37, 66–67, 93–99; Sheridan Bartlett, "A Colonial Tree: The Eucalyptus of the Palni Hills," *Landscape,* vol. 31, no. 1 (1993), 28–33.

34. Gayle M. Groenendaal, "*Eucalyptus* Helped Solve a Timber Problem: 1853 1880," in Richard B. Standiford and F. Thomas Ledig, eds., *Proceedings of a Workshop on* Eucalyptus *in California* (Berkeley: U.S. Department of Agriculture, 1983), 1–8; Viola L. Warren, "The Eucalyptus Crusade," *Southern California Quarterly,* vol. 44 (1962), 31–42.

35. Norman D. Ingham, *Eucalyptus in California* (Sacramento: California State Printing Office, 1908), 30–35; C. H. Sellers, *Eucalyptus: Its History, Growth, and Utilization* (Sacramento: A. J. Johnston, 1910); Warren, "Eucalyptus Crusade," 38.

36. Warren, "Eucalyptus Crusade," 39; Joseph R. Loftus Co., *What Eucalyptus Trees Will Do for You* (Los Angeles, n.d.); Jack London to Herbert Forder, 3 February 1911, in Earle Labor et al., eds., *The Letters of Jack London* (Stanford, Calif.: Stanford University Press, 1988), vol. 2, 978–79.

37. Warren, "Eucalyptus Crusade," 40–41; Penfold and Willis, *Eucalypts,* 6–7, 125–26.

38. Jacobs, *Growth Habits,* 68–81.

39. "The Fire Next Time," *Los Angeles Times,* 26 October 1991, B5; Pyne, *Burning Bush,* 24–25.

40. Nancy Vogel, "East Ball Hills Rebuild—with New Vigilance," *Sacramento Bee,* 1 December 1994, A13.

41. "The Fire Next Time"; Dan Morain and Jenifer Warren, " '70s Study Warned of Fire Threat," *Los Angeles Times,* 25 October 1991, A3; Robert Reinhold, "As Fire Toll Rises in Oakland Hills, So Do Questions," and Jane Gross, "Out of Ashes, Move to Ban Wood Roofs," both *New York*

Times, 23 October 1991, 19; Rick DelVecchio, "Some Plants May Be Banned in Fire Area," *San Francisco Chronicle,* 16 November 1992, A15.

42. DelVecchio, "Some Plants May Be Banned," 15; Rick DelVecchio, "Fire Survivors Rebuilding Bigger," *San Francisco Chronicle,* 13 October 1992, A1; Rick DelVecchio, "Keeping Oakland Hills Fire-Safe," *San Francisco Chronicle,* 2 November 1992, A12; David Darlington, "After the Firestorm," *Audubon,* March 1993, 72ff; Sharon Rummery, "A New Beginning," *San Francisco Chronicle,* 20 October 1993, Home, 1/Z1.

43. Kathleen A. Hughes, "Neighbors Seeking to Better Their Lot Are Often Up a Tree," *Wall Street Journal,* 15 April 1992, A9.

44. William D. Montalbano, "On Iberian Peninsula, a Protest Grows over Trees," *Los Angeles Times,* 4 December 1990, H2; Tony Smith, "Eucalyptus Trees Stir Iberian Ecology Alarm as They Soak Up Water," *Los Angeles Times,* 19 November 1989, A24; M. E. D. Poore and C. Fries, *The Ecological Effects of Eucalyptus,* FAO Forestry Paper 59 (Rome: Food and Agriculture Organization, 1985); Malcolm Smith, "Science: Live the High Life and Save the Wildlife," *Independent,* 30 May 1994, 19.

45. Montalbano, "Protest Grows over Trees"; Bartlett, "A Colonial Tree," 28–33.

46. Bartlett, "A Colonial Tree," 28–33.

47. Allen Lacy, "Reading Between Lines of Catalogue Fantasies," *New York Times,* 6 February 1992, C10; Allen Lacy, "Nature's Ribbons Cover an Industrial Wasteland," *New York Times,* 19 September 1991, C8.

8 The Computerized Office: The Revenge of the Body

1. See Asa Briggs, *Victorian Things* (Chicago: University of Chicago Press, 1988), 179; on the history of steel pens, 182–87.

2. On electronic system failures in transport, see "Computer Crash Freezes Train Traffic," RISKS electronic news group, 13 April 1995, from *Orange County Register,* 29 March 1995; and the posting of Klaus Brunnstein to the same list, "About the 'Altona Railway Software Glitch,' " 17 March 1995.

3. Jake Page, "Writing Got a Lot Easier When the Old 'Manual' Was New," *Smithsonian,* vol. 21, no. 9 (December 1990), 54–65.

4. *Fortune* Magazine, *The Fabulous Future: America in 1980* (New York: Dutton, 1956), 18, 37, 185, 201–3.

5. John Naisbitt, *Megatrends: Ten New Directions Transforming Our Lives* (New York: Warner, 1984), 5; Judith Davidsen, "Designing to Prevent RSI," *Interior Design,* 1 May 1991, 168ff.

6. Jon Jefferson, "Dying for Work," *ABA Journal,* vol. 79 (January 1993), 46–51; James C. Robinson, "The Rising Long-Term Trend in Occupational Injury Rates," *American Journal of Public Health,* vol. 78, no. 3 (March 1988), 276–81.

7. Robinson, "Rising Trend," 280–81.

8. Sally Squires, "Study Traces More Deaths to Working Than Driving," *Washington Post,* 31 August 1990, A4.

9. Paul Saffo, "A Conspiracy of Silence," *Byte,* July 1993, 278.

10. See Bogdan P. Radanov et al., "Role of Psychosocial Stress in Recovery from Common Whiplash," *Lancet,* vol. 338 (21 September 1991), 712–15, and related editorial, "Neck Injury and the Mind," 728–30. Neurosis does appear to affect the initial intensity of neck pain.

11. See the exchange of letters between Michael I. Weintraub and Richard A. Deyo in *Journal of the American Medical Association,* vol. 269, no. 3 (20 January 1993), 354–356.

12. Mariann Caprino, "From Heavy Lifting to Heavy Computing, Work Is Often a Pain in the Back," *Washington Post,* 13 February 1994, H2.

13. Fernand Braudel, *The Structures of Everyday Life: The Limits of the Possible,* vol. 1 of *Civilization and Capitalism: 15th–18th Century* (New York: Harper & Row, 1981), 287–90.

14. Jack London, *John Barleycorn,* ch. 23, quoted in Scott L. Malcolmson, "The Inevitable White Man: Jack London's Endless Journey," *Voice Literary Supplement,* February 1994, 10; Briggs, *Victorian Things,* 411.

15. Emilio Ambasz, personal communication, n.d.

16. Leonard S. Mark and Marvin J. Dainoff, "An Ecological Framework for Ergonomic Research," *Innovation,* vol. 7, no. 2 (Spring 1988), 8–11; Leonard S. Mark, Marvin J. Dainoff, et al., "An Ecological Framework for Ergonomic Research and Design," in R. R. Hoffman and D. A. Palermo, eds., *Cognition and the Symbolic Processes,* vol. 3 (Hillsdale, N.J.: Lawrence Erlbaum Associates, 1991), 477–505; Marvin J. Dainoff, "Reducing Health Complaints in the Computerized Workplace: The Role of Ergonomic Education," *Journal of Interior Design Education,* vol. 16, no. 2 (1990), 31–38.

17. Alix M. Freedman, "Today's Office Chair Promises Happiness, Has Lots of Knobs," *Wall Street Journal,* 18 June 1986, 1; Jon Van, "Furniture Is Adjusting to the Computerized Office," *Washington Post,* 16 June 1991,

H2; Chee Pearlman, "Made to Measure," *I.D.*, vol. 41, no. 5 (September–October 1994), 56–63.

18. On the higher stress levels in a legal office with document imaging, see Sandra Sugawara, "Cutting the Paper Chase," *Washington Post,* 17 August 1992, Washington Business, 19.

19. Jean Gottmann, "Urbanization and Employment: Towards a General Theory," in Jean Gottmann and Robert A. Harper, eds., *Since Megalopolis: The Urban Writings of Jean Gottmann* (Baltimore: Johns Hopkins University Press, 1990), 233.

20. National Institute for Occupational Safety and Health, Division of Standards and Technology Transfer, *NIOSH Publications on Video Display Terminals,* 2nd edn. (June 1991). This collection of testimony and reprints includes the final report of Richard Tell Associates, Inc., to NIOSH, 18 September 1990 (51–74), with data on radio signals and VDT-induced fields.

21. James E. Sheedy, "Vision Problems at Video Display Terminals: A Survey of Optometrists," *Journal of the American Optometric Association,* vol. 10, no. 10 (October 1992), 687–92.

22. Herman Miller Research & Design, *Issues Paper: Cumulative Trauma Disorders* (Zeeland, Mich.: Herman Miller, 1991), 2.

23. "Typing Pain: British Judge Dismisses It," *New York Times,* 29 October 1993, B9; Edward Felsenthal, "An Epidemic or a Fad? The Debate Heats Up over Repetitive Stress," *Wall Street Journal,* 14 July 1994, A7; Barnaby J. Feder, "A Spreading Pain, and Cries for Justice," *New York Times,* 5 June 1994, sec. 3, 1.

24. See Vern Putz-Anderson, ed., *Cumulative Trauma Disorders: A Manual for Musculoskeletal Diseases of the Upper Limbs* (London: Taylor & Francis, 1988), 11–13, 18.

25. John Napier, *Hands,* 2nd edn., rev. by Russell H. Tuttle (Princeton, N.J.: Princeton University Press, 1993), 70–71.

26. Tony Horwitz, "A Lack of Lancelots Fails to Discourage This Artisan's Work," *Wall Street Journal,* 2 March 1994, A1.

27. Adam Smith, *The Wealth of Nations,* ed. Edwin Cannon (London: Methuen, 1961), vol. 1, 11–12; "Very Debatable Units," *Economist,* vol. 316, no. 7670 (1 September 1990), 73–75.

28. "Cramp," *Encyclopaedia Britannica,* 11th edn. (1910), vol. 7, 363–64; Edwin Gabler, *The American Telegrapher: A Social History, 1860–1900* (New Brunswick, N.J.: Rutgers University Press, 1988), 83. Speedups, falling wages, and filthy workplaces appear to have been far more serious grievances for the telegraphers than "glass arm."

29. David Heilbroner, "Handling of an Epidemic: Repetitive Stress Injury," *Working Woman,* February 1993, 60ff; "The Strong Survive," *Back Letter,* vol. 7 (February 1992), 7.

30. Felsenthal, "Epidemic or Fad," A7; Linda Himelstein, "The Asbestos Case of the 1990s?" *Business Week,* 16 January 1995, 82–83.

31. Winn L. Rosch, "Does Your PC—or How You Use It—Cause Health Problems?" *PC Magazine,* 26 November 1991, 493; Janice M. Horowitz, "Crippled by Computers," *Time,* 12 October 1992, 70–72; Ronald E. Roel, "RSI: Little-Known Illness Is Big Hazard in the Workplace," *Los Angeles Times,* 11 June 1989, part 1, 3; John Ballard, "RSI on Trial," *New Scientist,* vol. 139, no. 1890 (11 September 1993), 24–26. On the ILO report: Shoshana Zuboff, *In the Age of the Smart Machine: The Future of Work and Power* (New York: Basic Books, 1988), 120–21.

32. Morton L. Kasdan et al., "Carpal Tunnel Syndrome: The Workup," *Patient Care,* vol. 27 (15 April 1993), 97–107.

33. Frank Swoboda, "Study Links Job Tension to VDT Physical Injuries," *Washington Post,* 21 June 1992, C1; Diana Hembree and Ricardo Sandoval, "RSI Has Become the Nation's Leading Work-Related Illness. How Are Reporters and Editors Coping with It?" *Columbia Journalism Review,* vol. 30, no. 2 (July–August 1991), 41–46; Jane E. Brody, "Epidemic at the Computer: Hand and Arm Injuries," *New York Times,* 3 March 1992, C10.

34. Herman Miller, *Cumulative Trauma Disorders,* 42–47.

35. Ibid., 26.

36. Lee A. Norman, "Mouse Joint—Another Manifestation of an Occupational Epidemic," *Western Journal of Medicine,* vol. 155 (October 1991), 413–15.

37. Frank Swoboda, "Study Links Job Tension to VDT Physical Injuries," *Washington Post,* 21 June 1992, C1; Hembree and Sandoval, "RSI Has Become the National's Leading Work-Related Illness," 41–46.

38. Mitch Betts, "Voice Strain Plagues Some PC Users," *Computerworld,* 24 April 1995, 1.

39. Gabriele Bammer and Brian Martin, "The Arguments About RSI: An Examination," *Community Health Studies,* vol. 12, no. 3 (1988), 348–58; Bammer and Martin, "Repetition Strain Injury in Australia: Knowledge, Social Movement, and De Facto Partisanship," *Social Problems,* vol. 39, no. 3 (August 1992), 219–37. North American ergonomists have avoided the phrase "repetitive strain injury" (RSI) partly because the Australian experience gave it such an ideological charge.

40. Herman Miller, *Cumulative Trauma Disorders,* 30; Bammer and Martin, "Repetition Strain Injury in Australia," 219–37.

41. Zuboff, *Smart Machine,* 141.

42. Ibid., 145.

43. D. Dainoff and J. Balliett, "Seated Posture and Workstation Configuration," in M. Kumashiro and E. D. Magaw, eds., *Towards Human Work: Solutions to Problems in Occupational Health and Safety* (London: Taylor & Francis, 1991), 156–63.

44. Dainoff, "Reducing Health Complaints," 37; Herman Miller, *Cumulative Trauma Disorders,* 36–37.

9 The Computerized Office: Productivity Puzzles

1. Julie Hart, "Buy Smart: Will Your Latest Purchase Hold Its Value?," *Computerworld,* 21 February 1994, 103.

2. Brooke Crothers and Rob Guth, "Strong Demand, Weak Dollar Squeeze DRAM," *InfoWorld,* 17 April 1995, 1.

3. See James R. Chiles, "Now That Everything's Portable, Getting Around Can Really Be a Drag," *Smithsonian,* January 1994, 110.

4. Rob Kling and Suzanne Iacono, "The Mobilization of Support for Computerization: The Role of Computerization Movements," *Social Problems,* vol. 35, no. 3 (June 1988), 226–43; Langdon Winner, "Mythinformation," *Whole Earth Review,* January 1985, 22–28, and reprinted in Langdon Winner, *The Whale and the Reactor: A Search for Limits in the New Age of High Technology* (Chicago: University of Chicago Press, 1986), 98–117. Rob Kling, ed., *Computerization and Controversy: Value Conflicts and Social Choices,* 2nd edn. (San Diego: Academic Press, 1995), is the most comprehensive anthology of reprinted and original papers on social aspects of computing.

5. The most recent statement is Paul A. David, "Computer and Dynamo: The Modern Productivity Paradox in a Not-Too-Distant Mirror," reprinted from *Technology and Productivity: The Challenge for Economic Policy* (Paris: OECD, 1991), 315–47.

6. Don L. Boroughs, "Desktop Dilemma," *U.S. News & World Report,* 24 December 1990, 46–48; Dean Foust, "Is the Computer Boost That Big?" *Business Week,* 16 January 1995, 24; Glenn Rifkin, "Heads That Roll If Computers Fail," *New York Times,* 14 May 1991, D1; Garry Ray, "The Productivity Chase," *Computerworld,* 27 December 1993–3 January 1994, 56. For the strongest defense of the computer industry's record, see U.S. Na-

tional Research Council, *Information Technology in the Service Society* (Washington: National Academy Press, 1994); for a powerful claim that poor usability is indeed undermining productivity, see Thomas K. Landauer, *The Trouble with Computers* (Cambridge, Mass.: MIT Press, 1995).

7. Lauren Ruth Wiener, *Digital Woes: Why We Should Not Depend on Software* (Reading, Mass.: Addison-Wesley, 1993), 10–13. See also Steven Casey, *Set Phasers on Stun and Other True Tales of Design, Technology, and Human Error* (Santa Barbara, Calif.: Aegean, 1993), 13–22, on how the medical technician's inability to see or hear the patient compounded the injury.

8. John Holusha, "The Painful Lessons of Disruption," *New York Times,* 17 March 1993, D1.

9. William M. Bulkeley, "To Read This, Give Us the Password . . . Oops! Try It Again," *Wall Street Journal,* 19 April 1995, A1; John Markoff, "Computer Viruses: Just Uncommon Colds After All?" *New York Times,* 1 November 1992; Darrel Ince, "Nasty Virus, but Not Fatal," *Independent,* 6 September 1993, 14; Christopher O'Malley, "Stalking Stealth Viruses," *Popular Science,* January 1992, 54ff; Mike Holderness, "On the Trail of the Cyberspace Pirates," *New Scientist,* vol. 137, no. 1860 (13 February 1993), 46–47.

10. See Andy Baird, "Surge 'Protectors'—Worse Than Useless?" *American Association of University Presses Computer Newsletter,* vol. 5, no. 1 (September 1990), 10–12.

11. Winn L. Rosch, "Cutting Surges Down to Size," *PC Magazine,* 13 April 1993, 261ff.

12. I am indebted to Michael S. Mahoney for identifying the substitution of programmers for accountants.

13. Kelly Conatser, "Where Has All the Documentation Gone?" *InfoWorld,* 17 April 1995, 1ff.

14. Randall Kennedy, "OS/2 Warp Goes Light Years Ahead of 2.1," *InfoWorld,* 14 November 1994, 167–70; Esther Schindler, "Third Time the Charm? OS/2 3 Goes to Warp Speed," *Computer Shopper,* January 1995, 586ff.

15. Neil Randall, "Roll Out Whatever Comes Next," *PC/Computing,* December 1994, 264–68. It is hard to understand why any computer manager would want to adopt Windows 95 after reading this article. See also John R. Wilke, "As Microsoft Adds Features to Windows, Other Software Makers Must Adapt or Die," *Wall Street Journal,* 20 January 1995, B1.

16. Charles R. Crawley, "From Charts to Glyphs: Rudolf Modley's Contribution to Visual Communication," *Technical Communication,* vol. 41, no. 1 (February 1944), 20ff.

17. Dave Kansas, "The Icon Crisis: Tiny Pictures Cause Confusion," *Wall Street Journal,* 17 November 1993, B1; Joe Lazzaro, "Adapting GUI Software for the Blind Is No Easy Task," *Byte,* May 1994, 33.

18. Laurie Hays, "Coloring Data Correctly Isn't a Black-and-White Issue," *Wall Street Journal,* 17 November 1993, B7.

19. Ibid.; Bill Dunne, "The Science of Vision and the Visualization of Science," *IBM Research Magazine,* n.v., no. 4 (1993), 8–11.

20. Mark Monmonier, *How to Lie with Maps* (Chicago: University of Chicago Press, 1991), 150, 156.

21. Landauer, *Trouble with Computers,* 210–20; Joel Garreau, "Labor-Saving Devices," *Washington Post,* 9 March 1994, C1; William M. Bulkeley, "Data Trap: How Using Your PC Can Be a Waste of Time, Money," *Wall Street Journal,* 4 January 1993, B5; Lamont Wood, "Office Futz Factor Is a Threat to PC Productivity," *Chicago Tribune,* 3 October 1993, Technology and the Workplace, 3.

22. Gary H. Anthes and William Braundel, "Quality Questioned," *Computerworld,* 24 April 1995, 1.

23. Nolan, Norton & Co., "Managing End-User Computing," research report (Boston: Nolan, Norton, 1992).

24. William M. Bulkeley, "Study Finds Hidden Costs of Computing," *Wall Street Journal,* 2 November 1992, B4.

25. Deborah Asbrand, "Lean Budgets Put the Squeeze on IS Departments," *InfoWorld,* 7 February 1994, 55.

26. David C. Churbuck, "Help! My PC Won't Work," *Forbes,* 13 March 1995, 103.

27. John S. Rigden, "The Lost Art of Oratory: Damn the Overhead Projector," *Physics Today,* vol. 43, no. 3 (March 1990), 73–74. " 'Eighty-seven years ago,' he began, as the image of his finger was seen to trace the coastline from North Carolina to Delaware, 'this was the new country that our forefathers brought to us: North Carolina, Virginia, Delaware, *et cetera.* The propositions on which they based their thinking are contained in this famous document.' The screen went brightly blank for a moment as the 13 colonies disappeared. Then a page of beautiful calligraphy starting with the words 'We hold these truths to be self-evident,' splendidly illuminated, came into view. . . ."

28. On the remarkable breeding of pigeons for long-distance flight, military and private, see Stephen J. Bodio, "Pigeon Racing: Homing in on an 'Invisible' Sport," *Smithsonian,* October 1990, 80–89.

29. The Spencerian revival coincided almost exactly with the rise of the

personal computer. See Wolf Von Eckardt, "Reforming with Zigs and Zags," *Time,* 21 March 1983, 86; Michael Kernan, "Cursives! A Lefty Remembers Palmer Penmanship," *Washington Post,* 17 April 1984, D1. On classic logotypes, see Hal Morgan, *Symbols of America* (New York: Penguin Books, 1987), 99, 117.

30. Laura Bird, "Marketers Sell Pen as Signature of Style," *Wall Street Journal,* 9 November 1993, B1. On other paradoxes of typography, see H. M. Collins, *Artificial Experts* (Cambridge, Mass.: MIT Press, 1990), 26–27.

31. Brian Livingston, "Unlocking TrueType Font Secrets," *PC/Computing,* March 1994, 204ff.

32. JoAnne Yates, electronic mail message, 5 November 1993.

33. See Gil Schwartz, "Mighty Multimedia Explodes from Little Presentation Packages," *PC/Computing,* February 1994, 102–4; Landauer, *Trouble with Computers,* 337, 332.

34. Jeffrey E. Kottemann, Fred D. Davis, and William E. Remus, "Computer-Assisted Decision Making: Performance, Beliefs, and the Illusion of Control," *Organizational Behavior and Human Decision Processes,* vol. 57, no. 1 (January 1994), 26ff.

35. Fred D. Davis, and Jeffrey E. Kottemann, "User Perceptions of Decision Support Effectiveness: Two Production Planning Experiments," *Decision Sciences,* vol. 25, no. 1 (January 1994), 57ff.

36. Arthur Howe, "No Calculator Required: Executives Flex Their Math Minds," *Philadelphia Inquirer,* 10 July 1986, 1A.

37. Raymond R. Panko and Richard P. Halverson, Jr., "Patterns of Errors in Spreadsheet Development," unpublished paper, December 1994.

38. William M. Bulkeley, "'Computerizing' Dull Meetings Is Touted as an Antidote to the Mouth That Bored," *Wall Street Journal,* 28 January 1992, B1.

39. Robert X. Cringely, *Accidental Empires* (New York: HarperBusiness, 1993), 115.

40. Susan Cohen, "White Collar Blues," *Washington Post Magazine,* 17 January 1993, 10–13, 24–27; Andrea Knox, "By Downsizing, Do Firms Ax Themselves in the Foot?" *Philadelphia Inquirer,* 17 February 1992.

41. Peter G. Sassone, "Survey Finds Low Office Productivity Linked to Staffing Imbalances," *National Productivity Review,* vol. 11, no. 2 (Spring 1992), 147–58.

42. Ibid., 154.

43. See, e.g., Herb Brody, "The Pleasure Machine," *Technology Review,* vol. 95, no. 3 (April 1992), 31–36; also a thoughtful and exuberant letter by

Andrew Paul Grell, "Life Is Sweeter on the Electronic Superhighway," *New York Times,* 13 July 1993, A18.

10 Sport: The Risks of Intensification

1. For a good popular survey, see Lionel Casson, "The First Olympics: Competing 'For the Greater Glory of Zeus,' " *Smithsonian,* June 1984, 64–73; on the modern Olympics as cultural dominance, see Allen Guttmann, *Games and Empires* (New York: Columbia University Press, 1994), 120–38, with the judo reference on 138.

2. Allen Guttmann, *From Ritual to Record: The Nature of Modern Sports* (New York: Columbia University Press, 1978), 15–55; and the response to his work in John Marshall Carter and Arnd Krüger, eds., *Ritual and Record: Sports Records and Quantification in Pre-Modern Societies* (New York: Greenwood Press, 1990), especially Dietrich Ramba, "Recordmania in Sports in Ancient Greece and Rome" (31–40).

3. John Rickards Betts, *America's Sporting Heritage: 1850–1950* (Reading, Mass.: Addison-Wesley, 1974), 31–34, 69–85. On the successful export of American methods of televising sport, see Neil Lyndon's marvelous diatribe "The Neutralising of a Generation," *Spectator,* vol. 274, no. 8703 (29 April 1995), 17–21.

4. Anson Rabinbach, *The Human Motor: Energy, Fatigue, and the Origins of Modernity* (New York: Basic Books, 1990), 224–25; Edward Tenner, "The Technological Imperative," *Wilson Quarterly,* vol. 19, no. 1 (Winter 1995), 26–34.

5. "Bicycle Helmets a Must," *Health News,* vol. 8, no. 5 (October 1990), 5ff; F. P. Rivara et al., "The Seattle Children's Bicycle Helmet Campaign: Changes in Helmet Use and Head Injury Admissions," *Pediatrics,* vol. 93, no. 4 (April 1994), 567–69.

6. Department of Transportation, U.S. Coast Guard, *Boating Statistics 1992,* COMDTPUB P16754.6 (June 1993), 6, 32–37.

7. "Dateline Norway; Lillehammer '94," *Edmonton Journal,* 16 February 1994, C1; Simon Barnes, "No Place for Boxing Among Exponents of Risk Game," *The Times* (London), 4 May 1994, 46; Jane E. Brody, "Fast Balls and Flying Elbows Mean Faces Need a Shield," *New York Times,* 14 April 1993, C12; Frederick P. Rivara and Abraham B. Bergman, "Strategies of a Successful Campaign to Promote the Use of Equestrian Helmets," *Public Health Reports,* vol. 108 (January–February 1993), 121–26; Jay Searcy, "The Fallen Jockeys," *Philadelphia Inquirer,* 2 May 1995, G5.

8. See Langdon Winner, "Do Artifacts Have Politics?" in *The Whale and the Reactor: The Search for Limits in an Age of High Technology* (Chicago: University of Chicago Press, 1986), 19–39.

9. Elliott J. Gorn, *The Manly Art: Bare-Knuckle Prize Fighting in America* (Ithaca, N.Y.: Cornell University Press, 1986), 220–24.

10. *The Boxing Debate* (London: British Medical Association, 1993), 1–30, 58–69; Luisa Dillner, "Boxing Should Be Counted Out, Says BMA Report," *British Medical Journal,* vol. 306, no. 6892 (12 June 1993), 1561–62; also Ira Berkow, "That 'Sinful Business' Remains in Business," *New York Times,* 17 August 1992, C5.

11. Christopher Lehmann-Haupt, "Woodpecker's Brain Survives, but Will Humans'?" *New York Times,* 1 October 1992, C17.

12. Kay Bartlett, "Athletes Find Ways to Stare Death in Face," *Los Angeles Times,* 13 February 1992, part a, 1; *Boxing Debate,* 24–28; Adele Lubell, "Chronic Brain Injury in Boxers: Is It Avoidable?" *Physician and Sportsmedicine,* vol. 17, no. 11 (November 1989), 126–32.

13. Schmid et al., "Experience with Headgear in Boxing," *Journal of Sports Medicine and Physical Fitness,* vol. 3 (1968), 171–76; Joyce Carol Oates, "The Cruelest Sport," *New York Review of Books,* vol. 39, no. 4 (13 February 1992), 3.

14. Brendan Kinney, "Wearing the Scars of Football," *New York Times,* 26 August 1990, sec. 8, 8.

15. This paragraph relies on English school rugby statistics as reported by George Hill, "Rugby Plays Safe over Its Dangers," *The Times* (London), 21 September 1992, Features section. On the relative risk of rugby and on concussions: Nathan Gibbs, "Injuries in Professional Rugby League: A Three-Year Prospective Study of the South Sydney Professional Rugby League Football Club," *American Journal of Sport Medicine,* vol. 21, no. 5 (September–October 1993), 696–700.

16. John S. Watterson, "Inventing Modern Football," *American Heritage,* September–October 1988, 102–13, with sidebar by James Weeks, "Football as a Metaphor for War," 113.

17. "The Safest Season," *Sports Illustrated,* 29 April 1991, 16; John Jeansonne, "The Ripple Effect: Football Changes, but Danger Factor Is Ever-Present," *New York Newsday,* 6 December 1992, Sports, 23; "Briefly," *Los Angeles Times,* 31 May 1988, part 4, 2; Jacob Weisman, "Pro Football—the Maiming Game," *Nation,* 27 January 1992, 84–87; James A. Nicholas et al., "A Historical Perspective of Injuries in Professional Football," *Journal of the American Medical Association,* vol. 260, no. 7 (19 August 1988), 939–44.

18. Robert Carmichael, "Is the Price Too High? Payment for Playing Football Can Be Debilitating Injuries, Lingering Physical Problems," *Los Angeles Times,* 1 January 1992, part C, 1; John W. Powell and Mario Schootman, "A Multivariate Risk Analysis of Selected Playing Surfaces in the National Football League: 1980 to 1989, an Epidemiologic Study of Knee Injuries," *American Journal of Sports Medicine,* vol. 20 (November–December 1992), 686–94.

19. Carmichael, "Is the Price Too High?" 1; Gerald Secor Couzens, "Football: A Painful Legacy for Players?" *Physician and Sportsmedicine,* vol. 20, no. 10 (1 October 1992), 146ff; Weisman, "Pro Football," 84; Peter Waldman, "Pro-Football Players Often Feel Pressure to Play When Hurt," *Wall Street Journal,* 16 September 1986, 1.

20. Ira Dreyfuss, "High Arch, Excessive Pronation, May Lead to Knee Injury in Runners," Associated Press article in Dow Jones News Retrieval, 11 March 1993, on the research of Benno Nigg, University of Calgary; Doug Thomas, "Runners Gain Without Pain? Researchers Disagree on Chronic Injury Risk," *Omaha World-Herald,* 28 March 1994, Living Today, 25ff; "Cushy Shoes Cause Sprains," *Edell Health Letter,* vol. 10 (September 1991), 1.

21. See R. Corrsin, "A More Explicit Estimate for the 'Implications of Athlete's Bradycardia on Lifespan,'" *Journal of Theoretical Biology,* vol. 96, no. 4 (21 June 1982), 683–88, replying to L. A. Kuehn, "Implications of Athlete's Bradycardia on Lifespan," *Journal of Theoretical Biology,* vol. 88, no. 7 (21 January 1981), 279–86. I am indebted to Joseph Keller for calling these papers to my attention in a public lecture and subsequent personal communications.

22. James F. Fixx, *Jim Fixx's Second Book of Running* (New York: Random House, 1980), 13–29; Barry A. Franklin et al., "Exercise and Cardiac Complications: Do the Benefits Outweigh the Risks?" *Physician and Sportsmedicine,* vol. 22, no. 2 (February 1994), 56ff; Marty Vogel, "Fitness After 50 from Walking to Swimming to Pumping Iron," *Washington Post,* 5 March 1986, T14. For a critique of the methods of Paffenbarger and others, see Henry A. Solomon, *The Exercise Myth* (New York: Harcourt Brace Jovanovich, 1984), 49–50, 55–56; Craig Sharp and Mark Parry-Billings, "Can Exercise Damage Your Health?" *New Scientist,* vol. 135, no. 1834 (15 August 1992), 33–37.

23. Timothy K. Smith, "Summer Olympics 1992: Quest for High-Tech Olympic Gear Poses an Ironic Challenge for U.S. Competitors," *Wall Street Journal,* 5 August 1992, A11.

24. Carl Potera, "Downhill Skiing: Is It Becoming Safer? Fractures Are Fewer, But Serious Knee Injuries Have More Than Doubled," *Washington Post,* 20 December 1988, Z16.

25. Christopher Clarey, "Alpine Skiing," *New York Times,* 6 February 1994, part 8a, 9; Tommy Hine, "Skiing's Real Down Side: Accidents," *Hartford Courant,* 5 February 1993, F1; Art Bentley, "Fatalities Don't Scare Extremists," *Los Angeles Daily News,* 24 March 1994, S14; Marj Charlier, "Resorts Go to Extremes to Attract Skiers," *Wall Street Journal,* 21 February 1992, B1; Alex Markels, "Going to Extremes: In Skiing, It Can Mean Death," *New York Times,* 18 April 1993, Sports, 5; Robert J. Johnson, "Skiing and Snowboarding Injuries: When Schussing Is a Pain," *Postgraduate Medicine,* vol. 88, no. 8 (December 1990), 36–50; Patrice Heinz Schelkun, "Cross-Country Skiing: Ski-Skating Brings Speed and New Injuries," *Physician and Sportsmedicine,* vol. 20, no. 2 (1 February 1992), 168–73.

26. Lex M. Bouter and Paul G. Knipschild, "Causes and Prevention of Injury in Downhill Skiing," *Physician and Sportsmedicine,* vol. 17, no. 11 (November 1989), 80–94; Carl Ettlinger and Robert Johnson, "Can Knee Injuries Be Prevented?," *Skiing,* March 1991, 120–23.

27. Johnson, "Skiing and Snowboarding Injuries," 42–43; "Black and Blue: Skiing Injuries," *Economist,* vol. 326, no. 7797 (6 February 1993), 88; John Underwood, "It's Pretty, It's Trendy, but Skiing Is Also Much Too Dangerous," *New York Times,* February 26, 1995, sec. 8, 9; George Anders, "Million-Dollar M.D.: A Top Physician Earns a Fortune Repairing Knees in Vail, Colo.," *Wall Street Journal,* 8 April 1993, A1.

28. Alvarez, "Feeding the Rat," *New Yorker,* 18 April 1988, 89ff; Jeremy Bernstein, *Mountain Passages* (Lincoln: University of Nebraska Press, 1978), 17–45; Ed Douglas, "The Mountain That Fell to Earth," *New Scientist,* vol. 138, no. 1875 (28 May 1993), 22–24; Eric Perlman, "A Mountaineer's Best Friends," in Eric W. Schrier and William F. Allman, eds., *Newton at the Bat* (New York: Charles Scribner's Sons, 1984), 71–76.

29. David G. Addiss and Sandra P. Baker, "Mountaineering and Rock-Climbing Injuries in U.S. National Parks," *Annals of Emergency Medicine,* vol. 18, no. 9 (September 1989), 975–79; Leonard Evans, *Traffic Safety and the Driver* (New York: Van Nostrand Reinhold, 1991), 134–35.

30. Mark Robinson, "Snap, Crackle, Pop: Climbing Injuries to Fingers and Forearms," *Climbing,* June–July 1993, 141–50.

31. Ibid., 142; Randall A. Lewis et al., "Acute Carpal Tunnel Syndrome: Wrist Stress During a Major Climb," *Physician and Sportsmedicine,* vol. 21, no. 7 (July 1993), 102–7; Will Gadd, "Casualties of Indoor Climb-

ing? Some Fear Indoor Trainers Are Ill-Prepared for Outside," *Seattle Times,* 14 October 1993, C7.

32. Robinson, "Snap, Crackle, Pop," 150.

33. Barry Meier, "With Rescue Costs Growing, U.S. Considers Billing the Rescued," *New York Times,* 28 March 1993, part 1, 18; Wallace Turner, "Season of Disasters Ends for Alpinists," *New York Times,* 21 October 1986, A18.

34. John Enders, "Call It What You Want, but McKinley Is a Killer," *Washington Post,* 5 June 1992, A2; Bill Richards, "Alaska's Denali Attracts the Hardy and the Foolhardy," *Wall Street Journal,* 27 June 1994, B1.

35. Ann Pomeroy, "Snow Avalanche Dangers Increase," *U.S. National Research Council NewsReport,* November 1990, 14–15.

36. *Snow Avalanche Hazards and Mitigation in the United States* (Washington, D.C.: National Academy Press, 1990), 6; Michael Romano, "Avalanches, Rugged Terrain Force Patrollers to Make Altitude Adjustments," *Rocky Mountain News,* Sunday Magazine, 27 February 1994, 8M.

37. Thomas Kleine-Brockhoff, " 'Ella Vegn! Sie Kommt!': Die Lawine—Höllensturz, Todeswalze, Naturschauspiel," *Die Zeit,* vol. 47, no. 8 (21 February 1992), 6–8.

38. John Meyer, "It Is Foolish to Ignore Dangers of Avalanche," *Rocky Mountain News,* 5 January 1994, 12B; William Oscar Johnson and Sally Guard, "Snow Business," *Sports Illustrated,* 8 March 1993, 18ff.

39. Susan Bell and Emma Wilkins, "Avalanche Survivor Says Off-Piste Skiers Should Not Be Afraid," *The Times* (London), 1 February 199.

40. "Friends Recall Sailor's Relentless Drive," *New York Newsday,* 27 November 1992, 148.

41. Phyllis Austin, "Is Technology Taming the Wilderness?," *AMC Outdoors,* vol. 60, no. 3 (April 1994), 23–25.

11 Sport: The Paradoxes of Improvement

1. Rick Reilly, "Sure Cures," *Sports Illustrated,* 9 July 1990, 56.

2. Brad Knickerbocker, "Person Power," *Christian Science Monitor,* 1 November 1993, Now, 12; Vic Sussman, "Laid-Back Bicycles," *U.S. News & World Report,* 6 July 1992, 72; Albert C. Gross et al., "The Aerodynamics of Human-Powered Land Vehicles," *Scientific American,* vol. 249, no. 142 (December 1983), 142ff.

3. Jeff Meyers, "High Adventure Lures Riders of the Wind," *Los Angeles Times,* 4 November 1993, J26.

4. David Bjerklie, "High-Tech Olympians," *Technology Review,* vol. 96, no. 1 (January 1993), 22–30.

5. Bernard Suits, "What Is a Game?" *Philosophy of Science,* vol. 34 (June 1967), 148–56. The examples are mine. I am indebted to J. Nadine Gelberg for calling this paper and other writings of Suits to my attention.

6. George F. Will, *Men at Work: The Craft of Baseball* (New York: HarperPerennial, 1991), 119.

7. James Cox, "Sponsoring the Games," *USA Today,* 21 February 1994, 1B.

8. Elliot J. Gorn, *The Manly Art: Bare-Knuckle Prize Fighting in America* (Ithaca, N.Y.: Cornell University Press, 1986), 47–55, 179–206.

9. Allen Guttmann, *From Ritual to Record: The Nature of Modern Sports* (New York: Columbia University Press, 1978), 53–54. I am indebted to Jonathan Rendall, "Sport and Values: Is Sport on the Wrong Track?," *Independent,* 3 February 1991, 22, for pointing this episode out, though, as Guttmann observes, the game did persist for another 150 years.

10. David Higdon, "13 Ways to Wake Up Pro Tennis," *Tennis,* May 1994, 42–46.

11. See Thomas George, "Instant Replay Goes the Way of the Single Wing," *New York Times,* 19 March 1992, B13; and Gerald Eskenazi, "Instant Replay Hardly Infallible," *New York Times,* 19 March 1992, B19.

12. Russell Davies, "Skiing: Split Seconds and Cold People Dull Interest," *Sunday Telegraph* (London), 28 November 1993, 3.

13. David Bjerklie, "High-Tech Olympians," *Technology Review,* vol. 96, no. 1 (January 1993), 30.

14. Carol A. Schwartz and Rebecca L. Turner, eds., *Encyclopedia of Associations,* 30th edn. (Detroit: Gale Research, Inc., 1995), vol. 1, part 2, 2611; Deborah M. Burek, ed., *Encyclopedia of Associations,* 26th edn. (Detroit: Gale Research, Inc., 1991), vol. 1, part 2, 2105.

15. Stephen Jay Gould, "Losing the Edge," in *The Flamingo's Smile: Reflections on Natural History* (New York: Norton, 1985), 215–29. Leo Torrey, *Stretching the Limits* (New York: Dodd, Mead, 1985), is still the best summary of the effects of technology on a full range of sports.

16. Elisabeth Geake, "A Record for Electronics," *New Scientist,* vol. 135, no. 1832 (1 August 1992), 36–37.

17. Stanley Mieses, "The Zen of the Game," *New York Newsday,* 4 June 1992, part II, 68; Gary Klein, "Aluminum Bat Cuts Cost, Raises Averages, Controversy," *Los Angeles Times,* 28 May 1987, part 9, 8.

18. Bjerklie, "High-Tech Olympians," 22; Charles Siebert, "Vaulting to New Heights," *New York Times Magazine,* 6 July 1986, 18–21.

19. Siebert, "Vaulting to New Heights," 20; Patrick Reusse, "Pole-Vault Backers Try to Take Sport to New Level," *Minneapolis Star-Tribune,* 24 March 1991, 3C.

20. Bjerklie, "High-Tech Olympians," 29–30; Kenny Moore, "Talk About a Change of Pace," *Sports Illustrated,* vol. 65, no. 4 (28 July 1986), 52–54; "Another Side of the Games," *Newsweek,* 20 August 1984, 32.

21. Simon Jones, "The Rackets Revolution: Power with Strings Attached," *Independent,* 9 June 1992, 30.

22. Anthony E. Foley, "Tennis Elbow," *American Family Physician,* vol. 48 (August 1993), 281–88; "Tennis Elbow: Better Answers," *Berkeley Wellness Letter,* vol. 8 (July 1992), 6.

23. Charles Arthur, "Anyone for Slower Tennis?" *New Scientist,* vol. 134, no. 1819 (2 May 1992), 24–28.

24. David Irvine, "Sampras Serves Up Short-Change Final," *Manchester Guardian Weekly,* vol. 151, no. 2 (10 July 1994), 32.

25. "A History of American Tennis Participation" (North Palm Beach, Fla.: Tennis Industry Association, n.d.), n.p.

26. "Tennis Racquets: 30 Years of Change" (North Palm Beach, Fla.: American Tennis Industry Federation, n.d.), n.p.

27. Richard Sandomir, "Off the Court, Tennis Is in a Slump," *New York Times,* 30 May 1993, 33.

28. Anne E. Platt, "Toxic Green: The Trouble with Golf," *World Watch,* vol. 7, no. 3 (May–June 1994), 27–32; Timothy Noah, "Golf Courses Denounced as Health Hazards," *Wall Street Journal,* 2 May 1994, B1; Sonni Efron, "Critics Take Swing at Japan's Golf Courses," *Los Angeles Times,* 7 July 1992, World Report, 6.

29. Frederick C. Klein, "Re-Staying the Course," *Wall Street Journal,* 7 June 1991, A11; *The Rules of Golf* (Far Hills, N.J.: U.S. Golf Association, 1994), 110–11. Actually the Overall Distance Standard for carry and roll is a maximum of "280 yards plus a tolerance of 6%," which "will be reduced to a minimum of 4% as test techniques are improved." The rules refer indirectly to the apparatus rather than give it its ubiquitous colloquial name, "Iron Byron," in honor of the flawless swing of Byron Nelson.

30. J. Cochran, "Science, Equipment Development and Standards," in A. J. Cochran, ed., *Science and Golf* (London: E. and F. N. Spon, 1990), 181.

31. *Rules of Golf,* 11, 40 (Rule 13).

32. Quoted in Lew Fishman, "Is Technology Really Making Golf Easier?" *Golf Digest,* May 1990, 56.

33. Cochran, "Science, Equipment Development and Standards," 179; Frank W. Thomas, "The State of the Game, Equipment and Science," paper presented at the Second World Scientific Congress of Golf, 1994. I am grateful to Dr. Thomas for a prepublication copy of the paper, and for an illuminating discussion and demonstration of his department's work.

34. Ibid.

35. Interview with Frank W. Thomas, June 1994.

36. Ed Weathers, "The Tests of Time," *Golf Digest,* June 1994, 158–69; Thomas, "State of the Game."

37. Thomas, "State of the Game," n.p.

38. Reilly, "Sure Cures," 57–58; Greg Logan, "The Keepers of the Game," *New York Newsday,* 8 June 1986, 11.

39. *Rules of Golf,* 105 10; Joe Strauss, "Square Groove Controversy: Ping Changes Face of Golf," *Atlanta Journal and Constitution,* 16 April 1993, E10.

40. Steven Pearlstein, "It Don't Mean a Thing if You Ain't Got That Swing," *Washington Post,* 1 September 1993, F1.

41. Strauss, "Square Groove Controversy"; James Willwerth, "Driving Reign," *Time,* 6 September 1993, 50.

42. Jaime Diaz, "Hard Times Land on Wealthy Greens," *New York Times,* 16 December 1991, C5; Michael Selz, "More Golfers Are Finding Their Home Is on the Range," *Wall Street Journal,* 7 October 1992, B2.

43. Hank Herman, "Some Sports Aren't a Pain in the Back," *New York Times,* 11 February 1991, C10; Joel Stashenko, "Oh, Those Aching Backs—They're Tour Pros' Most Painful Challenge," *Chicago Tribune,* 29 June 1994, special golf section, 10; J. B. S. Haldane, "What Is Fitness?" in *The Causes of Evolution* (London: Longmans, Green, 1932), 111–43. I am grateful to Peter Grant for calling the last to my attention.

12 Another Look Back, and a Look Ahead

1. John Tierney, "Betting on the Planet," *New York Times Magazine,* 2 December 1990, 52.

2. David S. Landes, *Revolution in Time: Clocks and the Making of the Modern World* (Cambridge, Mass.: Harvard University Press, 1983), 103–13;

Robert K. Merton, *Science, Technology, and Society in Seventeenth-Century England* (New York: Harper Torchbooks, 1970 [1938]), 167–77.

3. Derek Howse, *Greenwich Time and the Discovery of the Longitude* (Oxford: Oxford University Press, 1980), 44–72.

4. David S. Landes, *Revolution in Time: Clocks and the Making of the Modern World* (Cambridge, Mass.: Harvard University Press, 1983), 103–13, 144–57.

5. See Paul Slovic, "Perception of Risk," *Science*, vol. 236, no. 4799 (17 April 1987), 280–85; and Paul Slovic, "Perception of Risk: Reflections on the Psychometric Paradigm," in Sheldon Krimsky and Dominic Golding, eds., *Social Theories of Risk* (Westport, Conn.: Praeger, 1992), 117–52.

6. John P. Eaton and Charles A. Haas, *Titanic: Triumph and Tragedy* (New York: Norton, 1987), 310–11; Edward Bryant, *Natural Hazards* (Cambridge: Cambridge University Press, 1991), 68.

7. James T. Yenckel, "How Safe Is Cruising?," *Washington Post*, 11 August 1991, E6.

8. See Henry Petroski, *To Engineer Is Human: The Role of Failure in Successful Design* (New York: St. Martin's Press, 1985).

9. Mark F. Grady, "Torts: The Best Defense Against Regulation," *Wall Street Journal*, 3 September 1992, A11; Mark F. Grady, "Why Are People Negligent? Technology, Nondurable Precautions, and the Medical Malpractice Explosion," *Northwestern University Law Review*, vol. 82 (Winter 1988), 297–99, 312. See also Grady's review of Paul C. Weiler's *Medical Malpractice on Trial*, "Better Medicine Causes More Lawsuits, and New Administrative Courts Will Not Solve the Problem," *Northwestern University Law Review*, vol. 86 (Summer 1992), 1068–81.

10. On seventeenth-century navigation, see Landes, *Revolution in Time*, 105–11; Carla Rahn Phillips, *Six Galleons for the King of Spain* (Baltimore: Johns Hopkins University Press, 1986), 129–34. On Shovell's tomb: Margaret Whitney, *Sculpture in Britain, 1530 to 1830* (Baltimore: Penguin, 1964), 58. There is probably more to late-nineteenth- and early-twentieth-century litigation than Grady acknowledges: the deference shown by judges and juries of the time toward elite defendants in tort liability cases. To name just three other notorious cases, the industrialists' club that maintained the dam that broke and caused the Johnstown flood in 1889 (2,200 dead), the owners of the *General Slocum*, which caught fire in New York Harbor in 1904 (1,021 dead), and the proprietors of the Triangle Shirtwaist Factory, which burned with 145 dead, escaped civil and criminal action. Even with Astors, Wideners, and Guggenheims among the *Titanic* passen-

gers and the victims' families, the final settlement with the White Star Line was minuscule by late-twentieth-century standards: $663,000 (£136,701) on total claims of $16,804,112 (£3,464,765). See Eaton and Haas, *Titanic,* 277–79.

11. James C. Beniger, *The Control Revolution: Technological and Economic Origins of the Information Society* (Cambridge, Mass.: Harvard University Press, 1986), 221–26 and references.

12. Clay McShane, *Down the Asphalt Path: The Automobile and the American City* (New York: Columbia University Press, 1994), 42–45.

13. Christopher Gray, "Who Holds the Reins of Fate of a 1907 Horse-Auction Mart?" *New York Times,* 8 November 1987, Real Estate, 14; McShane, *Asphalt Path,* 51–54.

14. Daniel Pool, *What Jane Austen Ate and Charles Dickens Knew* (New York: Simon & Schuster, 1993), 250–51; McShane, *Asphalt Path,* 46–50; for a summary of modern fatalities, Leonard Evans, *Traffic Safety and the Driver* (New York: Van Nostrand Reinhold, 1991), 3.

15. Scott L. Bottles, *Los Angeles and the Automobile: The Making of the Modern City* (Berkeley: University of California Press, 1987), 22.

16. On powers and interests in highway building, see Mark H. Rose, *Interstate: Express Highway Politics, 1941–1956* (Lawrence: University Press of Kansas, 1979).

17. Mark S. Foster, *From Streetcar to Superhighway: American City Planners and Urban Transportation, 1900–1940* (Philadelphia: Temple University Press, 1981), 143–45; "Onwards and Outwards," *Economist,* vol. 333, no. 7885 (15 October 1994), 31.

18. Cited in Anatole Kopp, *Town and Revolution: Soviet Architecture and City Planning, 1917–1935* (New York: Braziller, 1970), 173. The Disurbanists (or Deurbanists) actually had in mind the relocation of urban functions along great highways, not the present American suburban pattern.

19. Kenneth Jackson, *Crabgrass Frontier: The Suburbanization of the United States* (New York: Oxford University Press, 1985), 249.

20. H. Schaeffer and Elliot Sclar, *Access for All: Transportation and Urban Growth* (Baltimore: Penguin, 1975), 40–44.

21. Ivan Illich, *Energy and Equity* (New York: Harper & Row, 1974), 18–19.

22. Ibid., 95–96; David Remnick, "Berserk on the Beltway," *Washington Post Magazine,* 7 September 1986, 66ff, 95; "Urban Freeways, Interstates in a Jam," *USA Today,* 18 September 1989, 10A; John F. Harris, "Auto Club, Citing Traffic, to Shut Fairfax Office," *Washington Post,* 1 October 1986.

23. Richard Arnott and Kenneth Small, "The Economics of Traffic Congestion," *American Scientist,* vol. 82, no. 5 (September–October 1994), 446–55; Bob Holmes, "When Shock Waves Hit Traffic," *New Scientist,* vol. 142, no. 1931 (25 June 1994), 36–40.

24. "Risk Homeostasis and the Purpose of Safety Regulation," *Ergonomics,* vol. 31, no. 4 (1988), 408–9.

25. Evans, *Traffic Safety,* 287–90; Haight remarks in telephone interview, October 1991.

26. J. Smeed and G. O. Jeffcoate, "Effects of Changes in Motorisation in Various Countries on the Number of Road Fatalities," *Traffic Engineering and Control,* vol. 12, no. 3 (July 1970), 150–51.

27. John Adams, "Smeed's Law and the Emperor's New Clothes," in Leonard Evans and Richard C. Schwing, eds., *Human Behavior and Traffic Safety* (New York: Plenum Press, 1985), 195–96, 235–37.

28. William K. Stevens, "When It Comes to Highway Chaos, India is No. 1," *New York Times,* 26 October 1983, A2; Steve Coll, " 'Road Kings' Truck Across India," *Washington Post,* 28 October 1989, A1; McShane, *Asphalt Path,* 174–77.

29. "Drivin' My Life Away," *Scientific American,* vol. 257, no. 2 (August 1987), 28, 30; Susan P. Baker, R. A. Whitefield, and Brian O'Neill, "Geographic Variations in Mortality from Motor Vehicle Crashes," *New England Journal of Medicine,* vol. 316, no. 22 (28 May 1987), 1384–87.

30. "In Portugal, Wheels of Misfortune," *New York Times,* 22 July 1990, Travel, 39.

31. Deborah Fallows, "Malaysia's Mad Motorists," *Washington Post,* 10 July 1988, C5.

32. James J. MacKenzie, Roger C. Dower, and Donald D. T. Chen, *The Going Rate: What It Really Costs to Drive* (Washington: World Resources Institute, 1992), 13.

33. Angus Kress Gillespie and Michael Aaron Rockland, *Looking for America on the New Jersey Turnpike* (New Brunswick, N.J.: Rutgers University Press, 1989), 114–15; Albert O. Hirschman, *The Strategy of Economic Development* (New Haven, Conn.: Yale University Press, 1958), 134, 143–45. Hirschman also points out that "a road that is not traveled is likely to deteriorate sooner than one that has to support heavy traffic: the former will surely be neglected whereas there is some hope that the latter will be maintained." Because bituminous surfaces show deterioration early, they may be more suitable for less-traveled roads in developing countries than gravel would be. They don't degrade gracefully, as electrical engineers put it; they demand attention.

34. "How Driving Under the Influence of Society Affects Traffic Deaths," *Washington Post,* 2 September 1991, A3.

35. Charles Stile, "N.J. Drivers Yielding to Safety," *Trenton Times,* 15 September 1991, A1.

36. Kenneth Grahame, *The Wind in the Willows* (New York: Charles Scribner's Sons, 1961), 121.

37. "Jay Gatsby Never Dreamed of Gridlock," *Trenton Times,* 19 November 1991, A18.

38. Michael S. Mahoney, personal communication; John Shore, "Why I Never Met a Programmer I Could Trust," *Communications of the ACM,* vol. 31, no. 4 (April 1988), 372; Wiener, *Digital Woes,* 99–100.

39. For the most useful recent summary of the vast literature on road issues, see the special issue of *CQ Researcher,* vol. 4, no. 17 (6 May 1994), 385–408.

40. N. Lee, "Has the Mortality of Male Doctors Improved with the Reductions in Their Cigarette Smoking?" *British Medical Journal,* 15 December 1979, 1538–40.

41. *Risk: Analysis, Perception, and Management* (London: Royal Society, 1992), 155–59, 138–42.

42. See Jonathan Beecher, *Charles Fourier: The Visionary and His World* (Berkeley: University of California Press, 1987), 338–41; Spencer Weart, "From the Nuclear Frying Pan into the Global Fire," *Bulletin of the Atomic Scientists,* vol. 48, no. 5 (June 1992), 18–27.

43. See Douglas R. Weiner, *Models of Nature* (Bloomington: University of Indiana Press, 1988), 195, for Engels's article "The Role of Labor in the Transformation from Ape to Man" as a rallying point for Soviet conservationists. Their opponents insisted that Engels meant only abusive capitalist development, not dialectically informed socialist intervention.

44. On early Soviet production quotas and the technological conservatism they encouraged, see Kendall E. Bailes, *Technology and Society Under Lenin and Stalin* (Princeton, N.J.: Princeton University Press, 1978), 350; Marshall I. Goldman, *What Went Wrong with Perestroika* (New York: Norton, 1991), 87; Marshall I. Goldman, *Gorbachev's Challenge* (New York: Norton, 1987), 123–24. Let those who have never used a Pentium computer to compose a yard-sale announcement cast the first stone.

45. The superiority of knowledge and the proper work ethic to wealth in resources became a watchword of 1980s reformers. Nathan Glazer's "Two Inspiring Lessons of the 1980s," *New York Times,* 24 December 1989, Review, 11, even suggests that some resources like the agricultural lands of Eu-

rope and Japan with their heavily subsidized surplus crops are becoming "a positive burden to economic success."

46. Matt Ridley, "Butterflies Fall Victim to Man's Interfering Hand," *Sunday Telegraph,* 17 July 1994, 32; Malcolm Smith, "Science: Live the High Life and Save the Wildlife," *Independent,* 30 May 1994, 19.

47. Robert Herman, Siamak A. Ardekani, and Jesse H. Ausubel, "Dematerialization," *Technological Forecasting and Social Change,* vol. 38 (1990), 333–47.

48. Lewis Carroll, *Through the Looking-Glass,* in *The Complete Works of Lewis Carroll* (New York: Modern Library, n.d. [1896 edn.]), 164; "How Much Land Can Ten Billion People Spare for Nature?," Council for Agricultural Science and Technology Task Force Report 121 (February 1994), 26.

INDEX